Strange Functions in Real Analysis

Second Edition

PURE AND APPLIED MATHEMATICS

A Program of Monographs, Textbooks, and Lecture Notes

MONOGRAPHS AND TEXTBOOKS IN PURE AND APPLIED MATHEMATICS

Recent Titles

Strange Functions
in Real Analysis
Second Edition

A. B. Kharazishvili

Tbilisi State University
Tbilisi, Georgia

CRC Press
Taylor & Francis Group
Boca Raton London New York

CRC Press is an imprint of the
Taylor & Francis Group, an **informa** business

A CHAPMAN & HALL BOOK

CRC Press
Taylor & Francis Group
6000 Broken Sound Parkway NW, Suite 300
Boca Raton, FL 33487-2742

First issued in paperback 2019

© 2006 by Taylor & Francis Group, LLC
CRC Press is an imprint of Taylor & Francis Group, an Informa business

No claim to original U.S. Government works

ISBN-13: 978-1-58488-582-5 (hbk)
ISBN-13: 978-0-367-39146-1 (pbk)

Library of Congress Cataloging-in-Publication Data

Catalog record is available from the Library of Congress

**Visit the Taylor & Francis Web site at
http://www.taylorandfrancis.com**

**and the CRC Press Web site at
http://www.crcpress.com**

Preface

At the present time, many strange (or singular) objects in various fields of mathematics are known and no working mathematician is greatly surprised if he meets some objects of this type during his investigations. In connection with strange (singular) objects, the classical mathematical analysis must be noticed especially. It is sufficient to recall here the well-known examples of continuous nowhere differentiable real-valued functions; examples of Lebesgue measurable real-valued functions nonintegrable on any nonempty open subinterval of the real line; examples of Lebesgue integrable real-valued functions with everywhere divergent Fourier series, and others.

There is a very powerful technique in modern mathematics by means of which we can obtain various kinds of strange objects. This is the so-called category method based on the classical Baire theorem from general topology. Obviously, this theorem plays one of the most important roles in mathematical analysis and its applications. Let us recall that, according to the Baire theorem, in any complete metric space E (also, in any locally compact topological space E) the complement of a first category subset of E is everywhere dense in E, and it often turns out that this complement consists precisely of strange (in a certain sense) elements. Many interesting applications of the category method are presented in the excellent textbook by Oxtoby [162] in which the deep analogy between measure and category is thoroughly discussed as well. In this connection, the monograph by Morgan [153] must also be pointed out where an abstract concept generalizing the notions of measure and category is introduced and investigated in detail.

Unfortunately, the category method does not always work and we sometimes need an essentially different approach to questions concerning the existence of singular objects.

This book is devoted to some strange functions in real analysis and their applications. Those functions can be frequently met during various studies in analysis and play an essential role there, especially as counterexamples to numerous statements that seem to be very natural but, finally, fail to be true in certain extraordinary situations (see, e.g., [64]). Another important role of strange functions, with respect to given concepts of analysis, is to show that those concepts are, in some sense, not satisfactory and hence have to be revised, generalized, or extended in an appropriate direction. In this context, we may say that strange functions stimulate and inspire the development of analysis.

The book deals with a number of important examples and constructions of strange functions. Primarily, we consider strange functions acting from the real line into itself. Notice that many such functions can be obtained by using the category method (for instance, a real-valued continuous function defined on the closed unit interval of the real line, which does not possess a finite derivative at each point of this interval). But, as mentioned above, there are some situations where the classical category method cannot be applied, and thus, in such a case, we have to appeal to the corresponding individual construction.

We begin with functions that can be constructed within the theory

ZF & DC,

where **ZF** denotes the Zermelo-Fraenkel set theory without the Axiom of Choice and **DC** denotes a certain weak form of this axiom: the so-called Axiom of Dependent Choices, which is enough for most domains of classical mathematics. Among strange functions whose existence can be established in **ZF & DC** the following ones are of primary interest: Cantor and Peano type functions, semicontinuous functions that are not countably continuous, singular monotone functions, everywhere differentiable nowhere monotone functions, and Jarnik's continuous nowhere approximately differentiable functions.

Then we examine various functions whose constructions need essentially noneffective methods, i.e., they need an uncountable form of the Axiom of Choice: functions nonmeasurable in the Lebesgue sense, functions without the Baire property, functions associated with a Hamel basis of the real line, Sierpiński-Zygmund functions that are discontinuous on each subset of the real line having the cardinality continuum, etc.

Finally, we consider a number of examples of functions whose existence cannot be established without the aid of additional set-theoretical axioms. However, it is demonstrated in the book that the existence of such functions follows from (or is equivalent to) certain widely known set-theoretical hypotheses (e.g., the Continuum Hypothesis).

Among other topics presented in this book and closely connected with strange functions in real analysis, we wish to point out the following ones: the construction (under Martin's Axiom) of absolutely nonmeasurable solutions of the Cauchy functional equation and their application to the measure extension problem raised by Banach; Egorov type theorems on the uniform convergence of sequences of measurable functions; some relationships between the classical Sierpiński partition of the Euclidean plane and Fubini type theorems; the existence of a function on a second category subset of a topological space, which cannot be extended to a function defined on the whole space and possessing the Baire property; sup-measurable and weakly sup-measurable functions with their applications in the theory of ordinary differential equations.

In the final chapter of our book, we consider the family of all continuous nondifferentiable functions from the points of view of category and measure. We present one general approach illuminating the basic reasons which necessarily imply that the above-mentioned family of functions has to be large in the sense of category or measure. Notice that, in connection with continuous nondifferentiable functions, a short scheme for constructing the classical Wiener measure is discussed in this chapter, too, and some simple but useful statements from the general theory of stochastic (random) processes are demonstrated.

This book is based on the course of lectures given by the author at I.Vekua Institute of Applied Mathematics of Tbilisi State University in the academic year 1997–1998, entitled:

Some Pathological Functions in Real Analysis.

These lectures (their role is played by the corresponding chapters of the book) are, in fact, mutually independent from the logical point of view but are strictly related from the point of view of the topics discussed and the methods applied (such as purely set-theoretical arguments and constructions, measure-theoretical methods, the Baire category method, and so on).

The material presented in the book is essentially self-contained and, consequently, is accessible to a wide audience of mathematicians (including graduate and postgraduate students). For the reader's convenience, the Introduction gives an overview of the subject. Here some preliminary notions and facts are presented that are useful in our further considerations. The reader can ignore this introductory chapter, returning to it if the need arises. In this connection, the standard graduate-level textbooks and well-known monographs (for instance, [20], [69], [74], [79], [84], [85], [119], [120], [124], [154], [162], [180]) should be pointed out containing all auxiliary notions and facts from set theory, general topology, classical descriptive set theory, integration theory and real analysis.

We begin with basic set-theoretical concepts such as: binary relations of special type (namely, equivalence relations, orderings, and functional graphs), ordinal numbers and cardinal numbers, the Axiom of Choice and the Zorn Lemma, some weak forms of the Axiom of Choice (especially, the countable form of **AC** and the Axiom of Dependent Choices), the Continuum Hypothesis, the Generalized Continuum Hypothesis, and Martin's Axiom as a set-theoretical assertion which is essentially weaker than the Continuum Hypothesis but rather helpful in various constructions of set theory, topology, measure theory and real analysis (cf. [71], [119]).

Then we briefly present some basic concepts of general topology and classical descriptive set theory, such as: the notion of a first category set in a topological space, the Baire property (and the Baire property in the restricted sense) of subsets of a topological space, the notion of a Polish space, Borel sets in a topological space, analytic (Suslin) subsets of a topological space, and the projective hierarchy of Luzin, which takes the Borel and analytic sets as the first two steps of this hierarchy. It is also stressed that Borel and analytic sets have a nice descriptive structure but this feature fails to be true for general projective sets (because, in certain models of set theory, there exist projective subsets of the real line that are not Lebesgue measurable and do not have the Baire property).

The final part of the Introduction is devoted to some classical facts and statements from real analysis. Namely, we recall here the notion of a real-valued lower (upper) semicontinuous function and demonstrate basic properties of such functions, formulate and prove the fundamental Vitali covering theorem, introduce the notion of a density point for a Lebesgue measurable set, and present the Lebesgue theorem on density points as a consequence of the above-mentioned Vitali theorem. In addition, we give here a short proof of the existence of a real-valued continuous nowhere differentiable function, starting with the well-known Kuratowski lemma on closed projections. Let us emphasize once more that the problem of the existence of real-valued continuous nondifferentiable functions, with respect to various concepts of generalized derivative, is one of the central questions in this book. We develop this topic gradually and, as mentioned earlier, investigate the question from different points of view. However, we hope that the reader will be able to see that the main kernel is contained in purely logical and set-theoretical aspects of the question.

The second edition of this book essentially differs from the first one. Five new chapters are added to the material of the previous edition and some old chapters are revised in details. Many new exercises (sometimes, rather difficult) are included in the text. They contain valuable information around the topics presented in the book and point out deep relationships between them. The list of references is also significantly expanded by adding recent works devoted to strange functions in real analysis and their applications.

A.B. Kharazishvili

Table of Contents

Introduction: basic concepts

In this chapter we fix the notation and present some auxiliary facts from set theory, general topology, classical descriptive set theory, measure theory and real analysis. We shall systematically utilize these facts in our further considerations.

The symbol **ZF** denotes the Zermelo–Fraenkel set theory, which is one of the most important formal systems of axioms for the whole of modern mathematics (in this connection, see [79], [119] and [124]; cf. also [20]). The basic notions of the Zermelo–Fraenkel system are sets and the membership relation ∈ between them. Of course, the system **ZF** consists of several axioms, which formalize various properties of sets in terms of the relation ∈. We do not present here a list of these axioms and, actually, we shall work in the so-called naive set theory.

The symbol **ZFC** denotes the Zermelo–Fraenkel theory with the Axiom of Choice. In other words, **ZFC** is the following theory:

$$\textbf{ZF} \; \& \; \textbf{AC}$$

where **AC** denotes, as usual, the Axiom of Choice (the precise formulation of **AC** will be given later with some of its equivalents).

At the present time, it is widely known that the theory **ZFC** is a basis of modern mathematics, i.e., almost all fields of mathematics can be developed by starting with **ZFC**. The Axiom of Choice is a very powerful set-theoretical assertion which implies many extraordinary and interesting consequences. Sometimes, in order to get a required result, we do not need the whole power of the Axiom of Choice. In such cases, it is sufficient to apply various weak forms of **AC**. Some of these forms are discussed below.

If x and X are any two sets, then the relation $x \in X$ means that x belongs to X. In this situation, we also say that x is an element of X.

One of the axioms of set theory implies that any set y is an element of some set Y (certainly, depending on y). Thus we see that the notion of an element is equivalent to the notion of a set.

The relation $X \subset Y$ means that a set X is a subset of a set Y, i.e., each element of X is also an element of Y.

1

If $X \subset Y$ and $X \neq Y$, then we say that X is a proper subset of Y.

If $R(x)$ is a relation depending on an element x (or, in other words, $R(x)$ is a property of an element x), then the symbol

$$\{x \ : \ R(x)\}$$

denotes the set (the family, the class) of all those elements x for which the relation $R(x)$ holds. In our further considerations we always suppose that $R(x)$ is such that the corresponding set $\{x \ : \ R(x)\}$ does exist. For example, a certain axiom of **ZF** states that there always exists a set of the type

$$\{x \ : \ x \in X \ \& \ S(x)\}$$

where X is an arbitrarily given set and $S(x)$ is an arbitrary relation. In this case we write

$$\{x \in X \ : \ S(x)\}$$

instead of $\{x \ : \ x \in X \ \& \ S(x)\}$. Also, if we have two relations $R_1(x)$ and $R_2(x)$, then we write

$$\{x \ : \ R_1(x), \ \ R_2(x)\}$$

instead of $\{x \ : \ R_1(x) \ \& \ R_2(x)\}$.

The symbol $\emptyset$ denotes, as usual, the empty set, i.e.,

$$\emptyset = \{x \ : \ x \neq x\}.$$

If X is any set, then the symbol $\mathcal{P}(X)$ denotes the family of all subsets of X, i.e., we have

$$\mathcal{P}(X) = \{Y \ : \ Y \subset X\}.$$

Note that the existence of $\mathcal{P}(X)$ is stated by one of the axioms of **ZF** (see, e.g., [119], [124]). The set $\mathcal{P}(X)$ is also called the power set of a given set X.

If x and y are any two elements, then the set

$$(x, y) = \{\{x\}, \{x, y\}\}$$

is called the ordered pair (or, simply, the pair) consisting of x and y. The reader can easily check that the implication

$$((x, y) = (x', y')) \Rightarrow (x = x' \ \& \ y = y')$$

is valid for all elements x, y, x', y'.

Let X and Y be any two sets. Then, as usual,

$X \cup Y$ denotes the union of X and Y;

$X \cap Y$ denotes the intersection of X and Y;

$X \setminus Y$ denotes the difference of X and Y;

$X \triangle Y$ denotes the symmetric difference of X and Y, i.e.,

$$X \triangle Y = (X \setminus Y) \cup (Y \setminus X).$$

We also put

$$X \times Y = \{(x, y) \ : \ x \in X, \ y \in Y\}.$$

The set $X \times Y$ is called the Cartesian product of the given sets X and Y. In a similar way, by recursion, we can define the Cartesian product

$$X_1 \times X_2 \times \ ... \ \times X_n$$

of a finite family $\{X_1, \ X_2, \ ... \ , \ X_n\}$ of arbitrary sets.

If X is a set, then the symbol $card(X)$ denotes the cardinality of X. Sometimes, $card(X)$ is also called the cardinal number of X.

ω is the first infinite cardinal (ordinal) number. In fact, ω is the cardinality of the set

$$\mathbf{N} = \{0, \ 1, \ 2, \ ... \ , \ n, \ ...\}$$

of all natural numbers. Sometimes, it is convenient to identify the sets ω and $\mathbf{N}$ and we always assume such an identification in our further considerations.

A set X is finite if $card(X) < \omega$. A set X is infinite if $card(X) \geq \omega$. A set X is (at most) countable if $card(X) \leq \omega$. Finally, a set X is uncountable if $card(X) > \omega$.

For an arbitrary set E, we put:

$[E]^{<\omega}$ = the family of all finite subsets of E;

$[E]^{\leq\omega}$ = the family of all countable subsets of E.

ω_1 is the first uncountable cardinal (ordinal) number. Notice that ω_1 is sometimes identified with the set of all countable ordinal numbers (countable ordinals).

Various ordinal numbers (ordinals) are denoted by symbols

$$\alpha, \ \beta, \ \gamma, \ \xi, \ \zeta, \ ... \ .$$

Let α be an ordinal number. We say that α is a limit ordinal if

$$\alpha = sup\{\beta \ : \ \beta < \alpha\}.$$

The cofinality of a limit ordinal α is the smallest ordinal ξ such that there exists a family $\{\alpha_\zeta \ : \ \zeta < \xi\}$ of ordinals satisfying the relations

$$(\forall \zeta < \xi)(\alpha_\zeta < \alpha), \qquad \alpha = sup\{\alpha_\zeta \ : \ \zeta < \xi\}.$$

The cofinality of a limit ordinal α is denoted by the symbol $cf(\alpha)$.

Clearly, we have the inequality $cf(\alpha) \leq \alpha$ for all limit ordinal numbers α.

A limit ordinal number α is called a regular ordinal if $cf(\alpha) = \alpha$.

A limit ordinal number α is called a singular ordinal if $cf(\alpha) < \alpha$.

Starting with the definitions of regular and singular limit ordinals, we can define, in the usual manner, regular infinite cardinals and singular infinite cardinals. For example, ω and ω_1 are regular ordinals (cardinals) and ω_ω is a singular ordinal (cardinal).

As mentioned above, in many considerations it is convenient to identify every ordinal α with the set of all those ordinals which are strictly less than α. Such an approach to the theory of ordinal numbers is due to von Neumann (see, e.g., [119], Chapter 1). It is also convenient to identify every cardinal κ with the smallest ordinal number α such that $card(\alpha) = \kappa$.

If κ is an arbitrary infinite cardinal number, then the symbol κ^+ denotes the smallest cardinal among all those cardinals which are strictly greater than κ. For example, we have

$$\omega^+ = \omega_1, \quad \omega_2 = (\omega_1)^+, \quad \dots .$$

The symbol $\mathbf{Z}$ denotes the set of all integers.

The symbol $\mathbf{Q}$ denotes the set of all rational numbers.

The symbol $\mathbf{R}$ denotes the set of all real numbers.

If the set $\mathbf{R}$ is equipped with its standard structures (order structure, algebraic structure, topological structure), then $\mathbf{R}$ is usually called the real line. This object is basic for classical mathematical analysis.

The symbol $\mathbf{c}$ denotes the cardinality of the continuum, i.e., we have

$$\mathbf{c} = 2^\omega = card(\mathbf{R}).$$

The Continuum Hypothesis (**CH**) is the assertion

$$\mathbf{c} = \omega_1.$$

The Generalized Continuum Hypothesis (**GCH**) asserts that

$$2^{\omega_\alpha} = \omega_{\alpha+1}$$

for all ordinals α.

At the present time, it is well known that the theory **ZFC** is consistent if and only if both theories

$$\mathbf{ZFC} \ \& \ (the \ Continuum \ Hypothesis),$$

ZFC & (*the negation of the Continuum Hypothesis*)

are consistent (see, e.g., [119], Chapter 7). Moreover, it is also well known that the theory **ZFC** is consistent if and only if the theory

ZFC & (**c** *is a singular cardinal*)

is consistent. More precisely, it was established that, for an infinite cardinal number ω_α satisfying the relation $cf(\omega_\alpha) > \omega$, there exists a model of **ZFC** in which we have the equality

$$\mathbf{c} = \omega_\alpha.$$

Actually, if we start with an arbitrary countable transitive model for **ZFC** (strictly speaking, for a relevant fragment of **ZFC**) satisfying the Generalized Continuum Hypothesis, then the above-mentioned equality is true in a certain Cohen model for **ZFC** extending the original model (for details, see [119], Chapter 7).

The Generalized Continuum Hypothesis holds in a special model of set theory, first constructed by Gödel. This model is called the Constructible Universe of Gödel and usually denoted by **L**. Actually, the Constructible Universe **L** is a subclass of the well-known von Neumann Universe which is a natural model of set theory. Various facts and statements concerning **L** are discussed in Chapter 6 of [119] (see also [71], [72] and [79]). It is reasonable to note here that, in **L**, some naturally defined subsets of the real line are bad from the point of view of Lebesgue measure and Baire property (i.e., they are not measurable in the Lebesgue sense and do not have the Baire property).

Let n be a fixed natural number. The symbol $\mathbf{R}^n$ denotes, as usual, the n-dimensional Euclidean space. If $n = 0$, then $\mathbf{R}^n$ is the one-element set consisting of zero only. If $n > 0$, then it is sometimes convenient to consider $\mathbf{R}^n$ as a vector space V over the field $\mathbf{Q}$ of all rational numbers. According to a fundamental statement of the theory of vector spaces (over arbitrary fields), there exists a basis in the space V (see, e.g., [42] where much more general assertions are discussed for universal algebras). This basis is usually called a Hamel basis of V. Obviously, the cardinality of any Hamel basis of V is equal to the cardinality of the continuum. Notice also that the existence of a Hamel basis of V cannot be established without the aid of uncountable forms of the Axiom of Choice because the existence of such a basis immediately implies the existence of a subset of the real line $\mathbf{R}$, nonmeasurable with respect to the standard Lebesgue measure λ given on $\mathbf{R}$. Some nontrivial applications of Hamel bases will be discussed in Chapter 9 of this book.

Let X and Y be any two sets. A binary relation between X and Y is an arbitrary subset G of the Cartesian product of X and Y, i.e.,

$$G \subset X \times Y.$$

In particular, if we have $X = Y$, then we say that G is a binary relation on the basic set X.

For any binary relation $G \subset X \times Y$, we put

$$pr_1(G) = \{x \ : \ (\exists y)((x,y) \in G)\};$$

$$pr_2(G) = \{y \ : \ (\exists x)((x,y) \in G)\}.$$

It is clear that

$$G \subset pr_1(G) \times pr_2(G).$$

The Axiom of Dependent Choices is the following set-theoretical statement:

If G is a binary relation on a nonempty set X and, for each element $x \in X$, there exists an element $y \in X$ such that $(x,y) \in G$, then there exists a sequence $(x_0, x_1, ..., x_n, ...)$ of elements of X, such that

$$(\forall n \in \mathbf{N})((x_n, x_{n+1}) \in G).$$

The Axiom of Dependent Choices is usually denoted by **DC**. Actually, the statement **DC** is a weak form of the Axiom of Choice. This form is completely sufficient for most fields of classical mathematics: geometry of a finite-dimensional Euclidean space, mathematical analysis on the real line, Lebesgue measure theory, etc. We shall deal with the axiom **DC** many times in our further considerations and discuss some interesting applications of this axiom.

It was established by Blair that, in the theory **ZF**, the following two assertions are equivalent:

(a) the Axiom of Dependent Choices;

(b) no nonempty complete metric space is of first category on itself (the classical theorem of Baire).

Exercise 1. Prove the logical equivalence of assertions (a) and (b) in the theory **ZF**. Note that the implication (a) $\Rightarrow$ (b) is widely known in analysis. In order to establish the converse implication, equip a nonempty set X with the discrete topology and consider the complete metric space X^ω. Further, by starting with a given binary relation G on X satisfying

$$(\forall x \in X)(\exists y \in X)((x,y) \in G),$$

define a certain countable family of everywhere dense open subsets of X^ω and obtain with the aid of this family the desired sequence of elements from X.

Exercise 2. Show that if E is a complete separable metric space, then the Baire theorem for E is valid within **ZF**. Deduce from this fact (again within **ZF**) that the real line **R** cannot be represented as the union of countably many finite sets. In particular, we have (in **ZF**) the classical theorem of Cantor saying that **R** is not countable.

In connection with the previous exercise, let us remark that there are models of **ZF** in which **R** is expressible in the form of countable union of countable sets (for more details, see [79] and [119]).

Exercise 3. Show that the following two assertions are equivalent in **ZF**:

(a) there exists a subset of **R** of cardinality ω_1;

(b) there exists a function f acting from $[\mathbf{R}]^{\leq\omega}$ into **R** and satisfying the relation $f(D) \notin D$ for each countable set $D \subset \mathbf{R}$.

Note that none of these two assertions is provable within the theory **ZF** & **DC** (see [169] and [182]; cf. also Chapter 8 of this book).

Let X be an arbitrary set. A binary relation $G \subset X \times X$ is called an equivalence relation on the set X if the following three conditions hold:

(1) $(x, x) \in G$ for all elements $x \in X$;

(2) $(x, y) \in G$ and $(y, z) \in G$ imply $(x, z) \in G$;

(3) $(x, y) \in G$ implies $(y, x) \in G$.

If G is an equivalence relation on X, then the pair (X, G) is called a set equipped with an equivalence relation. In this case, the set X is also called the basic set for the given equivalence relation G.

Obviously, if G is an equivalence relation on X, then we have a partition of X canonically associated with G. This partition consists of the sets

$$G(x) \quad (x \in X),$$

where $G(x)$ denotes the section of G corresponding to an element $x \in X$; in other words, $G(x)$ is defined by the formula

$$G(x) = \{y \; : \; (x, y) \in G\}.$$

Conversely, every partition of a set X canonically determines an equivalence relation G on X. Namely, we put $(x, y) \in G$ if and only if x and y belong to the same element of the partition.

Let X be an arbitrary set and let G be a binary relation on X. We say that G is a partial order on X if the following three conditions hold:

(1) $(x, x) \in G$ for each element $x \in X$;
(2) $(x, y) \in G$ and $(y, z) \in G$ imply $(x, z) \in G$;
(3) $(x, y) \in G$ and $(y, x) \in G$ imply $x = y$.
Suppose that G is a partial order on a set X. As usual, we write

$$x \leq y \quad iff \quad (x, y) \in G.$$

The pair $(X, \leq)$ is called a set equipped with a partial order (or, simply, a partially ordered set). The set X is called the basic set for the given partial order $\leq$.

Let $(X, \leq)$ be a partially ordered set. Let x and y be any two elements of the basic set X. We say that x and y are comparable (with respect to $\leq$) if

$$x \leq y \ \lor \ y \leq x.$$

According to this definition, we say that x and y are incomparable if

$$x \not\leq y \ \& \ y \not\leq x.$$

Further, we say that elements x and y of the set X are consistent (in X) if there exists an element z of X such that $z \leq x$ and $z \leq y$. Now, it is clear that elements x and y of the set X are inconsistent (in X) if there does not exist an element z of X having the above property.

Obviously, if x and y are inconsistent, then they are incomparable. The converse assertion is not true in general.

A subset Y of the set X is called a chain in X (or a subchain of X) if any two elements of Y are comparable with respect to $\leq$.

A subset Y of the set X is called an antichain in X if any two distinct elements of Y are incomparable with respect to $\leq$. In this case, Y is also called a free subset of X with respect to $\leq$.

A subset Y of the set X is called consistent if, for any two elements y_1 and y_2 of Y, there exists an element y of Y such that $y \leq y_1$ and $y \leq y_2$. From this definition it immediately follows, by induction, that if a nonempty set $Y \subset X$ is consistent, then, for every finite set $\{y_1, y_2, \dots, y_n\} \subset Y$, there exists an element y of Y satisfying the relations

$$y \leq y_1, \ y \leq y_2, \ \dots, \ y \leq y_n.$$

A subset Y of the set X is called totally inconsistent if any two distinct elements of Y are inconsistent in X.

The following definition is very important for modern set theory and its various applications to general topology and measure theory (see, e.g., [119]).

Let $(X, \leq)$ be a partially ordered set. We say that $(X, \leq)$ satisfies the countable chain condition (or the Suslin condition) if every totally inconsistent subset of X is at most countable.

Let $(X, \leq)$ be a partially ordered set and let Y be a subset of X. We say that Y is coinitial in X if, for each element x of X, there exists an element y of Y such that $y \leq x$.

Martin's Axiom is the following set-theoretical statement:

If $(X, \leq)$ is a partially ordered set satisfying the countable chain condition and $\mathcal{F}$ is a family of coinitial subsets of X, such that $card(\mathcal{F}) < \mathbf{c}$, then there exists a consistent subset Y of X which intersects each set from the family $\mathcal{F}$, i.e.,

$$(\forall Z \in \mathcal{F})(Y \cap Z \neq \emptyset).$$

Martin's Axiom is usually denoted by **MA**. One can easily show that the Continuum Hypothesis implies Martin's Axiom.

Exercise 4. Prove in **ZF** & **DC** the implication **CH** $\Rightarrow$ **MA**.

On the other hand, it was established (by Martin and Solovay) that if the theory **ZFC** is consistent, then the theory

ZFC & **MA** & (*the negation of the Continuum Hypothesis*)

is consistent, too (see, e.g., [119], Chapter 8). Moreover, it was established that Martin's Axiom is a set-theoretical statement much weaker than the Continuum Hypothesis, because it does not bound from above the size of the continuum. At the present time, many applications of Martin's Axiom to the theory of infinite groups, to general topology, to measure theory and real analysis are known (see, e.g., [10], [38], [68], [71], [72], [89], [105], [119]). Some nontrivial applications of **MA** will be discussed below. Note, in addition, that **MA** can also be formulated in purely topological terms (see [71], [72], [119]).

Let $(X, \leq)$ be again a partially ordered set.

We say that $(X, \leq)$ is a linearly ordered set if any two elements of X are comparable with respect to $\leq$. A linearly ordered set is also called a chain.

We say that an element x of a partially ordered set $(X, \leq)$ is maximal (in X) if, for each element y of X, we have the implication

$$x \leq y \Rightarrow x = y.$$

In a similar way one can define a minimal element of a partially ordered set $(X, \leq)$.

The next set-theoretical statement is a well-known equivalent of the Axiom of Choice, formulated in terms of partially ordered sets. This statement is usually called the Zorn Lemma.

Let $(X, \leq)$ be a partially ordered set such that each subchain of X is bounded from above. Then there exists at least one maximal element in X. Moreover, if x is an arbitrary element of X, then there exists at least one maximal element y in X satisfying the inequality $x \leq y$.

Sometimes, it is convenient to apply the Zorn Lemma instead of the Axiom of Choice whose standard formulation looks as follows:

For any family $\mathcal{Z}$ of nonempty sets which pairwise have no common elements, there exists a set Z such that each set from $\mathcal{Z}$ has one and only one common element with Z.

The proof of the fact that **AC** is equivalent to the Zorn Lemma (in the theory **ZF**) is not difficult and can be found in many books (see, e.g., [85] or [124]).

Let $(X, \leq)$ be a partially ordered set and let $x \in X$. We say that x is a smallest (first, least) element of X if $x \leq y$ for all elements $y \in X$. In a similar way we can define a largest (last, greatest) element of X. It is easy to check that a smallest (respectively, largest) element of X is unique. Moreover, the smallest (respectively, largest) element of the set X is a unique minimal (respectively, unique maximal) element of X.

We say that a partially ordered set $(X, \leq)$ is well ordered if any nonempty subset of X, equipped with the induced order, has a smallest element.

Obviously, every well ordered set is linearly ordered. Well ordered sets are very important in the class of all partially ordered sets because one can directly apply the principle of transfinite induction and the method of transfinite recursion to well ordered sets.

It is known that, for every well ordered set $(X, \leq)$, there exists a unique ordinal number α such that $(X, \leq)$ and α are isomorphic as partially ordered sets. Thus, without loss of generality, well ordered sets can be identified with corresponding ordinal numbers.

Let $(X, \leq)$ be an arbitrary linearly ordered set. Then it is not difficult to prove (by using the Zorn Lemma) that there exists a subset Y of X satisfying the following two relations:

(1) Y is a well ordered set with respect to the induced order;

(2) Y is a cofinal subset of X, i.e., for each element x of X, there exists an element y of Y such that $x \leq y$.

This fact immediately implies that the Zorn Lemma can be formulated in the weaker version:

Let $(X, \leq)$ be a partially ordered set such that every well ordered subset of X is bounded from above. Then there exists a maximal element in X. Moreover, if x is an arbitrary element of X, then there exists a maximal element y in X such that $x \leq y$.

Now, we shall consider some simple facts concerning the fundamental notion of a partial function (partial mapping).

Let X and Y be any two sets. Suppose that G is a binary relation between X and Y, i.e., $G \subset X \times Y$. We say that G is a functional graph if the implication

$$((x, y) \in G \ \& \ (x, y') \in G) \Rightarrow (y = y')$$

holds for all elements x, y, y'. It is easy to see that a binary relation $G \subset X \times Y$ is a functional graph if and only if

$$(\forall x)(card(G(x)) \leq 1).$$

We say that a triple

$$g = (G, X, Y)$$

is a partial function (or a partial mapping) acting from X into Y if G is a functional graph and $G \subset X \times Y$. In this case, we also say that the set G is the graph of a partial function g. Sometimes, it is convenient to identify a partial function with its graph.

Furthermore, we say that a triple $g = (G, X, Y)$ is a function (mapping) acting from X into Y if g is a partial function acting from X into Y and $X = pr_1(G)$. In this case, we also write

$$g \ : \ X \to Y.$$

If x is an arbitrary element of X, then the symbol $g(x)$ denotes the unique element y of Y for which $(x, y) \in G$. The element $g(x)$ is called the value of g at x. Hence we can write

$$g \ : \ x \to g(x) \qquad (x \in X, \ g(x) \in Y).$$

We may use a similar notation for any partial function $g = (G, X, Y)$, too. For example, it is sometimes rather convenient to write $g \ : \ X \to Y$ for this partial function. But we wish to emphasize that the symbol $g(x)$ can be applied only in the case when $x \in pr_1(G)$.

Let $g = (G, X, Y)$ be again a partial mapping acting from X into Y. If A is a subset of the set X, then we put

$$g(A) = \{g(x) \ : \ x \in pr_1(G) \cap A\}.$$

The set $g(A)$ is usually called the image of A under (with respect to) g. Obviously, one can introduce, by the same definition, the set $g(A)$ for an arbitrary set A.

If B is a subset of the set Y, then we put

$$g^{-1}(B) = \{x \ : \ x \in pr_1(G), \ g(x) \in B\}.$$

The set $g^{-1}(B)$ is usually called the preimage of B under (with respect to) g. Clearly, one can introduce, by the same definition, the set $g^{-1}(B)$ for any set B.

If A is a subset of X, then the symbol $g|A$ denotes the restriction of g to this subset, i.e., we put

$$g|A = (G \cap (A \times Y), A, Y).$$

Evidently, the same definition can be applied to an arbitrary set A.

We say that a partial function g is an extension of a partial function f if f is a restriction of g.

Let g be a partial function whose graph is G.

The set $pr_1(G)$ is called the domain of g. It is denoted by $dom(g)$.

The set $pr_2(G)$ is called the range of g. It is denoted by $ran(g)$.

Obviously, we have the equality

$$ran(g) = g(dom(g)).$$

We say that a partial function $g = (G, X, Y)$ is an injective partial function (or, simply, an injection) if the implication

$$g(x) = g(x') \Rightarrow x = x'$$

holds for all elements x, x' from the domain of g.

If $g = (G, X, Y)$ is injective, then we can consider a partial function

$$g^{-1} \ : \ Y \to X$$

whose graph coincides with the set

$$G^{-1} = \{(y, x) \ : \ (x, y) \in G\}.$$

This partial function is called the partial function inverse to g.

We say that $g = (G, X, Y)$ is a surjective partial function (or, simply, a surjection) if the equality $ran(g) = Y$ is fulfilled.

Finally, we say that a function $g = (G, X, Y)$ is a bijective function (or, simply, a bijection) if g is an injection and a surjection simultaneously. In this case, we also say that g is a one–to–one correspondence between the sets X and Y.

A transformation of a set X is an arbitrary bijection acting from X onto X. The set of all transformations of a set X becomes a group with respect to the natural operation $\circ$ of composition of transformations. This group is called the symmetric group of X and denoted by the symbol $Sym(X)$.

The group $Sym(X)$ is universal in the following sense: if $(\Gamma, \cdot)$ is an abstract group such that

$$card(\Gamma) \leq card(X),$$

then there exists a subgroup of $Sym(X)$ isomorphic to $(\Gamma, \cdot)$.

Suppose that I is a set and g is a function with $dom(g) = I$. Then we say that

$$(g(i))_{i \in I} \quad (or \; \{g(i) \; : \; i \in I\})$$

is a family of elements indexed by I. In this case, we also say that I is the set of indices of the family mentioned above.

Moreover, suppose that E is a fixed set and, for each index $i \in I$, the element $g(i)$ coincides with a subset F_i of E. Then we say that $\{F_i \; : \; i \in I\}$ is an indexed family of subsets of E. Actually, in such a case, we have a certain mapping

$$F \; : \; I \to \mathcal{P}(E)$$

where $\mathcal{P}(E)$ denotes the family of all subsets of E. A mapping of this type is usually called a set–valued mapping (or a multi–valued mapping). As we know, the graph of F is the set

$$\{(i, F(i)) \; : \; i \in I\} \subset I \times \mathcal{P}(E).$$

But if we treat F as a set–valued mapping, then it is sometimes useful to consider another notion of the graph of F. Namely, the graph of a set–valued mapping F is the following set:

$$\{(i, e) \in I \times E \; : \; e \in F(i)\}.$$

Obviously, the concept of a set-valued mapping is more general (in some sense) than the concept of an ordinary mapping. Indeed, every ordinary mapping

$$f \; : \; X \to Y$$

can be regarded as a set-valued mapping

$$F_f \; : \; X \to \mathcal{P}(Y)$$

of a special type; namely, for any element $x \in X$, we put $F_f(x) = \{f(x)\}$. In this way we come to a canonical one-to-one correspondence between ordinary mappings f acting from X into Y and set-valued mappings

$$F : X \to \mathcal{P}(Y)$$

satisfying the condition

$$(\forall x \in X)(card(F(x)) = 1).$$

Evidently, there are various set-valued mappings canonically associated with the given ordinary mapping $f : X \to Y$. For instance, we can define a set-valued mapping

$$F^f : Y \to \mathcal{P}(X)$$

by the following formula:

$$F^f(y) = f^{-1}(y) \qquad (y \in Y).$$

Let $\{X_i : i \in I\}$ be a family of sets. In the usual way we define the union

$$\cup\{X_i : i \in I\} = \{x : (\exists i \in I)(x \in X_i)\}$$

of this family. If $I \neq \emptyset$, then we may define the intersection

$$\cap\{X_i : i \in I\} = \{x : (\forall i \in I)(x \in X_i)\}$$

of this family. Further, if J is an arbitrary subset of I, then $\{X_i : i \in J\}$ is called a subfamily of the family $\{X_i : i \in I\}$ (in fact, a subfamily of a given family is some restriction of the function which determines this family).

We recall that a family of sets $\{X_i : i \in I\}$ is disjoint if the equality

$$X_i \cap X_j = \emptyset$$

holds for all indices $i \in I$, $j \in I$, $i \neq j$.

Exercise 5. Show that there is a canonical one-to-one correspondence between ordinary mappings

$$f : X \to Y$$

and set-valued mappings

$$F : Y \to \mathcal{P}(X)$$

satisfying the condition that the family $\{F(y) : y \in Y\}$ forms a disjoint covering of X.

We say that a family of elements $\{x_i \ : \ i \in I\}$ is a selector of a family of sets $\{X_i \ : \ i \in I\}$ if the relation

$$(\forall i \in I)(x_i \in X_i)$$

holds. A selector $\{x_i \ : \ i \in I\}$ is called injective if the corresponding function

$$i \to x_i \quad (i \in I, \ x_i \in X_i)$$

is injective.

The set of all selectors of a given family $\{X_i \ : \ i \in I\}$ is called the Cartesian product of this family and denoted by the symbol

$$\prod\{X_i \ : \ i \in I\}.$$

Let X and Y be any two sets. Then the symbol Y^X denotes the set of all mappings acting from X into Y. Obviously, the set Y^X can be regarded as a particular case of the Cartesian product of a family of sets.

In fact, the Axiom of Choice states that the relation

$$(\forall i \in I)(X_i \neq \emptyset)$$

implies the relation

$$\prod\{X_i \ : \ i \in I\} \neq \emptyset.$$

Kelley showed that, in the theory **ZF**, the following two statements are equivalent:

(a) the Axiom of Choice;

(b) the product space of an arbitrary family of quasicompact topological spaces is quasicompact.

Exercise 6. Prove in the theory **ZF** that statements (a) and (b) formulated above are equivalent (we recall that a topological space E is quasicompact if any open covering of E contains a finite subcovering of E). Note that the implication (a) $\Rightarrow$ (b) is widely known in general topology. In order to establish the converse implication, consider a family $(X_i)_{i \in I}$ of nonempty sets and take any element x such that

$$x \notin \cup\{X_i \ : \ i \in I\}.$$

Further, for each index $i \in I$, define

$$X_i' = X_i \cup \{x\}$$

and equip X_i' with the topology $\mathcal{T}_i$ where

$$\mathcal{T}_i = \{\emptyset, \ X_i', \ \{x\}\}.$$

In this way, all X_i' become quasicompact spaces. Finally, apply (b) to the product space of the family $(X_i')_{i \in I}$.

Remark 1. A topological space is compact if it is quasicompact and Hausdorff simultaneously. At the present time, it is known that the following two assertions are not equivalent in **ZF**:

(a) the Axiom of Choice;

(b') the product space of an arbitrary family of compact spaces is compact.

However, Sierpiński showed in the theory **ZF** & **DC** that the compactness of the product space $\{0,1\}^{\mathbf{R}}$, where $\{0,1\}$ is equipped with the discrete topology, implies the existence of a subset of $\mathbf{R}$ nonmeasurable in the Lebesgue sense (cf. [192], [107]).

The countable form of the Axiom of Choice is the restriction of this axiom only to countable families of sets:

If $\{X_n : n \in \omega\}$ is an arbitrary countable family of nonempty pairwise disjoint sets, then there exists a selector of $\{X_n : n \in \omega\}$.

Obviously, the countable form of the Axiom of Choice is a rather weak version of this axiom. It is not difficult to show, in the theory **ZF**, that:

(1) **DC** implies the countable form of **AC**;

(2) the countable form of **AC** is sufficient to prove the equivalence of the Cauchy and Heine definitions of the continuity of a partial function $f : \mathbf{R} \to \mathbf{R}$ at a point $t \in dom(f)$;

(3) the countable form of **AC** implies that the union of a countable family of countable sets is a countable set.

Exercise 7. Prove the above-mentioned assertions (1), (2) and (3) in the theory **ZF**.

We say that a set X is infinite if, for every natural number n, we have $card(X) \neq n$.

Verify that in **ZF** the countable form of **AC** implies the following statement: a set X is infinite if and only if there exists an injection from ω into X.

Actually, the countable form of the Axiom of Choice is completely sufficient for classical mathematical analysis, classical Euclidean geometry and even for the elementary topology of point sets.

Exercise 8. Let E be a metric space, E' be another metric space and let $f : E \to E'$ be a function. Suppose that E contains an everywhere dense well ordered subset. Show that the following two assertions are equivalent within the theory **ZF**:

(a) f is continuous (on E) in the Cauchy sense;

(b) f is continuous (on E) in the Heine sense.

Deduce from this fact that if E is separable (in particular, if $E = \mathbf{R}$), then the equivalence of (a) and (b) does not rely on the Axiom of Choice, i.e., this equivalence is provable in the theory **ZF**.

Observe that the implication (a) $\Rightarrow$ (b) is also provable in **ZF**.

Exercise 9. Verify that the following two assertions are valid in **ZF**:

(a) every infinite bounded subset of $\mathbf{R}$ has at least one accumulation point;

(b) every continuous (in the Cauchy sense) function

$$f : [t_1, t_2] \to \mathbf{R}$$

is uniformly continuous (where $[t_1, t_2]$ is an arbitrary closed bounded subinterval of $\mathbf{R}$).

Infer from (b) that f is bounded on $[t_1, t_2]$.

Now, we are going to present some notions and facts from general topology (these facts will be needed for our further purposes).

We recall that a topological space is any pair $(E, \mathcal{T})$ where E is a basic set and $\mathcal{T}$ is a topology (or a topological structure) defined on E. If $\mathcal{T}$ is fixed in our considerations, then we simply say that E is a topological space.

Let E be a topological space and let X be a subset of E. We put:

$cl(X) = $ the closure of the set X;

$int(X) = $ the interior of the set X;

$bd(X) = $ the boundary of the set X.

Consequently, closed subsets of E are all those sets $X \subset E$ for which we have $cl(X) = X$, and open subsets of E are all those sets $X \subset E$ for which we have $int(X) = X$.

We say that a set $X \subset E$ is an F_σ–subset of E if X can be represented as the union of a countable family of closed subsets of E.

We say that a set $Y \subset E$ is a G_δ–subset of E if Y can be represented as the intersection of a countable family of open subsets of E.

The Borel σ–algebra of a space E is the σ–algebra of subsets of E, generated by the family of all open subsets of E. This σ–algebra is denoted by the symbol $\mathcal{B}(E)$. Obviously, we may say that $\mathcal{B}(E)$ is generated by the family of all closed subsets of E. Elements of $\mathcal{B}(E)$ are called Borel subsets of a space E.

In many cases below we assume (without special pointing out) that, for topological spaces E which are under consideration, the following property is valid: all singletons in E are closed (hence, Borel in E).

Quasicompact (and compact) spaces were already introduced in this chapter. We also recall that a topological space E is Lindelöf if any open covering of E contains a countable subcovering. Evidently, every quasicompact space is Lindelöf. In addition, any topological space with a countable base is Lindelöf (and, moreover, is hereditarily Lindelöf).

Exercise 10. Let E be a topological space and let

$$G = \{g_i : i \in I\}$$

be a family of mappings acting from E into E. We say that a set $A \subset E$ is G-invariant in E if

$$A = cl(\cup\{g_i(A) : i \in I\}).$$

In particular, such an A is always closed in E.

By using the Zorn Lemma or the method of transfinite induction, show that if E is nonempty and quasicompact, then there exists a subset X of E satisfying the following relations:

(a) X is nonempty and G-invariant;

(b) for any G-invariant set $Y \subset E$, we have either $X \subset Y$ or $X \cap Y = \emptyset$.

Let us underline that in this exercise the given mappings g_i $(i \in I)$ are not assumed to be continuous.

We say that a topological space E is Polish if E is homeomorphic to a complete separable metric space.

We have the following topological characterization of all Polish spaces: a topological space E is Polish if and only if it is homeomorphic to a G_δ–subset of the Hilbert cube $[0,1]^\omega$ (see, e.g., [120]).

We say that a metrizable topological space E is an analytic (or Suslin) space if E can be represented as a continuous image of a Polish space. There is also another definition of analytic spaces starting with the so-called (A)–operation applied to the family of all closed subsets of a Polish space (for the concept of (A)-operation and its basic properties, see [120] or [124]).

Let X be a Polish topological space. The family of all analytic subsets of X is denoted by the symbol $\mathcal{A}(X)$. This family is closed under countable unions and countable intersections. Moreover, we have the inclusion

$$\mathcal{B}(X) \subset \mathcal{A}(X).$$

If a given Polish space X is uncountable, then the inclusion mentioned above is proper. This classical result is due to Suslin (for the proof, see [120] or Chapter 1 of the present book).

Another important result, due to Alexandrov and Hausdorff, states that any uncountable analytic set $A \subset X$ contains a subset which is homeomorphic to the Cantor discontinuum $\{0,1\}^\omega$. Consequently, the equality $card(A) = \mathbf{c}$ holds. In particular, for each uncountable Borel set $B \subset X$, we also have the equality $card(B) = \mathbf{c}$ (the proofs of these statements can be found in the same book [120]).

Let E be a topological space and let μ be a measure given on E. We recall that μ is a Borel measure (on E) if

$$dom(\mu) = \mathcal{B}(E).$$

We also say that μ is a Radon measure on a Hausdorff topological space E if μ is σ-finite, $dom(\mu) = \mathcal{B}(E)$ and, for each Borel subset B of E, the relation

$$\mu(B) = sup\{\mu(K) \ : \ K \subset B, \ K \ is \ compact\}$$

holds true.

Finally, we say that a Hausdorff topological space E is a Radon space if every σ-finite Borel measure on E is Radon.

According to the well-known result from topological measure theory, all Polish spaces turn out to be Radon spaces. More generally, any analytic space is Radon. In addition, if E is a Polish space and A is an arbitrary analytic subset of E, then the space $E \setminus A$ is Radon, too. In other words, all co-analytic subsets of a Polish space are Radon. For more information about Radon measures and Radon spaces, see, e.g., [21]. These measures and spaces play an important role in various questions of analysis and probability theory.

Let X and Y be any two topological spaces and let f be a mapping acting from X into Y. We say that f is a Borel mapping if, for each Borel subset B of Y, the preimage $f^{-1}(B)$ is a Borel subset of X.

Clearly, every continuous mapping acting from X into Y is a Borel mapping. Also, the composition of Borel mappings is a Borel mapping.

We say that a mapping $f : X \to Y$ is a Borel isomorphism from X onto Y if f is a bijection and both mappings f and f^{-1} are Borel. In this case, we say that the spaces X and Y are Borel isomorphic.

Obviously, if two topological spaces are homeomorphic, then they are Borel isomorphic. The converse assertion is not true in general.

Let Z be a Borel subset of a Polish topological space, let Y be a metrizable topological space and let f be an injective Borel mapping acting from Z into Y. Then the image $f(Z)$ is a Borel subset of Y. In particular, we see that the family of all Borel subsets of a Polish space E is invariant under the family of all injective Borel mappings acting from E into E.

More generally, let Z be a Borel subset of a Polish space, Y be a metrizable topological space and let

$$f : Z \to Y$$

be a Borel mapping such that

$$(\forall y \in Y)(card(f^{-1}(y)) \leq \omega).$$

Then the set $f(Z)$ is Borel in Y.

The last fact will be essentially applied in Chapter 14 of this book, which is devoted to bad functions given on second category sets.

The above-mentioned facts are rather deep theorems of classical descriptive set theory and their proofs are heavily based on the so-called separation principle of analytic sets (for details, see [84], [120], or [133]).

We also have the following important result (see, e.g., [120]). Let X and Y be any two uncountable Borel subsets of a Polish topological space E. Then there exists a Borel isomorphism from X onto Y.

Because all infinite countable subsets of a Polish space E are Borel sets, trivially being Borel isomorphic, we can easily deduce from the result mentioned above that, for any Borel subsets X and Y of E, these two conditions are equivalent:

(1) $card(X) = card(Y)$;

(2) X and Y are Borel isomorphic.

Unfortunately, we do not have (in the theory **ZFC**) an analogous nice equivalence for analytic subsets of Polish topological spaces.

Notice also that if Z is a Borel subset of a Polish topological space and f is a Borel mapping acting from Z into a Polish space E, then the image $f(Z)$ is an analytic subset of E.

The classical theory of Borel subsets and analytic subsets of Polish topological spaces is considered in detail in the fundamental monograph by Kuratowski [120] (see also [71], [72], [84], [124] and [133]).

Let E be an arbitrary Polish topological space. We define the classes

$$\mathcal{P}r_0(E), \quad \mathcal{P}r_1(E), \quad ..., \quad \mathcal{P}r_n(E), \quad ...$$

of subsets of E by recursion. Namely, first of all, we put

$$\mathcal{P}r_0(E) = \mathcal{B}(E).$$

Suppose now that, for a natural number $n > 0$, the class $\mathcal{P}r_{n-1}(E)$ has already been defined. If n is an odd number, then, by definition, $\mathcal{P}r_n(E)$ is the class of all continuous images (in E) of sets from the class $\mathcal{P}r_{n-1}(E)$. If n is an even number, then, by definition, $\mathcal{P}r_n(E)$ is the class of all complements of the sets from the class $\mathcal{P}r_{n-1}(E)$.

Finally, we put

$$\mathcal{P}r(E) = \cup\{\mathcal{P}r_n(E) \ : \ n < \omega\}.$$

Sets from the class $\mathcal{P}r(E)$ are called projective subsets of a space E. The notion of a projective set was introduced by Luzin and, independently, by Sierpiński. At the present time, there are many remarkable works devoted to the theory of projective sets. Elements of this theory are presented in the monograph by Kuratowski mentioned above (a more detailed discussion of this subject can be found in [71], [72], [84], [124] and [133]).

Thus we conclude that Borel subsets of E (i.e., sets from the class $\mathcal{P}r_0(E)$) and analytic subsets of E (i.e., sets from the class $\mathcal{P}r_1(E)$) are very particular cases of projective sets.

Note that many natural problems concerning projective sets cannot be solved in the theory **ZFC**. For example, the following statements are true:

(a) it cannot be proved, in **ZFC**, that each uncountable set from the class $\mathcal{P}r_2(\mathbf{R})$ contains a subset homeomorphic to the Cantor discontinuum;

(b) it cannot be proved, in **ZFC**, that each set from the class $\mathcal{P}r_3(\mathbf{R})$ is measurable in the Lebesgue sense.

Note that the statement analogous to (b) and concerning the Baire property of sets from the class $\mathcal{P}r_3(\mathbf{R})$ is true, too.

Now, let us recall some elementary facts about the Baire property of subsets of general topological spaces.

Let E be an arbitrary topological space.

We say that a set $X \subset E$ is nowhere dense (in E) if

$$int(cl(X)) = \emptyset.$$

For example, if V is an open subset of E, then the set $bd(V)$ is nowhere dense in E.

We say that a set $X \subset E$ is a first category subset of E (or X is a meager subset of E) if X can be represented in the form

$$X = \cup\{X_n \ : \ n \in \omega\},$$

where all X_n $(n \in \omega)$ are nowhere dense subsets of E.

We say that a set $X \subset E$ is of second category in E if X is not of first category in E.

Finally, we say that a set $X \subset E$ is residual (co-meager) in E if the set $E \setminus X$ is of first category in E.

Exercise 11. Let E be a topological space and let X and Y be two subsets of E. Show that if $X \cap cl(Y)$ is nowhere dense in $cl(Y)$, then $X \cap Y$ is nowhere dense in Y.

Deduce from this fact that if $X \cap cl(Y)$ is of first category in $cl(Y)$, then $X \cap Y$ is of first category in Y.

The family of all first category subsets of E is denoted by the symbol $\mathcal{K}(E)$. If E is not a first category space, then $\mathcal{K}(E)$ forms a certain σ–ideal of subsets of E, which plays an important role in many questions of analysis and topology.

We say that a set $X \subset E$ has the Baire property (in E) if X can be represented in the form

$$X = (U \cup Y) \setminus Z,$$

where U is an open subset of E and both Y and Z are first category subsets of E. It is easy to check that a set $X \subset E$ has the Baire property if and only if X can be represented in the form $X = V \triangle P$ where V is an open subset of E and P is a first category subset of E.

The family of all subsets of a space E, having the Baire property (in E), is denoted by the symbol $\mathcal{B}a(E)$. Obviously, $\mathcal{B}a(E)$ coincides with the σ–algebra of subsets of E generated by the family $\mathcal{T}(E) \cup \mathcal{K}(E)$, where $\mathcal{T}(E)$ is the topology of E (i.e., the family of all open subsets of E). Hence we have the inclusion

$$\mathcal{B}(E) \subset \mathcal{B}a(E).$$

As a rule, this inclusion is proper. But there are some interesting examples of topological spaces E for which this inclusion becomes the equality. For instance, if E is a classical Luzin subset of the real line $\mathbf{R}$, everywhere dense in $\mathbf{R}$, then we have

$$\mathcal{K}(E) = [E]^{\leq \omega}.$$

Consequently, in this case, we easily obtain $\mathcal{B}(E) = \mathcal{B}a(E)$. Extensive information on Luzin subsets of the real line is contained in [119], [120], [149] and [162]. We shall deal with Luzin sets in the subsequent sections of our book. At this moment, we only wish to notice that the existence of Luzin subsets of $\mathbf{R}$ cannot be proved in the theory **ZFC**. On the other hand, the existence of such subsets of $\mathbf{R}$ easily follows from the Continuum Hypothesis (see Chapter 10).

Another interesting example (in the theory **ZFC**) of a topological space E, for which the equality $\mathcal{B}(E) = \mathcal{B}a(E)$ holds, can be obtained if one takes the set of all real numbers equipped with the so-called density topology (see information on this topology in [162] and [219]). We shall consider below some elementary properties of the density topology.

It can be proved that, under some natural assumptions, the Baire property and the measurability of sets are preserved under the (A)-operation (see, for instance, [120]). This gives, in particular, that all analytic subsets of a Polish space E possess the Baire property and are universally measurable in E (the latter means that they are measurable with respect to the completion of any σ-finite Borel measure on E).

Concerning measure and category, let us mention two important consequences of Martin's Axiom that will be often applied in this book. Namely, under this axiom, the σ-ideal $\mathcal{K}(\mathbf{R})$ of all first category subsets of $\mathbf{R}$ and the σ-ideal $\mathcal{I}(\lambda)$ of all Lebesgue measure zero subsets of $\mathbf{R}$ are $\mathbf{c}$-additive. In other words, the union of strictly less than continuumly many members of $\mathcal{K}(\mathbf{R})$ (respectively, of $\mathcal{I}(\lambda)$) is again a member of $\mathcal{K}(\mathbf{R})$ (respectively, of $\mathcal{I}(\lambda)$).

This also yields, under the same axiom, that the σ-algebra $\mathcal{B}a(\mathbf{R})$ of subsets of $\mathbf{R}$ with the Baire property and the σ-algebra $dom(\lambda)$ of all Lebesgue measurable subsets of $\mathbf{R}$ are $\mathbf{c}$-additive (for more details, see [119]).

Let E be again an arbitrary topological space and let X be a subset of E. We say that X has the Baire property in the restricted sense if, for each subspace Y of E, the set $X \cap Y$ has the Baire property in the space Y.

Clearly, the family of all subsets of a space E, having the Baire property in the restricted sense, is a σ–algebra of subsets of E. We denote this σ–algebra by $\mathcal{B}ar(E)$. Obviously, we have the inclusion

$$\mathcal{B}ar(E) \subset \mathcal{B}a(E).$$

It is also easy to check that $\mathcal{B}(E) \subset \mathcal{B}ar(E)$. Moreover, it can be shown that

$$\mathcal{A}(E) \subset \mathcal{B}ar(E),$$

i.e., all analytic subsets of E have the Baire property in the restricted sense (see, for instance, [120]).

Exercise 12. Let E be a topological space and let X be a subset of E. Prove that the following three assertions are equivalent:

(a) X has the Baire property in the restricted sense;

(b) for any closed set $F \subset E$, the set $F \cap X$ has the Baire property in F;

(c) for any perfect set $P \subset E$, the set $P \cap X$ has the Baire property in P.

Note that, for the class of all complete metric spaces, the Baire property in the restricted sense is a topological invariant, i.e., if E and E' are two complete metric spaces, $X \subset E$ has the Baire property in the restricted sense and $Y \subset E'$ is a homeomorphic image of X, then Y also has the Baire property in the restricted sense (see [120]).

Let X and Y be any two topological spaces and let f be a mapping acting from X into Y. We say that f has the Baire property if, for each Borel subset B of Y, the set $f^{-1}(B)$ has the Baire property in X.

Evidently, every Borel mapping acting from X into Y has the Baire property.

The composition of two mappings, each of which has the Baire property, can be a mapping without the Baire property. Moreover, it is not difficult to give an example of two functions

$$f : \mathbf{R} \to \mathbf{R}, \qquad g : \mathbf{R} \to \mathbf{R},$$

each of which has the Baire property, but their composition $g \circ f$ does not possess this property (an analogous phenomenon can be observed for the composition of two Lebesgue measurable functions acting from $\mathbf{R}$ into $\mathbf{R}$).

However, if X, Y, Z are three topological spaces, $f : X \to Y$ has the Baire property and $g : Y \to Z$ is a Borel mapping, then $g \circ f$ has the Baire property, too.

In a similar way we can define a mapping with the Baire property in the restricted sense. Namely, we say that $f : X \to Y$ has the Baire property in the restricted sense if, for each Borel subset B of Y, the set $f^{-1}(B)$ is a subset of X having the Baire property in the restricted sense.

It is easy to see that all Borel mappings have the Baire property in the restricted sense.

Exercise 13. Let X be a topological space, Y be a topological space with a countable base and let

$$f : X \to Y$$

be a function. Show that the following two assertions are equivalent:

(a) f possesses the Baire property;

(b) there exists a first category set $A \subset X$ such that the restricted function $f|(X \setminus A)$ is continuous.

Let X and Y be any two topological spaces and let

$$F : X \to \mathcal{P}(Y)$$

be a set-valued mapping. We say that F has closed graph if

$$G_F = \{(x, y) \in X \times Y : y \in F(x)\}$$

is a closed subset of the product space $X \times Y$.

It is clear that if the given set-valued mapping F has closed graph, then, for each element $x \in X$, the set $F(x)$ is a closed subset of the space Y. The converse assertion is not true in general.

Set-valued mappings with closed graphs are important in different domains of mathematics, especially, in those questions which concern the existence of fixed points of set-valued mappings (we recall that an element $x \in dom(F)$ is a fixed point for the set-valued mapping F if $x \in F(x)$).

Note that theorems on the existence of fixed points for set-valued mappings found many interesting applications (see, e.g., [51] or [213]).

Let X and Y be again two topological spaces and let

$$F : X \to \mathcal{P}(Y)$$

be a set-valued mapping. We say that F is lower semicontinuous if the following conditions hold:

(1) for each point $x \in X$, the set $F(x)$ is closed in Y;

(2) for each open subset V of Y, the set

$$F^{-1}(V) = \{x \in X \ : \ F(x) \cap V \neq \emptyset\}$$

is open in X.

There are certain similarities between set-valued mappings with closed graphs and lower semicontinuous set-valued mappings (cf., for example, Theorem 1 below).

Exercise 14. Let X be a topological space and let f be a real-valued function on X. We recall that f is lower (respectively, upper) semicontinuous if, for any real number t, the set $\{x \in X \ : \ f(x) > t\}$ (respectively, the set $\{x \in X \ : \ f(x) < t\}$) is open in X. Show that:

(a) f is lower semicontinuous if and only if $-f$ is upper semicontinuous;

(b) f is lower semicontinuous if and only if the set

$$G^*(f) = \{(x,t) \in X \times \mathbf{R} \ : \ f(x) \leq t\}$$

is closed in the product space $X \times \mathbf{R}$;

(c) f is lower semicontinuous if and only if, for each $x_0 \in X$, we have the equality

$$liminf_{x \to x_0} f(x) = f(x_0);$$

(d) f is continuous if and only if it is lower and upper semicontinuous;

(e) if f is lower semicontinuous, g is continuous and $ran(g) \subset X$, then $f \circ g$ is lower semicontinuous on $dom(g)$;

(f) if $k \geq 0$ and f, g are lower semicontinuous on X, then kf and $f + g$ are lower semicontinuous on X.

Exercise 15. Let X be a set and let f be a real-valued function on X. We introduce two set-valued mappings

$$F_{1,f} : X \to \mathcal{P}(\mathbf{R}),$$

$$F_{2,f} : X \to \mathcal{P}(\mathbf{R})$$

by the following formulas:

$$F_{1,f}(x) = \ ]-\infty, f(x)] \quad (x \in X),$$

$$F_{2,f}(x) = [f(x), +\infty[\quad (x \in X).$$

Suppose now that X is a topological space. Show that the following two assertions are equivalent:

(a) f is lower semicontinuous (as an ordinary function);
(b) $F_{1,f}$ is lower semicontinuous (as a set-valued function).
Show that the next two assertions are also equivalent:
(c) f is upper semicontinuous (as an ordinary function);
(d) $F_{2,f}$ is lower semicontinuous (as a set-valued function).

Exercise 16. Let X be a nonempty quasicompact topological space and let

$$f : X \to \mathbf{R}$$

be a lower semicontinuous function. Show that there exists a point $x_0 \in X$ satisfying the relation

$$f(x_0) = inf_{x \in X} f(x).$$

Formulate and prove an analogous result for upper semicontinuous real-valued functions defined on X.

Exercise 17. Let X and Y be any two topological spaces and let

$$f : X \to \mathbf{R}, \quad g : Y \to \mathbf{R}$$

be any two lower (respectively, upper) semicontinuous functions such that

$$f(x) \geq 0, \quad g(y) \geq 0 \quad (x \in X, \ y \in Y).$$

Define a function

$$h : X \times Y \to \mathbf{R}$$

by the formula

$$h(x, y) = f(x) \cdot g(y) \qquad (x \in X, \; y \in Y).$$

Show that h is also lower (respectively, upper) semicontinuous.

Exercise 18. Let X be a completely regular topological space (for the definition, see, e.g., [120]) and let

$$f \; : \; X \to \mathbf{R}$$

be a lower semicontinuous function such that $f(x) \geq 0$ for all $x \in X$. Show that there exists a family $(f_i)_{i \in I}$ of functions acting from X into $\mathbf{R}$ and satisfying the following conditions:

(a) for each $i \in I$, the function f_i is continuous;

(b) for all $i \in I$ and for all $x \in X$, we have $f_i(x) \geq 0$;

(c) $f = sup_{i \in I} f_i$;

(d) $card(I) \leq w(X) + \omega$, where $w(X)$ denotes the topological weight of X (i.e., $w(X)$ is the smallest cardinality of a base of X).

In particular, if X has a countable base (i.e., X is metrizable), then f can be represented as a pointwise limit of an increasing sequence of positive continuous real-valued functions defined on X.

Exercise 19. Let $[a, b]$ be a closed subinterval of $\mathbf{R}$ and let

$$f \; : \; [a, b] \to \mathbf{R}, \qquad g \; : \; [a, b] \to \mathbf{R}$$

be two functions such that $f \geq g$. Suppose, in addition, that f is lower semicontinuous and g is upper semicontinuous. Demonstrate that there exists a continuous function

$$h \; : \; [a, b] \to \mathbf{R}$$

satisfying the inequalities

$$g \leq h \leq f.$$

This simple result admits a number of generalizations and, actually, is a direct consequence of the well-known Michael theorem on continuous selectors (see [145], [146], or [173]).

Exercise 20. Let X be a second category topological space, $\{f_i \; : \; i \in I\}$ be a family of real-valued lower semicontinuous functions on X and suppose that, for each point $x \in X$, the set $\{f_i(x) \; : \; i \in I\}$ is bounded from above. Show that there exists a nonempty open set $V \subset X$ for which the set

$$\cup \{f_i(V) \; : \; i \in I\}$$

is bounded from above, too.

Formulate and prove an analogous result for upper semicontinuous functions.

Exercise 21. Let $(G, +)$ be a second category topological group and let $\{f_i : i \in I\}$ be a family of real-valued lower semicontinuous functions on G. Suppose that the following conditions hold:

(a) for any index $i \in I$, the function f_i is subadditive, i.e., we have

$$f_i(x + y) \leq f_i(x) + f_i(y) \qquad (x \in G, \ y \in G);$$

(b) for each point $x \in G$, the set $\{f_i(x) : i \in I\}$ is bounded from above.

Show that the given family $\{f_i : i \in I\}$ is locally bounded from above. This means that, for any point $x \in G$, there exists its neighborhood $V(x)$ for which the set

$$\cup \{f_i(V(x)) : i \in I\}$$

is bounded from above.

Formulate and prove an analogous statement for upper semicontinuous functions.

Notice that the result presented in Exercise 21 easily implies the well-known Banach-Steinhaus theorem (see [14] or [81]).

Exercise 22. Let E be a topological space, let

$$f : E \to \mathbf{R}$$

be a partial function bounded from above (respectively, from below) and suppose that $dom(f) \neq \emptyset$. For any point $x \in cl(dom(f))$, let us put

$$f^*(x) = limsup_{y \to x, y \in dom(f)} f(y),$$

and, respectively,

$$f_*(x) = liminf_{y \to x, y \in dom(f)} f(y).$$

Verify that:

(a) the function f^* is upper semicontinuous on $cl(dom(f))$;

(b) the function f_* is lower semicontinuous on $cl(dom(f))$.

Infer from (a) and (b) that:

(c) f is upper semicontinuous on $dom(f)$ if and only if f^* extends f;

(d) f is lower semicontinuous on $dom(f)$ if and only if f_* extends f.

Observe also that f admits a continuous extension on $cl(dom(f))$ if and only if the equality $f^* = f_*$ is valid.

Exercise 23. Let $[0,1]^\omega$ denote the Hilbert cube and let us introduce two mappings:

$$f_1 : [0,1]^\omega \to [0,1],$$

$$f_2 : [0,1]^\omega \to [0,1]$$

by putting

$$f_1(x) = inf(x_n)_{n<\omega}, \quad f_2(x) = sup(x_n)_{n<\omega} \quad (x = (x_n)_{n<\omega} \in [0,1]^\omega).$$

Verify that:

(a) f_1 is upper semicontinuous;

(b) f_2 is lower semicontinuous.

Let us remark that, as shown by van Mill and Pol, none of these two functions is countably continuous (see [148] and Chapter 3 of this book).

For our further considerations, we need one auxiliary proposition on closed projections. This proposition is due to Kuratowski (see, for instance, [120]) and has numerous applications in topology and analysis.

Lemma 1. *Let X be a topological space, let Y be a quasicompact space and let pr_1 denote the canonical projection from $X \times Y$ into X, i.e., the mapping*

$$pr_1 : X \times Y \to X$$

is defined by the formula

$$pr_1((x,y)) = x \quad ((x,y) \in X \times Y).$$

Then pr_1 is a closed mapping, i.e., for each closed subset A of $X \times Y$, the image $pr_1(A)$ is closed in X.

Proof. Take any point $x \in X$ such that $U(x) \cap pr_1(A) \neq \emptyset$ for all neighborhoods $U(x)$ of x. We are going to show that $x \in pr_1(A)$. For this purpose, it is sufficient to establish that

$$(\{x\} \times Y) \cap A \neq \emptyset.$$

Suppose otherwise, i.e.,

$$(\{x\} \times Y) \cap A = \emptyset.$$

Then, for each point $y \in Y$, there exists an open neighborhood $W(x,y)$ of the point (x,y), satisfying the relation

$$W(x,y) \cap A = \emptyset.$$

We may assume, without loss of generality, that

$$W(x, y) = U(x) \times V(y),$$

where $U(x)$ is an open neighborhood of x and $V(y)$ is an open neighborhood of y. Because the space $\{x\} \times Y$ is quasicompact, there exists a finite sequence

$$(x, y_1), \ (x, y_2), \ \dots, \ (x, y_n)$$

of points from $\{x\} \times Y$, such that $\{W(x, y_i) \ : \ 1 \leq i \leq n\}$ is a finite covering of $\{x\} \times Y$. Now, let us put

$$U_i(x) = pr_1(W(x, y_i)),$$

$$O(x) = \cap\{U_i(x) \ : \ 1 \leq i \leq n\}.$$

Then it is easy to check that $O(x)$ is a neighborhood of the point x, satisfying the equality

$$O(x) \cap pr_1(A) = \emptyset.$$

But this is impossible. So we get a contradiction, and the Kuratowski lemma is proved.

Exercise 24. Show, e.g., for a metrizable topological space Y, that the property of Y described in Lemma 1 is equivalent to the compactness of Y. More precisely, prove that, for a given metric space Y, the following two assertions are equivalent:

(a) Y is compact;

(b) the canonical projection $pr_1 \ : \ \mathbf{R} \times Y \to \mathbf{R}$ is a closed mapping.

Now, we want to give some applications of Lemma 1 to set-valued mappings. For the sake of simplicity (and motivated by the aims of mathematical analysis), we restrict our further considerations to the class of all metric spaces, but it is not difficult to see that the results presented below remain true in more general situations.

Theorem 1. *Let X be a metric space, let Y be a compact metric space and let $F \ : \ X \to \mathcal{P}(Y)$ be a set-valued mapping. Then the following two assertions are equivalent:*

(1) F has closed graph;

(2) for each point $x \in X$, the set $F(x)$ is closed in Y and, for each closed subset A of Y, the set

$$F^{-1}(A) = \{x \in X \ : \ F(x) \cap A \neq \emptyset\}$$

is closed in X.

Proof. Suppose that assertion (1) is true. Then, obviously, for any element $x \in X$, the set $F(x)$ is closed in Y. Now, let A be an arbitrary closed subset of Y. It is easy to see that

$$F^{-1}(A) = pr_1((X \times A) \cap G_F),$$

where G_F denotes the graph of the given set-valued mapping F. Clearly, the set $(X \times A) \cap G_F$ is closed in the product space $X \times Y$, as the intersection of two closed subsets of this space. Therefore, by the Kuratowski lemma proved above, we obtain that $F^{-1}(A)$ is a closed subset of the space X. Thus, assertion (2) is true and the implication (1) $\Rightarrow$ (2) is established.

Now, assume that assertion (2) holds. Let us prove that G_F is a closed subset of the product space $X \times Y$. For this purpose, take an arbitrary sequence $\{(x_i, y_i) \ : \ i \in \mathbf{N}\}$ of points of the graph G_F, such that

$$lim_{i \to +\infty} \ (x_i, y_i) = (x, y) \in X \times Y.$$

Let us show that the point (x, y) also belongs to G_F. Suppose otherwise, i.e. $(x, y) \notin G_F$. This means, by the definition, that $y \notin F(x)$. Since $F(x)$ is a closed set in Y, there exists a neighborhood $V(y)$ of the point y, satisfying the relation

$$V(y) \cap F(x) = \emptyset.$$

Furthermore, we have

$$lim_{i \to +\infty} \ x_i = x, \quad lim_{i \to +\infty} \ y_i = y.$$

Without loss of generality, we may assume that $y_i \in V(y)$ for all indices $i \in \mathbf{N}$. Let us put

$$A = \{y\} \cup \{y_i \ : \ i \in \mathbf{N}\}.$$

Evidently, A is a closed subset of Y. It is also clear that $A \cap F(x) = \emptyset$. On the other hand, we have

$$y_i \in F(x_i), \quad x_i \in F^{-1}(A) \quad (i \in \mathbf{N}).$$

Taking into account that $lim_{i \to +\infty} \ x_i = x$, and that the set $F^{-1}(A)$ is closed in X, we get

$$x \in F^{-1}(A), \quad F(x) \cap A \neq \emptyset,$$

which yields a contradiction. It shows us that our set-valued mapping F has closed graph. Thus, the implication (2) $\Rightarrow$ (1) is established and the proof of Theorem 1 is completed.

Remark 2. One can easily see that, in the proof of Theorem 1, only the implication $(1) \Rightarrow (2)$ relies essentially on the Kuratowski lemma on closed projections. The converse implication $(2) \Rightarrow (1)$ does not need this lemma. So we can conclude the following fact.

Suppose that X and Y are arbitrary metric spaces and

$$F : X \to \mathcal{P}(Y)$$

is a set-valued mapping satisfying the conditions:

(a) $F(x)$ is a closed subset of Y for each point $x \in X$;

(b) $F^{-1}(A)$ is a closed subset of X for each closed set $A \subset Y$.

Then the set-valued mapping F has closed graph.

A particular case of this fact is the following one.

Let X and Y be any two metric spaces and let

$$f : X \to Y$$

be a continuous mapping. Then the graph

$$G_f = \{(x, y) \in X \times Y : y = f(x)\}$$

of this mapping is a closed subset of the product space $X \times Y$.

Notice that the converse assertion is not true in general. Indeed, it is not difficult to construct an example of a function $g : \mathbf{R} \to \mathbf{R}$ that is discontinuous but has closed graph. Moreover, there exists a function $g : \mathbf{R} \to \mathbf{R}$ with closed graph, such that the set of all points of discontinuity of g is a nonempty perfect subset of $\mathbf{R}$ (the function g with this property can be constructed by starting with the classical Cantor set on the real line). On the other hand, it is reasonable to notice here that any function $f : \mathbf{R} \to \mathbf{R}$ having closed graph belongs to the first Baire class (see Chapter 2). Therefore, according to a well-known Baire theorem (see, e.g., [6], [120], [154], [162] or Chapter 2), for each nonempty perfect set $A \subset \mathbf{R}$, there exists a point of A at which the function $f|A$ is continuous.

Exercise 25. Let X be a metric space, let Y be a compact metric space and let f be a mapping acting from X into Y. Deduce from Theorem 1 that the following two conditions are equivalent:

(a) f is a continuous mapping;

(b) the graph G_f is a closed subset of the product space $X \times Y$.

Exercise 26. Let X and Y be two compact metric spaces and let f be a mapping acting from X into Y. Deduce from the result of the previous exercise that these two conditions are equivalent:

(a) f is a continuous mapping;

(b) the graph G_f is a compact subset of the product space $X \times Y$.

Let us now present a typical application of the Kuratowski lemma to the classical theory of real functions. We mean here the existence of continuous nowhere differentiable functions. Let $C[0,1]$ denote, as usual, the family of all continuous real-valued functions defined on the unit segment $[0,1]$. This family becomes a separable Banach space with respect to the standard sup-norm.

The following famous result is due to Banach and Mazurkiewicz (see [12] and [143]).

Theorem 2. *The family of all those functions from the Banach space $C[0,1]$, which are nowhere differentiable on $[0,1]$, is the complement of a first category set.*

Proof. Let h be a nonzero rational number such that $|h| < 1$. For every natural number $n > 0$, consider the set

$$\Phi_{h,n} = \{f \in C[0,1] \ : \ (\exists x \in [0,1])(\forall \delta)(0 < |\delta| < |h| \Rightarrow$$

$$|(f(x+\delta) - f(x))/\delta| \leq n)\}.$$

It is not hard to check that $\Phi_{h,n}$ is a closed subset of the space $C[0,1]$. Indeed, let us put

$$Z_{h,n} = \{(f,x) \in C[0,1] \times [0,1] \ : \ (\forall \delta)(0 < |\delta| < |h| \Rightarrow$$

$$|(f(x+\delta) - f(x))/\delta| \leq n)\}.$$

Then $Z_{h,n}$ is a closed subset of the product space $C[0,1] \times [0,1]$ and

$$\Phi_{h,n} = pr_1(Z_{h,n}),$$

where

$$pr_1 \ : \ C[0,1] \times [0,1] \to C[0,1]$$

denotes the canonical projection onto $C[0,1]$. Taking account of the compactness of the unit segment $[0,1]$ and applying Lemma 1, we immediately obtain that the set $\Phi_{h,n}$ is closed in $C[0,1]$. Simultaneously, $\Phi_{h,n}$ is nowhere dense in $C[0,1]$ (the latter fact is almost trivial from the geometrical point of view). Consequently, the set

$$D = \cup\{\Phi_{h,n} \ : \ h \in \mathbf{Q} \setminus \{0\}, \ |h| < 1, \ n \in \mathbf{N} \setminus \{0\}\}$$

is of first category in $C[0,1]$. Now, it is clear that any function belonging to the set $C[0,1] \setminus D$ is nowhere differentiable on $[0,1]$. This completes the proof of Theorem 2.

Exercise 27. Let f be a function acting from $\mathbf{R}$ into $\mathbf{R}$ and let x be a point of $\mathbf{R}$. Recall that f possesses a symmetric derivative at the point x if there exists a (finite) limit

$$lim_{h \to 0, h \neq 0} \frac{f(x+h) - f(x-h)}{2h}.$$

In such a case, this limit is called the symmetric derivative of f at x (denoted by the symbol $f'_s(x)$).

Demonstrate that f can possess a symmetric derivative at a point x being even discontinuous at this point.

Show that if f is differentiable (in the usual sense) at a point x, then there exists a symmetric derivative $f'_s(x)$ and the equality $f'(x) = f'_s(x)$ is fulfilled. Show also that the converse assertion is not true in general.

In addition, investigate the question whether a direct analogue of Theorem 2 holds for the symmetric derivative (instead of the derivative in the usual sense).

The notion of a symmetric derivative of a function can be regarded as a simple example of the concept of a generalized derivative. In the subsequent sections of the book we shall discuss some other types of a generalized derivative. The notion of an approximate derivative (introduced by Khinchin in 1914) is of special interest and will be defined and discussed in Chapter 6. It is well known that this notion plays an important role in various questions of real analysis (for instance, in the theory of generalized integrals). The definition of an approximate derivative relies on the concept of a density point for a given Lebesgue measurable subset of $\mathbf{R}$. Let λ denote the standard Lebesgue measure on $\mathbf{R}$ and let X be an arbitrary λ-measurable subset of $\mathbf{R}$. We say that $x \in \mathbf{R}$ is a density point for X if

$$lim_{h \to 0, h > 0} \frac{\lambda(X \cap [x-h, x+h])}{2h} = 1.$$

The classical theorem of real analysis, due to Lebesgue, states that almost all (with respect to λ) points of X are its density points. In order to establish this fact, we need the concept of a Vitali covering of a set lying in $\mathbf{R}$, and the important result of Vitali concerning such coverings. For the sake of completeness, we formulate and prove this result.

Let $\{D_i \ : \ i \in I\}$ be a family of nondegenerate segments on $\mathbf{R}$ and let Z be a subset of $\mathbf{R}$. We say that this family is a Vitali covering of Z if, for each point $z \in Z$, we have

$$inf\{\lambda(D_i) \ : \ i \in I, \ z \in D_i\} = 0.$$

The following fundamental result was obtained by Vitali (cf., e.g., [154], [162] or [180]).

Theorem 3. *If Z is a subset of $\mathbf{R}$ and $\{D_i : i \in I\}$ is a Vitali covering of Z, then there exists a countable set $J \subset I$ such that the partial family $\{D_j : j \in J\}$ is disjoint and*

$$\lambda(Z \setminus \cup \{D_j : j \in J\}) = 0.$$

Proof. Without loss of generality we may assume that Z is bounded. Let U be an open bounded set in $\mathbf{R}$ containing Z. We may also assume that $D_i \subset U$ for each index $i \in I$. Define by recursion a disjoint countable subfamily of segments

$$D_{i(0)}, \; D_{i(1)}, \; \dots, \; D_{i(k)}, \; \dots \; .$$

Take $D_{i(0)}$ arbitrarily. Suppose that $D_{i(0)}$, $D_{i(1)}$, $\dots$, $D_{i(k)}$ have already been defined. Put

$$t(k) = sup\{\lambda(D_i) : D_i \subset U \setminus (D_{i(0)} \cup \dots \cup D_{i(k)})\}.$$

Let $D_{i(k+1)}$ be a segment from $\{D_i : i \in I\}$ such that

$$D_{i(k+1)} \subset U \setminus (D_{i(0)} \cup \dots \cup D_{i(k)}), \quad \lambda(D_{i(k+1)}) \geq t(k)/2.$$

In this way we obtain the desired disjoint sequence $\{D_{i(k)} : k \in \mathbf{N}\}$. Note that

$$\sum_{k \in \mathbf{N}} \lambda(D_{i(k)}) \leq \lambda(U) < +\infty,$$

so we have the equality

$$lim_{k \to +\infty} \lambda(D_{i(k)}) = 0.$$

We are going to show that

$$\{D_j : j \in J\} = \{D_{i(k)} : k \in \mathbf{N}\}$$

is the required subfamily. For this purpose, denote by $D'_{i(k)}$ the segment in $\mathbf{R}$ whose center coincides with the center of $D_{i(k)}$ and for which

$$\lambda(D'_{i(k)}) = 5\lambda(D_{i(k)}).$$

Let us demonstrate that, for each natural number n, the inclusion

$$Z \setminus \cup \{D_{i(k)} : k \in \mathbf{N}\} \subset \cup \{D'_{i(k)} : k \in \mathbf{N}, \; k > n\}$$

holds true. Indeed, let z be an arbitrary point from $Z \setminus \cup \{D_{i(k)} : k \in \mathbf{N}\}$. Then, in particular,

$$z \in Z \setminus (D_{i(0)} \cup ... \cup D_{i(n)}).$$

Because $\{D_i : i \in I\}$ is a Vitali covering of Z, there exists a segment D_i for which

$$z \in D_i, \quad D_i \cap (D_{i(0)} \cup ... \cup D_{i(n)}) = \emptyset.$$

Obviously, we have $\lambda(D_i) > 0$. At the same time, as mentioned above, the relation $lim_{k \to +\infty} \lambda(D_{i(k)}) = 0$ is valid. So, for some natural numbers k, we must have

$$D_i \cap D_{i(k+1)} \neq \emptyset.$$

Let k be the smallest natural number with this property. Evidently, $k \geq n$. Thus we get

$$D_i \cap D_{i(k+1)} \neq \emptyset, \quad D_i \cap (D_{i(0)} \cup ... \cup D_{i(k)}) = \emptyset.$$

In addition,

$$\lambda(D_i) \leq 2\lambda(D_{i(k+1)}),$$

which immediately implies (in view of the definition of $D'_{i(k+1)}$) the inclusion $D_i \subset D'_{i(k+1)}$. Consequently,

$$z \in D_i \subset D'_{i(k+1)} \subset \cup\{D'_{i(m)} : m \in \mathbf{N}, \ m > n\}.$$

Finally, for every natural number n, we have the inequality

$$\lambda(\cup\{D'_{i(m)} : m > n\}) \leq 5 \sum_{m>n} \lambda(D_{i(m)}),$$

from which we can conclude that

$$\lambda(Z \setminus \cup\{D_j : j \in J\}) = 0,$$

and the theorem is proved.

We shall present some standard applications of Theorem 3 in the subsequent sections of the book. Here we only want to recall how the above-mentioned Lebesgue result on density points of λ-measurable sets can be easily derived from Theorem 3.

Theorem 4. *Let X be an arbitrary λ-measurable set on $\mathbf{R}$ and let*

$$d(X) = \{x \in \mathbf{R} : x \ is \ a \ density \ point \ of \ X\}.$$

Then we have the equality

$$\lambda(X \setminus (X \cap d(X))) = 0.$$

Proof. We may assume, without loss of generality, that X is bounded. For any natural number $n > 0$, let us define

$$X_n = \{x \in X \; : \; liminf_{h \to 0, h > 0} \lambda(X \cap [x - h, x + h])/2h \; < 1 - 1/n\}.$$

Clearly, it suffices to show that $\lambda^*(X_n) = 0$ where λ^* denotes the outer measure associated with λ. For this purpose, fix $\varepsilon > 0$. Let U be an open subset of $\mathbf{R}$ such that

$$X_n \subset U, \quad \lambda(U) \leq \lambda^*(X_n) + \varepsilon.$$

In virtue of the definition of X_n, there exists a Vitali covering $\{D_i \; : \; i \in I\}$ of X_n such that

$$(\forall i \in I)(\frac{\lambda(D_i \cap X)}{\lambda(D_i)} < 1 - 1/n).$$

Obviously, we may suppose that $D_i \subset U$ for each index $i \in I$. According to Theorem 3, there exists a disjoint countable subfamily $\{D_j \; : \; j \in J\}$ of this covering, for which we have

$$\lambda(X_n \setminus \cup \{D_j \; : \; j \in J\}) = 0.$$

Then we can write

$$\lambda^*(X_n) \leq \sum_{j \in J} \lambda(X \cap D_j) \leq (1 - 1/n) \sum_{j \in J} \lambda(D_j)$$

$$\leq (1 - 1/n)\lambda(U) \leq (1 - 1/n)(\lambda^*(X_n) + \varepsilon).$$

Because $\varepsilon > 0$ was taken arbitrarily, we obtain

$$\lambda^*(X_n) \leq (1 - 1/n)\lambda^*(X_n).$$

Finally, in view of the inequality $\lambda^*(X_n) < +\infty$, we conclude that

$$\lambda^*(X_n) = 0,$$

and the theorem is proved.

Exercise 28. For any λ-measurable set $X \subset \mathbf{R}$, show that the set $d(X)$ is Borel in $\mathbf{R}$.

Exercise 29. Demonstrate that if $z \in \mathbf{R}$ is a density point of any two λ-measurable sets X and Y, then z is a density point of the set $Z = X \cap Y$.

This fact is important for introducing the so-called density topology on $\mathbf{R}$ which will be discussed in our further considerations (see, e.g., Exercise 9 from Chapter 6). Here we only wish to mention that the density topology on $\mathbf{R}$ can be defined within the theory **ZF** & **DC** and is a very particular case of a von Neumann topology, which is associated with an arbitrary nonzero σ-finite complete measure μ. The theorem on the existence of a von Neumann topology for such a μ is one of the most deep results in measure theory and relies on uncountable forms of the Axiom of Choice (cf. [138], [222]).

Exercise 30. Let E be a topological space and let $\{U_i : i \in I\}$ be a family of open subsets of E such that

$$(\forall i \in I)(U_i \ \textit{is a first category set in} \ E).$$

By using the Zorn Lemma, prove that the open set $U = \cup\{U_i : i \in I\}$ is also of first category in E.

This important statement is due to Banach (see, e.g., [120] or [162]). Let us underline the circumstance that the cardinality of a set I can be arbitrarily large here (in particular, $card(I)$ can be uncountable).

Deduce from the above statement that if a set $X \subset E$ is locally of first category in E, then X is of first category in E.

1. Cantor and Peano type functions

It is well known that one of the first mathematical results of Cantor (which turned out to be rather surprising to him) was the discovery of the existence of a bijection between the set $\mathbf{R}$ of all real numbers and the corresponding product set $\mathbf{R}^2 = \mathbf{R} \times \mathbf{R}$ (i.e., the Euclidean plane). For a time, Cantor did not believe that such a bijection exists and even wrote to Dedekind about his doubts in this connection. Of course, Cantor already knew of the existence of a bijection between the set $\mathbf{N}$ of all natural numbers and the product set $\mathbf{N} \times \mathbf{N}$. A simple way to construct such a bijection is the following one. First, we observe that a function

$$f \; : \; \mathbf{N} \to \mathbf{N} \setminus \{0\}$$

defined by the formula

$$f(n) = n + 1 \qquad (n \in \mathbf{N})$$

is a bijection between $\mathbf{N}$ and the set of all strictly positive natural numbers. Then, for each integer $n > 0$, we have a unique representation of n in the form

$$n = 2^k(2l + 1)$$

where k and l are some natural numbers. Now, define a function

$$g \; : \; \mathbf{N} \setminus \{0\} \to \mathbf{N} \times \mathbf{N}$$

by the formula
$$g(n) = (k, l) \qquad (n \in \mathbf{N} \setminus \{0\}).$$

One can immediately check that g is a bijection, which also yields the corresponding bijection between $\mathbf{N}$ and $\mathbf{N} \times \mathbf{N}$.

By starting with the latter bijection, it is not hard to establish a one-to-one correspondence between the real line and the Euclidean plane (respectively, between the unit segment $[0, 1]$ and the unit square $[0, 1]^2$). Indeed, a simple argument (within the theory $\mathbf{ZF}$) shows that the sets

$$\mathbf{R}, \quad [0, 1], \quad 2^{\mathbf{N}}$$

are equivalent, i.e., there exists a bijective mapping from each of them to any other one. So we only have to verify that the sets

$$2^{\mathbf{N}}, \quad 2^{\mathbf{N}} \times 2^{\mathbf{N}}$$

are equivalent, too. But this is obvious since the product set $2^{\mathbf{N}} \times 2^{\mathbf{N}}$ is equivalent with the set $2^{\mathbf{N} \times \mathbf{N}}$, and the latter set is equivalent with $2^{\mathbf{N}}$ because of the existence of a bijection between $\mathbf{N}$ and $\mathbf{N} \times \mathbf{N}$.

Keeping in mind these simple constructions, it is reasonable to introduce the following definition.

We say that a mapping f acting from $\mathbf{R}$ into $\mathbf{R}^2$ (respectively, from $[0,1]$ into $[0,1]^2$) is a Cantor type function if f is a bijection.

As mentioned above, Cantor type functions do exist.

Exercise 1. Prove that if $\mathbf{R} = X \cup Y$, then $card(X) = \mathbf{c}$ or $card(Y) = \mathbf{c}$ (use the existence of a bijection between $\mathbf{R}$ and $\mathbf{R}^2$). Analogously, by using the existence of a bijection between $\mathbf{R}$ and $\mathbf{R}^{\omega}$, prove that if

$$\mathbf{R} = \cup \{X_n : n < \omega\},$$

then there is an index $n_0 < \omega$ such that $card(X_{n_0}) = \mathbf{c}$.

Observe that both these facts need some form of the Axiom of Choice. On the other hand, demonstrate in **ZF** that

$$\mathbf{c} \neq \omega_{\omega}.$$

Remark 1. As pointed out earlier, one-to-one correspondences between $\mathbf{N}$ and $\mathbf{N} \times \mathbf{N}$ (respectively, between $\mathbf{R}$ and $\mathbf{R} \times \mathbf{R}$ or between $[0,1]$ and $[0,1]^2$) can be constructed effectively, i.e., without the aid of the Axiom of Choice. In this connection, let us recall that, for an arbitrary infinite set X, we also have a bijection between X and $X \times X$, but the existence of such a bijection needs the whole power of the Axiom of Choice. More precisely, according to the classical result of Tarski (cf. [124], [190]), the following two assertions are equivalent in the theory **ZF**:

(1) the Axiom of Choice;

(2) for any infinite set X, there exists a bijection from X onto $X \times X$.

Exercise 2. Let X be an arbitrary set. Show, in the theory **ZF**, that there exists a well ordered set Y such that there is no injection from Y into X. We may suppose, without loss of generality, that $X \cap Y = \emptyset$. Demonstrate (in the same theory) that if

$$card(X \times Y) \leq card(X \cup Y),$$

then there exists an injection from X into Y and, consequently, X can be well ordered. Show also (in **ZF**) that the relation

$$card((X \cup Y)^2) \leq card(X \cup Y)$$

implies the inequality $card(X \times Y) \leq card(X \cup Y)$.

Deduce from these results that, in the theory **ZF**, the following two assertions are equivalent:

(a) the Axiom of Choice;

(b) for any infinite set X, the equality $card(X \times X) = card(X)$ is satisfied.

Now, let f be an arbitrary Cantor type function acting, for example, from **R** onto **R**^{2}. It is well known that, in such a case, f cannot be continuous. Indeed, suppose for a moment that f is continuous. Then we may write

$$\mathbf{R}^2 = \cup\{f([-n, n]) \; : \; n \in \mathbf{N}\},$$

where each set $f([-n, n])$ $(n \in \mathbf{N})$ is compact (hence closed) in **R**^{2}. In accordance with the classical Baire theorem, at least one of these sets has a nonempty interior. Let k be a natural number such that

$$int(f([-k, k])) \neq \emptyset.$$

Then we have a bijective continuous mapping

$$f|[-k, k] \; : \; [-k, k] \to f([-k, k])$$

that is obviously a homeomorphism between $[-k, k]$ and $f([-k, k])$. But this is impossible since $[-k, k]$ is a one-dimensional space and $f([-k, k])$ is a two-dimensional one. If we want to avoid an argument based on the notion of a dimension of a topological space (and it is reasonable to avoid such an argument because we do not discuss this important notion in our book), we can argue in the following manner.

Consider the function

$$f^{-1}|f([-k, k]) \; : \; f([-k, k]) \to [-k, k]$$

that is also a homeomorphism. Let L denote any circumference contained in the set $f([-k, k])$, i.e., let L be a subset of $f([-k, k])$ isometric to

$$\{(x, y) \in \mathbf{R}^2 \; : \; x^2 + y^2 = r\},$$

where r is some strictly positive real (the existence of L is evident since $f([-k,k])$ has a nonempty interior). We thus see that the function

$$f^{-1}|L \; : \; L \to [-k,k]$$

is injective and continuous. This immediately yields a contradiction since there is no injective continuous function acting from a circumference into the real line (cf. the next exercise).

Exercise 3. Let L be an arbitrary circumference on the plane and let $g \; : \; L \to \mathbf{R}$ be a continuous mapping. By using the classical Cauchy theorem on intermediate values for continuous functions, prove that there exist two points $z \in L$ and $z' \in L$ satisfying the relations:

(a) $g(z) = g(z')$;

(b) z and z' are antipodal in L, i.e., the linear segment $[z, z']$ in $\mathbf{R}^2$ is a diameter of L.

Conclude that the mapping g cannot be an injection.

This simple result admits an important generalization to the case of an n-dimensional sphere (instead of L) and of an n-dimensional Euclidean space (instead of $\mathbf{R}$). The corresponding statement is known as the Borsuk-Ulam theorem on antipodes and plays an essential role in algebraic topology (see, for example, [121]). In particular, this theorem shows that there are no injective continuous mappings acting from the sphere $\mathbf{S}^n$ into the space $\mathbf{R}^n$.

The following statement is also of some interest in connection with Cantor type functions (see, e.g., [190]).

Theorem 1. *Let f be a function from $\mathbf{R}^2 = \mathbf{R} \times \mathbf{R}$ into $\mathbf{R}$ continuous with respect to each of the variables $x \in \mathbf{R}$ and $y \in \mathbf{R}$ (separately). Then f is not an injection.*

Proof. Suppose otherwise, i.e., suppose that our f is injective. Denote

$$\phi(x) = f(x,0) \qquad (x \in \mathbf{R}).$$

Then, according to the assumption of the theorem, ϕ is a continuous function acting from $\mathbf{R}$ into $\mathbf{R}$. Let us put

$$\phi(0) = a, \qquad \phi(1) = b.$$

Because f is injective, we have $a \neq b$. Consequently, either $a < b$ or $b < a$. We may assume, without loss of generality, that $a < b$. The function ϕ, being continuous on the segment $[0, 1]$, takes all values from the segment

$$[\phi(0), \phi(1)] = [a, b].$$

In particular, there exists at least one point $x_0 \in]0,1[$ such that

$$\phi(x_0) = (a+b)/2.$$

Further, let us define

$$\psi(y) = f(x_0, y) \qquad (y \in \mathbf{R}).$$

Then ψ is a continuous function, too, and

$$\psi(0) = f(x_0, 0) = \phi(x_0) = (a+b)/2.$$

Hence we get the inequalities

$$a < \psi(0) < b$$

that imply the existence of a neighborhood $U(0)$ of the point 0, such that

$$(\forall y \in U(0))(a < \psi(y) < b)$$

or, equivalently,

$$(\forall y \in U(0))(a < f(x_0, y) < b).$$

Thus, on the one hand, we have the inclusion

$$\{f(x_0, y) \ : \ y \in U(0) \setminus \{0\}\} \subset \ ]a, b[.$$

On the other hand, we have the relation

$$\{f(x, 0) \ : \ 0 \le x < 1\} \supset [a, b[.$$

Therefore, for some reals $y_0 \ne 0$ and x_1, we get

$$f(x_0, y_0) = f(x_1, 0),$$

which contradicts the injectivity of f. The contradiction obtained finishes the proof of Theorem 1.

Exercise 4. Answer the following question: does there exist an injective mapping

$$f \ : \ \mathbf{R}^2 \to \mathbf{R}$$

continuous with respect to one of the variables $x \in \mathbf{R}$ and $y \in \mathbf{R}$?

Exercise 5. Show that there exists a bijection

$$f \ : \ [0,1] \to [0,1]^2$$

such that the function $pr_1 \circ f$ is continuous, where

$$pr_1 \; : \; [0,1]^2 \to [0,1]$$

denotes, as usual, the first canonical projection from $[0,1]^2$ onto $[0,1]$.
   More generally, let $f_1 : [0,1] \to [0,1]$ be a function satisfying the relation

$$(\forall x \in [0,1])(card(f_1^{-1}(x)) = \mathbf{c}).$$

Show that there exists a function $f_2 : [0,1] \to [0,1]$ such that the mapping $f = (f_1, f_2)$ is a bijection between $[0,1]$ and $[0,1]^2$.

   We thus see that Cantor type functions cannot be continuous. In this connection, it is reasonable to ask whether there exist continuous surjections from $\mathbf{R}$ onto $\mathbf{R}^2$ or from $[0,1]$ onto $[0,1]^2$. It turned out that such surjections do exist and the first example of the corresponding function acting from $[0,1]$ onto $[0,1]^2$ was constructed by Peano.
   Hence the following definition seems to be natural. Let

$$f \; : \; [0,1] \to [0,1]^2.$$

We shall say that f is a Peano type function if f is continuous and surjective.
   In order to demonstrate the existence of Peano type functions, we recall the classical Cantor construction of his famous discontinuum. Take the unit segment $[0,1]$ on the real line $\mathbf{R}$. The first step of Cantor's construction is to remove from this segment the open interval $]1/3, 2/3[$ whose center coincides with the center of $[0,1]$ and whose length is equal to the one-third of the length of our segment. After this step we obtain the two segments without common points. Then we apply the same operation to each of these two segments, etc. After ω-many steps we come to the subset C of $[0,1]$ which is called the Cantor discontinuum (or the Cantor space). The set C is closed (because we removed open intervals from $[0,1]$) and, in addition, C is perfect because the removed intervals are disjoint and pairwise have no common end-points. Moreover, the sum of lengths of the removed intervals is equal to 1 (which can easily be checked). So we infer that C is nowhere dense in $\mathbf{R}$ and its Lebesgue measure equals zero. Consequently, C is a small subset of $\mathbf{R}$ from the point of view of the Baire category and from the point of view of the standard Lebesgue measure λ on $\mathbf{R}$. The geometric construction of C described above and due to Cantor himself is rather visual but, sometimes, other constructions and characterizations of C are needed in order to formulate the corresponding results in a more general form. We present some of such constructions and characterizations of Cantor's discontinuum in the next two exercises.

Exercise 6. Take the two-element set $2 = \{0,1\}$ and equip this set with the discrete topology. Equip also the Cartesian product 2^ω with the product topology. Demonstrate that 2^ω is homeomorphic to the classical Cantor discontinuum C.

Exercise 7. Let E be a topological space. Show that E is homeomorphic to C if and only if the conjunction of the following four relations holds:

(a) E is nonempty and compact;

(b) E has a countable base;

(c) there are no isolated points in E;

(d) E is zero-dimensional, i.e., for each point $e \in E$ and for any neighborhood $U(e)$ of e, there exists a neighborhood $V(e)$ of e such that

$$V(e) \subset U(e), \quad bd(V(e)) = \emptyset,$$

where the symbol $bd(V(e))$ denotes the boundary of $V(e)$.

Actually, the last relation means that the family of all clopen subsets of E forms a base for E.

The abstract characterization of the Cantor space, given in Exercise 7, implies many useful consequences. For instance, by using this characterization, it is not difficult to show that, for each natural number $k \geq 2$, the product space k^ω is homeomorphic to the Cantor discontinuum (of course, here k is equipped with the discrete topology).

Exercise 8. Demonstrate that k^ω is homeomorphic to C. Verify also that ω^ω is homeomorphic to the space of all irrational real numbers (where ω is equipped with the discrete topology).

Naturally, the Cantor discontinuum has numerous applications in various branches of mathematics (especially, in topology and analysis). The next exercise presents a typical application of C in real analysis.

Exercise 9. Construct a set C' on the segment $[0,1]$ such that:

(a) C' is the image of C under some homeomorphism of $[0,1]$ onto itself;

(b) the Lebesgue measure of C' is strictly positive.

Deduce from relations (a) and (b) that the Lebesgue measure λ is not quasiinvariant with respect to the group of all homeomorphisms of $\mathbf{R}$, i.e., this group does not preserve the σ-ideal of all λ-measure zero sets.

It immediately follows from the construction of C that, for each clopen set $X \subset C$ and for any $\varepsilon > 0$, there exists a finite partition of X consisting of clopen subsets of X, each of which has diameter strictly less than ε.

It is also easy to check that every zero-dimensional compact metric space possesses an analogous property. At the same time, if E is an arbitrary compact metric space, then, for each closed set $X \subset E$ and for any $\varepsilon > 0$, there exists a finite covering of X consisting of closed subsets of X, each of which has diameter strictly less than ε.

These simple observations lead to the following important statement due to Alexandrov (see, e.g., [120]).

Theorem 2. *Let E be an arbitrary nonempty compact metric space. Then there exists a continuous surjection from the Cantor space C onto E.*

Proof. Taking account of the preceding remarks, we can recursively define two sequences

$$\{(X_{n,k})_{1 \leq k \leq m(n)} \ : \ n \in \omega\},$$

$$\{(Y_{n,k})_{1 \leq k \leq m(n)} \ : \ n \in \omega\},$$

satisfying the conditions:
(1) for any $n \in \omega$, the finite family

$$(X_{n,k})_{1 \leq k \leq m(n)}$$

is a partition of C consisting of clopen subsets of C each of which has diameter strictly less than $1/(n+1)$;
(2) for any $n \in \omega$, the finite family

$$(Y_{n,k})_{1 \leq k \leq m(n)}$$

is a covering of E by nonempty closed sets each of which has diameter strictly less than $1/(n+1)$;
(3) for any $n \in \omega$, the family $(X_{n+1,k})_{1 \leq k \leq m(n+1)}$ (respectively, the family $(Y_{n+1,k})_{1 \leq k \leq m(n+1)}$) is inscribed in the family $(X_{n,k})_{1 \leq k \leq m(n)}$ (respectively, in the family $(Y_{n,k})_{1 \leq k \leq m(n)}$), i.e., each set of the first family is contained in some set of the second family;
(4) for all $n \in \omega$ and for any natural numbers k and l such that

$$1 \leq k \leq m(n), \quad 1 \leq l \leq m(n+1),$$

we have

$$X_{n+1,l} \subset X_{n,k} \Rightarrow Y_{n+1,l} \subset Y_{n,k}.$$

Let now x be an arbitrary point of C. Then x uniquely determines the sequences

$$\{k(n) \ : \ n \in \omega\}, \quad \{X_{n,k(n)} \ : \ n \in \omega\},$$

such that
$$(\forall n \in \omega)(1 \leq k(n) \leq m(n) \,\&\, x \in X_{n,k(n)}).$$

Consider the corresponding sequence $\{Y_{n,k(n)} \; : \; n \in \omega\}$. Obviously, we have
$$(\forall n \in \omega)(Y_{n+1,k(n+1)} \subset Y_{n,k(n)}),$$
$$lim_{n \to +\infty} \; diam(Y_{n,k(n)}) = 0.$$

Hence there exists one and only one point y belonging to all the sets $Y_{n,k(n)}$ $(n \in \omega)$. Let us put $y = f(x)$. In this way we obtain the mapping
$$f \; : \; C \to E.$$

By starting with the definition of f, it is not hard to verify that f is continuous and surjective. Theorem 2 has thus been proved.

Exercise 10. Let X be a nonempty closed subset of C. Show that there exists a continuous mapping
$$f \; : \; C \to X$$

satisfying the relation
$$(\forall x \in X)(f(x) = x).$$

In other words, each nonempty closed subset of C is a retract of C.

Notice that this simple but useful result follows directly from the well-known Michael theorem concerning the existence of continuous selectors for lower semicontinuous set-valued mappings defined on zero-dimensional paracompact spaces (in this context, see [145], [146] or [173]).

Exercise 11. Using the result of Exercise 10, give another proof of Alexandrov's theorem. Namely, starting with a canonical continuous surjection
$$h \; : \; 2^{\omega} \to [0,1],$$

show that there exists a continuous surjection
$$h_1 \; : \; 2^{\omega} \to [0,1]^{\omega}.$$

Then, for each closed subset Y of $[0,1]^{\omega}$, show that there exists a closed subset X of C such that $h_1(X) = Y$. Finally, apply the classical theorem of Urysohn stating that every compact metric space can be realized as a closed subset of the space $[0,1]^{\omega}$ (see, e.g., [85]).

Theorem 2 immediately implies the existence of Peano type functions.

Theorem 3. *There exists a Peano type function, i.e., there exists a continuous surjection from* $[0,1]$ *onto* $[0,1]^2$.

Proof. It follows from Theorem 2 that there is a continuous surjection

$$g \ : \ C \to [0,1]^2.$$

Consider any open interval U that was removed from $[0,1]$ during the construction of C. We may add U to the domain of g and extend g in such a way that the extended function will be affine on U and will coincide with g on the end-points of U. Doing this for all removed intervals simultaneously, we come to the function f defined on the whole segment $[0,1]$. The construction of f immediately implies that f is continuous. In addition, since f is an extension of g, we conclude that f is a surjection as well. This ends the proof of Theorem 3.

Furthermore, let us note that the existence of a Peano type function from $[0,1]$ onto $[0,1]^2$ implies at once the existence of a continuous surjection

$$h \ : \ \mathbf{R} \to \mathbf{R}^2.$$

Such a function h can also be regarded as a Peano type function (acting from $\mathbf{R}$ onto $\mathbf{R}^2$).

For many interesting properties of Peano type functions, we refer the reader to [64] and [202]. Exercises presented below are also closely connected with the existence of Peano type functions.

Exercise 12. Let P be a nonempty perfect subset of the real line. Demonstrate that there exists a continuous surjection from P onto the square $[0,1]^2$. Infer from this result that there exists a disjoint family

$$\{P_j \ : \ j \in J\} \subset \mathcal{P}(P)$$

of nonempty perfect subsets of $\mathbf{R}$, such that $card(J) = \mathbf{c}$.

Exercise 13. Prove that, for any Peano type function

$$f = (f_1, f_2) \ : \ \mathbf{R} \to \mathbf{R}^2,$$

the relation

$$\{x \in \mathbf{R} \ : \ f_1'(x) \ exists\} \cup \{x \in \mathbf{R} \ : \ f_2'(x) \ exists\} \neq \mathbf{R}$$

holds true (cf. Exercise 7 from Chapter 13).

Exercise 14. By starting with the existence of a Peano type function acting from the segment $[0,1]$ onto the square $[0,1]^2$, demonstrate that there exists an injective continuous mapping

$$g : [0,1] \to \mathbf{R}^3 = \mathbf{R} \times \mathbf{R} \times \mathbf{R}$$

satisfying the following relations:

(a) the orthogonal projection of $g([0,1])$ on the plane $\mathbf{R} \times \mathbf{R} \times \{0\}$ coincides with $[0,1]^2$;

(b) every plane in $\mathbf{R}^3$ parallel to $\mathbf{R} \times \mathbf{R} \times \{0\}$ has at most one common point with $g([0,1])$.

As shown by Theorem 2, any nonempty compact metric space is a continuous image of C or, equivalently, of 2^ω. A natural question arises concerning the description of all those Hausdorff topological spaces which are continuous images of C. It turns out that only nonempty compact metrizable spaces are such images.

The following exercise provides the corresponding result.

Exercise 15. Let E be a topological space. We say that a family S of subsets of E is a net in E (in the sense of Archangelskii) if, for any open set $U \subset E$, there exists a subfamily of S whose union coincides with U. Clearly, every base of E is a net but the converse assertion does not hold in general.

Now, let E be a compact space and let κ be an infinite cardinal number. Show that if E admits a net of cardinality κ, then E possesses also a base of cardinality κ. Deduce from this fact that if E' and E'' are two compact spaces and

$$g : E' \to E''$$

is a continuous surjection, then the topological weight of E'' is less than or equal to the topological weight of E'. In particular, if E' has a countable base, then E'' has a countable base, too.

Conclude from this result that if a Hausdorff topological space Y is a continuous image of C, then Y is compact and metrizable.

Exercise 16. Let κ be an arbitrary infinite cardinal number. Equip the set 2^κ with the product topology (where the two-element set $2 = \{0,1\}$ is endowed with the discrete topology). The space 2^κ is usually called the generalized Cantor discontinuum (of weight κ). This space can also be regarded as a commutative compact topological group with respect to the addition operation modulo 2. Hence there exists a Haar probability measure μ on 2^κ. By starting with the fact that $\mu(U) > 0$ for each nonempty open subset U of 2^κ, show that 2^κ satisfies the so-called Suslin condition or countable chain condition, i.e., any disjoint family of nonempty open subsets of 2^κ is at most countable. Deduce from this fact that if a topological space E is a continuous image of 2^κ, then E satisfies the countable chain condition, too. Conclude that there exists a nonempty compact topological space X such that, for all $\kappa \geq \omega$, there is no continuous surjection from 2^κ onto X.

On the other hand, show that, for any compact topological space Y of weight κ, there exists a closed subset Z of 2^κ such that Y is a continuous image of Z.

Note that the results presented in Exercise 16 are essentially due to Szpilrajn (Marczewski).

As mentioned earlier, the Cantor set C is small from the topological point of view (i.e., C is a nowhere dense subset of the real line $\mathbf{R}$) and from the measure-theoretical point of view (i.e., C is of Lebesgue measure zero). On the other hand, the operation of vector sum yields the set

$$C + C = \{x + y \ : \ x \in C, \ y \in C\},$$

which is not small at all. Indeed, it can easily be seen that this set contains in itself a nonempty open interval. More precisely, we have the equality

$$C + C = [0, 2].$$

Exercise 17. Prove that $C + C = [0, 2]$. Moreover, give a simple geometrical interpretation of this equality.

The result of Exercise 17 shows, in particular, that the operation of vector sum does not preserve the two classical σ-ideals on the real line: the σ-ideal $\mathcal{K}(\mathbf{R})$ of all first category subsets of $\mathbf{R}$ and the σ-ideal $\mathcal{I}(\lambda)$ of all Lebesgue measure zero subsets of $\mathbf{R}$. Let us note that, at the same time, there exist many σ-ideals on the real line, which are (under some additional set-theoretical axioms) isomorphic to $\mathcal{K}(\mathbf{R})$ and $\mathcal{I}(\lambda)$, are invariant with respect to the group of all translations of $\mathbf{R}$ and also invariant with respect to the operation of vector sum of sets.

Exercise 18. Consider on the Euclidean plane $\mathbf{R}^2$ the family of all straight lines parallel to the line $\{0\} \times \mathbf{R}$. Let $\mathcal{J}$ denote the σ-ideal of subsets of the plane, generated by this family. Check that:

(a) $\mathcal{J}$ is invariant under the group of all translations of $\mathbf{R}^2$;

(b) $\mathcal{J}$ is invariant under the operation of vector sum of subsets of $\mathbf{R}^2$.

In addition, note that, by assuming the Continuum Hypothesis (or, more generally, Martin's Axiom), it can be shown that $\mathcal{J}$ is isomorphic to each of the σ-ideals $\mathcal{K}(\mathbf{R}^2)$ and $\mathcal{I}(\lambda_2)$ where λ_2 denotes the standard two-dimensional Lebesgue measure on $\mathbf{R}^2$. This fact follows directly from the Sierpiński-Erdős Duality Principle (see, for instance, [34], [153] or [162]). Let us also remark that the σ-ideals $\mathcal{K}(\mathbf{R})$ and $\mathcal{K}(\mathbf{R}^2)$ (respectively, $\mathcal{I}(\lambda)$ and $\mathcal{I}(\lambda_2)$) are isomorphic to each other, and the corresponding isomorphisms can be constructed within the theory **ZF** & **DC** (see, e.g., [34]).

Now, by taking account of the fact that the additive group of $\mathbf{R}$ is isomorphic to the additive group of $\mathbf{R}^2$ (see, e.g., Chapter 9 of this book), demonstrate that there exists a σ-ideal $\mathcal{I}$ of subsets of $\mathbf{R}$, satisfying the relations:

(c) $\mathcal{I}$ is invariant with respect to the group of all translations of $\mathbf{R}$;

(d) $\mathcal{I}$ is invariant with respect to the operation of vector sum of subsets of $\mathbf{R}$;

(e) under the Continuum Hypothesis (or, more generally, under Martin's Axiom), $\mathcal{I}$ is isomorphic to each of the σ-ideals $\mathcal{K}(\mathbf{R})$ and $\mathcal{I}(\lambda)$.

The preceding considerations show us that the operation of vector sum is rather bad from the point of view of preserving σ-ideals on $\mathbf{R}$. Analogously, this operation is bad from the point of view of descriptive set theory. The latter fact may be illustrated, for instance, by various kinds of examples of two G_δ-subsets of $\mathbf{R}$ whose vector sum is not a Borel subset of $\mathbf{R}$ (on the other hand, such a vector sum is always an analytic subset of $\mathbf{R}$). A similar phenomenon can be observed when dealing with the distance set and with the difference set of a given point set lying on the real line or in a finite-dimensional Euclidean space.

We recall that the distance set of a set X lying in a metric space (E, d) is the set of all distances $d(x, y)$ where x and y range over X.

The difference set of a set X lying in a commutative group $(\Gamma, +)$ is the set of all elements $x - y$ where x and y range over X.

Sierpiński was the first mathematician who gave an example of a G_δ-subset of the Euclidean plane, whose distance set is not Borel (see [197]). Much later, several authors (see, e.g., [174], [207] and [208]) constructed two G_δ-subsets of the real line, whose vector sum is not Borel. Moreover, Erdős and Stone constructed in [55] a compact subset of $\mathbf{R}$ and a G_δ-subset of $\mathbf{R}$, such that their vector sum is not a Borel set in $\mathbf{R}$.

Here we would like to present an example of a G_δ-subset of $\mathbf{R}$ whose difference set is not Borel. This example is due to Rogers (see [174]).

First of all, we need to recall the notion of the Hausdorff metric. Let (E, d) be an arbitrary metric space. Denote by the symbol $\mathcal{F}(E)$ the family of all nonempty closed bounded subsets of E. For any two sets $X \in \mathcal{F}(E)$

and $Y \in \mathcal{F}(E)$, we define

$$d'(X, Y) = inf\{\varepsilon > 0 \; : \; X \subset V_\varepsilon(Y) \; \& \; Y \subset V_\varepsilon(X)\}$$

where $V_\varepsilon(X)$ (respectively, $V_\varepsilon(Y)$) denotes the ε-neighborhood of X (respectively, of Y). It is easy to see that d' is a metric on $\mathcal{F}(E)$, so we get the metric space $(\mathcal{F}(E), d')$. The metric d' is usually called the Hausdorff metric associated with d. If the original metric space (E, d) has good properties, then, sometimes, those properties can be transferred to the space $(\mathcal{F}(E), d')$. For instance, the following assertions are valid:

(1) if (E, d) is complete, then $(\mathcal{F}(E), d')$ is complete, too;

(2) if (E, d) is compact, then $(\mathcal{F}(E), d')$ is compact, too.

Assertions (1) and (2) can be established without any difficulties. On the other hand, let us remark that if the original space (E, d) is separable, then, in general, $(\mathcal{F}(E), d')$ need not be separable. Moreover, it can easily be observed that if (E, d) is bounded but not totally bounded, then $(\mathcal{F}(E), d')$ is necessarily nonseparable.

Exercise 19. Let $\lambda : \mathcal{F}([0, 1]) \to [0, 1]$ denote the standard Lebesgue measure restricted to the family of all nonempty closed subsets of $[0, 1]$. Equipping this family with the Hausdorff metric, we obtain the compact metric space $\mathcal{F}([0, 1])$. Show that the function λ is upper semicontinuous on this space.

The Hausdorff metric can be successfully applied in establishing the classical result of Suslin which states the existence of analytic non-Borel subsets of uncountable Polish spaces (see, for instance, [84], [120] or [133]). For the sake of completeness, we shall give a short proof of this result, especially taking account of the fact that the proof is essentially based on Theorem 2 which is the main tool for constructing various Peano type functions.

Theorem 4. *Let E be an arbitrary uncountable Polish topological space (or, more generally, an uncountable Borel subset of a Polish space). Then there exists an analytic subset of E which is not Borel.*

Proof. Because all uncountable Borel subsets of Polish spaces are Borel isomorphic, it suffices to show that, in the standard Cantor space $C = 2^\omega$, there exists an analytic subset which is not Borel. In order to do this, let us take the product space

$$W = 2^\omega \times [0, 1]$$

and observe that it is compact. Denote by $\mathcal{F}(W)$ the family of all nonempty closed subsets of W and equip $\mathcal{F}(W)$ with the Hausdorff metric (or, equivalently, with the Vietoris topology that is metrizable by this metric). It is easy to verify that, in such a way, $\mathcal{F}(W)$ becomes a compact metric space (see assertion (2) formulated above). According to Theorem 2, there exists a surjective continuous mapping

$$h \; : \; 2^\omega \to \mathcal{F}(W).$$

Further, let Z denote the set of all irrational points of the segment $[0, 1]$. We recall that Z is homeomorphic to the canonical Baire space ω^ω where ω is equipped with the discrete topology. Now, we define a set-valued mapping

$$\Phi \; : \; 2^\omega \to \mathcal{P}(2^\omega)$$

by the following formula:

$$\Phi(t) = pr_1((2^\omega \times Z) \cap h(t)) \qquad (t \in 2^\omega).$$

It is clear that $(2^\omega \times Z) \cap h(t)$ ranges over the family of all closed subsets of the space $2^\omega \times Z$ as t ranges over 2^ω. By starting with this fact, it is not difficult to claim that

$$ran(\Phi) = \mathcal{A}(2^\omega),$$

where $\mathcal{A}(2^\omega)$ denotes the family of all analytic subsets of 2^ω. Let us put

$$X = \{t \in 2^\omega \; : \; t \notin \Phi(t)\}$$

and establish that X is not analytic in 2^ω. Suppose, for a while, that X is analytic. Then, for some $t_0 \in 2^\omega$, we must have the equality

$$X = \Phi(t_0).$$

But, according to the definition of X, we get

$$t_0 \in X \Leftrightarrow t_0 \notin X,$$

which obviously yields a contradiction. Consequently, X is not an analytic subset of the Cantor space. On the other hand, let us verify that the set

$$2^\omega \setminus X = \{t \in 2^\omega \; : \; t \in \Phi(t)\}$$

is an analytic subset of the Cantor space. Indeed, we may write

$$2^\omega \setminus X = pr_1(D)$$

where the set D is defined by the formula

$$D = \{(t, y) \in W \; : \; (t, y) \in h(t) \; \& \; (t, y) \in 2^\omega \times Z\}.$$

Because our mapping h is continuous, the set

$$D' = \{(t, y) \in W \ : \ (t, y) \in h(t)\}$$

is closed in W and, therefore, is a G_δ-subset of W. Also, the set

$$D'' = \{(t, y) \in W \ : \ (t, y) \in 2^\omega \times Z\}$$

is a G_δ-subset of W. Hence the intersection

$$D' \cap D'' = D$$

is a G_δ-subset of W, too, and $pr_1(D)$ is an analytic set in the Cantor space 2^ω. Finally, we easily infer that $pr_1(D)$ is not Borel because, as has been shown above, the set

$$X = 2^\omega \setminus pr_1(D)$$

is not analytic. This ends the proof of Theorem 4.

Example 1. In connection with Theorem 4, it is reasonable to point out that, in classical mathematical analysis, there are many interesting concrete sets of functions, analytic but not Borel (in an appropriate Polish space). For instance, Mauldin established in [142] that the set of all continuous nowhere differentiable real-valued functions defined on the unit segment $[0, 1]$ is an analytic but not Borel subset of the separable Banach space $C[0, 1]$. Many other such sets can be constructed in the theory of trigonometric series (for more information, see especially [84] and the references therein).

Now, keeping in mind Theorem 4, we are going to prove the following statement due to Rogers [174].

Theorem 5. *There exists a G_δ-subset B of $\mathbf{R}$ such that its difference set $B - B$ is not Borel.*

Proof. We begin with some simple observations. First of all, let us introduce two subsets C_1 and C_2 of the unit segment $[0, 1]$. Namely, the set C_1 consists of all real numbers x with decimal expansions

$$x = 0, t_1 t_2 t_3 ...$$

where $t_n = 0$ or $t_n = 1$ for every natural index $n \geq 1$; analogously, the set C_2 consists of all real numbers y with decimal expansions

$$y = 0, t_1 t_2 t_3 ...$$

where $t_n = 0$ or $t_n = 2$ for every natural index $n \geq 1$.

Obviously, C_1 and C_2 are uncountable closed subsets of the segment $[0, 1]$. Consequently, applying Theorem 4 proved above, we may choose an analytic subset A_1 of C_1 that is not Borel. Also, we can represent the set A_1 as the projection (on the x-axis) of a certain G_δ-subset Z_1 of the Cartesian product $C_1 \times C_2 \subset \mathbf{R}^2$.

Now, let us consider a mapping

$$\phi \; : \; \mathbf{R}^2 \to \mathbf{R}$$

defined by the formula

$$\phi(x, y) = x + y \qquad ((x, y) \in \mathbf{R}^2).$$

This mapping is continuous. Moreover, by taking account of the definitions of C_1 and C_2, it can easily be checked that the restriction of ϕ to the product set $C_1 \times C_2$ is a homeomorphism between $C_1 \times C_2$ and $\phi(C_1 \times C_2)$. Hence

$$Z_0 = \phi(Z_1) = \{x + y \; : \; (x, y) \in Z_1\}$$

is a G_δ-subset of the compact set $\phi(C_1 \times C_2)$. This also implies that Z_0 is a G_δ-subset of the real line. Now, let us put

$$B = Z_0 \cup (C_2 - 3).$$

Evidently, B is a G_δ-subset of $\mathbf{R}$. We are going to show that the difference set of B is not Borel in $\mathbf{R}$. Let us denote this difference set by

$$D = \{x - y \; : \; x \in B, \; y \in B\}.$$

It is enough to establish that the set $D \cap (C_1 + 3)$ is not Borel. First, let us note that the following inclusions are fulfilled:

$$Z_0 \subset [0, 2], \quad C_2 - 3 \subset [-3, -2], \quad C_1 + 3 \subset [3, 4].$$

Further, it is not difficult to verify that each point of the set $D \cap (C_1 + 3)$ is of the form $u - v$ where

$$u - v \in C_1 + 3, \quad u \in Z_0, \quad v \in C_2 - 3.$$

But the relation $u \in Z_0$ is true if and only if

$$u = x + y$$

where
$$x \in C_1, \quad y \in C_2, \quad (x, y) \in Z_1.$$
Consequently, $D \cap (C_1 + 3)$ is the set of points of the form $x + y - v$ where
$$x + y - v - 3 \in C_1,$$
$$x \in C_1, \quad y \in C_2, \quad v + 3 \in C_2, \quad (x, y) \in Z_1.$$
Now, let us remark that if the points x and y are fixed and $(x, y) \in Z_1$, then there exists one and only one point v satisfying the relations
$$x + y - v - 3 \in C_1, \quad v + 3 \in C_2.$$
Namely, such a point is $v = y - 3$. Hence $D \cap (C_1 + 3)$ coincides with the following set:
$$\{x + 3 \ : \ (\exists y)((x, y) \in Z_1)\}.$$
In other words, we have
$$D \cap (C_1 + 3) = A_1 + 3.$$
Because A_1 is not Borel, the set $D \cap (C_1 + 3)$ is not Borel, either. This also shows that the difference set D is not Borel in $\mathbf{R}$.

Theorem 5 has thus been proved.

Furthermore, putting
$$X = B, \quad Y = -B,$$
we get two G_δ-subsets X and Y of $\mathbf{R}$ for which the vector sum $X + Y$ is not Borel in $\mathbf{R}$.

We now wish to present another application of Theorem 2. Namely, we are going to prove the classical Banach-Mazur theorem on the universality of the space $C[0, 1]$ for the class of all separable metric spaces.

Let $\mathcal{M}$ be a class of metric spaces and let X be some space from this class. We shall say that X is universal for $\mathcal{M}$ if, for any space Y belonging to $\mathcal{M}$, there exists an isometric embedding of Y into X. In other words, X is universal for $\mathcal{M}$ if all spaces from $\mathcal{M}$ can be realized as isometric copies of corresponding subsets of X.

Let $\mathcal{M}_s$ denote the class of all separable metric spaces. The first example of a space universal for $\mathcal{M}_s$ was constructed by Urysohn. Later, Banach and Mazur discovered that the classical function space $C[0, 1]$ is universal for $\mathcal{M}_s$, too. In order to establish this fact, we need two lemmas.

Lemma 1. *For any metric space* (X, d), *there exists a Banach space* E *such that:*

(1) X *can be isometrically embedded in* E;

(2) the weight of E *is equal to the weight of* X *(in particular, if* X *is separable, then* E *is separable, too).*

Proof. We may assume, without loss of generality, that $X \neq \emptyset$. Let us fix a point $t \in X$. Further, for any point $x \in X$, let us define a function

$$f_x \ : \ X \to \mathbf{R}$$

by the formula

$$f_x(y) = d(x, y) - d(t, y) \qquad (y \in X).$$

Obviously, f_x is continuous, and the relation

$$|f_x(y)| \leq |d(x, y) - d(t, y)| \leq d(x, t)$$

indicates directly that f_x is bounded.

Now, let the symbol $C_b(X)$ denote the Banach space (with respect to the standard sup-norm) of all real-valued bounded continuous functions defined on X. We introduce a mapping

$$\phi \ : \ X \to C_b(X)$$

by the following formula:

$$\phi(x) = f_x \qquad (x \in X).$$

Let us check that ϕ is an isometric embedding of X into $C_b(X)$. Indeed, for any two elements $x \in X$ and $x' \in X$, we may write

$$||f_x - f_{x'}|| = sup_{y \in X} \, |f_x(y) - f_{x'}(y)| = sup_{y \in X} \, |d(x, y) - d(x', y)|.$$

But it can easily be observed that

$$sup_{y \in X} \, |d(x, y) - d(x', y)| = d(x, x').$$

Consequently, we get

$$||\phi(x) - \phi(x')|| = ||f_x - f_{x'}|| = d(x, x'),$$

which shows that ϕ is an isometric embedding. Let us define E as the closed vector subspace of $C_b(X)$ generated by the set $\phi(X)$. Then E is obviously a Banach space whose weight is equal to the weight of $\phi(X)$ or, equivalently, to the weight of X. We thus conclude that E is the required Banach space.

Before the formulation of the next lemma, we need some auxiliary notions from the theory of topological vector spaces (below, we restrict ourselves to the class of topological vector spaces over the field $\mathbf{R}$). Let E and E' be any two vector spaces over $\mathbf{R}$. We say that these spaces are in a duality if a bilinear function

$$\Phi \ : \ E \times E' \to \mathbf{R}$$

is given such that:

(a) for each $x \neq 0$ from E, the partial linear functional

$$y \to \Phi(x, y) \quad (y \in E')$$

is not identically equal to zero;

(b) for each $y \neq 0$ from E', the partial linear functional

$$x \to \Phi(x, y) \quad (x \in E)$$

is not identically equal to zero.

In such a situation, it is usually said that Φ establishes a duality between the given spaces E and E'.

In particular, let E be an arbitrary Banach space (or, more generally, normed vector space). As usual, we denote by E^* the vector space of all continuous linear functionals on E. Evidently, a bilinear function

$$\Phi \ : \ E^* \times E \to \mathbf{R}$$

defined by the formula

$$\Phi(u, x) = u(x) \quad (u \in E^*, \ x \in E)$$

establishes a duality between E^* and E.

Let X be a subset of E. We equip E^* with the weakest topology $\sigma(E^*, X)$ for which all linear functionals from the family $(\Phi(\cdot, x))_{x \in X}$ are continuous. Clearly, the pair $(E^*, \sigma(E^*, X))$ is a topological vector space.

Exercise 20. Let X be an everywhere dense subset of a normed vector space E. Denote by B the closed unit ball in E^*, i.e., put

$$B = \{u \in E^* \ : \ ||u|| \leq 1\}.$$

Verify that the topologies on B induced by $\sigma(E^*, X)$ and $\sigma(E^*, E)$, respectively, are identical.

Lemma 2. *Let E be a separable normed vector space and let B denote the closed unit ball in E^*. Then B is a compact metric space with respect to the topology induced by $\sigma(E^*, E)$.*

Proof. According to Exercise 20, the topology $\sigma(E^*, E)$ restricted to B coincides with the topology

$$\sigma(E^*, \{x_n \ : \ n \in \omega\})$$

restricted to B, where $\{x_n \ : \ n \in \omega\}$ is a countable everywhere dense subset of E. Let us define a mapping

$$h \ : \ B \to \mathbf{R}^\omega$$

by the formula
$$h(u) = (u(x_n))_{n \in \omega} \qquad (u \in B).$$
Actually, h maps B into the product space

$$\prod_{n \in \omega} [-\|x_n\|, \|x_n\|],$$

which obviously is compact and metrizable. Now, by starting with the definition of $\sigma(E^*, \{x_n \ : \ n \in \omega\})$, it is not hard to demonstrate that B is homeomorphic to some closed subset of the above-mentioned product space and, consequently, B is compact and metrizable as well. Lemma 2 has thus been proved.

We are ready to formulate and prove the following classical result of Banach and Mazur.

Theorem 6. *The space $C[0, 1]$ is universal for the class $\mathcal{M}_s$.*

Proof. Let X be an arbitrary separable metric space. According to Lemma 1, there exists a separable Banach space E containing an isometric copy of X. Hence it suffices to show that E can be isometrically embedded in $C[0, 1]$. Let us denote by B the closed unit ball in E^*. Because E is separable, the ball B equipped with the topology induced by $\sigma(E^*, E)$ is compact and metrizable (see Lemma 2). According to Theorem 2, there exists a continuous surjection g from the Cantor space $C \subset [0, 1]$ onto B. It is easy to see that g can be extended to a continuous mapping acting from $[0, 1]$ onto B (cf. the proof of Theorem 3). For the sake of simplicity, the extended mapping will be denoted by the same symbol g. Now, let us take an arbitrary element x from E and define a function

$$f_x \ : \ [0,1] \to \mathbf{R}$$

by the formula

$$f_x(t) = g(t)(x) \qquad (t \in [0,1]).$$

Because g is continuous, f_x is continuous, too. Moreover, $g(t) \in B$ for each $t \in [0,1]$ and we get

$$|f_x(t)| \leq ||x|| \qquad (t \in [0,1]),$$

i.e., f_x is also bounded. If y is another element from E, then, for any $t \in [0,1]$, we may write

$$|f_x(t) - f_y(t)| = |g(t)(x-y)| \leq ||x-y||.$$

On the other hand, a simple consequence of the Hahn-Banach theorem says that if $x \neq y$, then there exists a continuous linear functional

$$u \ : \ E \to \mathbf{R}$$

satisfying the relations

$$||u|| = 1, \quad u(x-y) = ||x-y||.$$

In particular, $u \in B$ and, because g is a surjection from $[0,1]$ onto B, there exists a point $t_0 \in [0,1]$ such that $u = g(t_0)$. Then, for the point t_0, we have

$$|f_x(t_0) - f_y(t_0)| = |u(x-y)| = ||x-y||,$$

which shows us that

$$||f_x - f_y|| = ||x-y||$$

and, consequently, the mapping

$$x \to f_x \qquad (x \in E, \ f_x \in C[0,1])$$

is an isometric embedding. This completes the proof of Theorem 6.

Remark 2. According to the preceding result, the space $C[0,1]$ is universal for the class $\mathcal{M}_s$. At the same time, it is obvious that $C[0,1]$ cannot be universal for the class $\mathcal{M}(\mathbf{c})$ consisting of all those metric spaces whose cardinalities are less than or equal to the cardinality of the continuum (clearly, we have the proper inclusion $\mathcal{M}_s \subset \mathcal{M}(\mathbf{c})$). The problem of the existence of a metric space $X \in \mathcal{M}(\mathbf{c})$ universal for $\mathcal{M}(\mathbf{c})$ was investigated, with related problems for other infinite cardinals, by Sierpiński (see [191] and [193]).

Evidently, analogous questions about the existence of universal objects can be posed for various mathematical structures: for algebraic structures (e.g., groups), binary relations, topological structures and so on. There are some important results in this direction.

For instance, let us recall the well-known theorem of Cantor stating that the set $\mathbf{Q}$ equipped with its standard order is universal for the class of all countable linearly ordered sets, i.e., every countable linearly ordered set can be isomorphically embedded in $\mathbf{Q}$. The following example illustrates the situation for linearly ordered sets whose cardinalities are less than or equal to $\mathbf{c}$.

Example 2. Suppose that the Continuum Hypothesis holds. Then there exists a linearly ordered set $(X, \leq)$ with $card(X) = \mathbf{c}$ such that, for any linearly ordered set $(Y, \leq)$ with $card(Y) \leq \mathbf{c}$, there is a monomorphism from Y into X. For the proof, see, e.g., [190] where related results for other infinite cardinals are also discussed.

Exercise 21. Starting with the fact that $(\mathbf{Q}, \leq)$ is universal for the class of all countable linearly ordered sets, prove in the theory **ZF** that there exists a partition $\{A_\xi : \xi < \omega_1\}$ of $\mathbf{R}$.

This classical result is due to Lebesgue. As shown by Luzin and Sierpiński [137], in the same **ZF** there are partitions $\{B_\xi : \xi < \omega_1\}$ of $\mathbf{R}$ such that all sets B_ξ $(\xi < \omega_1)$ are Borel in $\mathbf{R}$. Moreover, it is known that any uncountable analytic (co-analytic) set $X \subset \mathbf{R}$ admits a canonical representation

$$X = \cup\{X_\xi : \xi < \omega_1\},$$

where all sets X_ξ $(\xi < \omega_1)$ are nonempty, pairwise disjoint and Borel in $\mathbf{R}$ (for details, see [120]).

Apply the existence of $\{A_\xi : \xi < \omega_1\}$ and prove within **ZF** that the set $\mathcal{P}(\mathbf{R})$ cannot be represented as the union of a countable family of countable sets.

The last result is due to Tarski. It is useful to compare this result with the theorem of **ZF** stating that $\mathbf{R}$ cannot be represented as the union of a countable family of finite sets (see Exercise 2 from the Introduction).

In addition, deduce from the existence of $\{A_\xi : \xi < \omega_1\}$ that the inequality

$$2^{\omega_1} \leq 2^{\mathbf{c}}$$

holds true within **ZF**. Consequently, we have in **ZF** the relation

$$\omega_1 < 2^{\omega_1} \leq 2^{\mathbf{c}}.$$

Note that the stronger inequality $\omega_1 \leq \mathbf{c}$ cannot be established within **ZF** & **DC** since it implies the existence of a subset of **R** nonmeasurable in the Lebesgue sense (the result of Shelah [182] and Raisonnier [169]).

2. Functions of first Baire class

Let E be a topological space and let $f : E \to \mathbf{R}$ be a function.

We shall say that f is of Baire zero class if f is continuous at all points of E (i.e., f is continuous on E).

The family of all continuous functions acting from E into $\mathbf{R}$ is usually denoted by the symbol $C(E, \mathbf{R})$. In accordance with the definition above, we will also use the notation $Ba_0(E, \mathbf{R})$ for the same family of functions. Thus, we have

$$Ba_0(E, \mathbf{R}) = C(E, \mathbf{R}).$$

By the standard definition due to Baire (see [5], [6]), a given function

$$f : E \to \mathbf{R}$$

is of first Baire class if there exists a sequence $\{f_n : n < \omega\}$ of functions from $Ba_0(E, \mathbf{R})$ such that

$$lim_{n \to +\infty} f_n(x) = f(x)$$

for each point $x \in E$. In other words, $f : E \to \mathbf{R}$ belongs to the first Baire class if and only if f can be represented as a pointwise limit of a sequence of functions belonging to $Ba_0(E, \mathbf{R})$.

It is well known that functions of first Baire class play a significant role in various topics of real analysis. The following simple but important example emphasizes this circumstance.

Example 1. Let $E = \mathbf{R}$ and let $f : E \to \mathbf{R}$ be a derivative. Then there exists a continuous function $g : E \to \mathbf{R}$ such that

$$g'(x) = f(x) \quad (x \in E).$$

Define a sequence $\{g_n : 1 \leq n < \omega\}$ of real-valued functions on $\mathbf{R}$ by the formula

$$g_n(x) = n(g(x + 1/n) - g(x)) \quad (x \in \mathbf{R}, \ n = 1, 2, ...).$$

63

Obviously, we have

$$lim_{n \to +\infty} g_n(x) = f(x) \qquad (x \in \mathbf{R}).$$

Because each g_n is a continuous function on $\mathbf{R}$, we conclude that f belongs to the first Baire class.

For any topological space E, the family of all functions $f : E \to \mathbf{R}$ belonging to the first Baire class will be denoted by $Ba_1(E, \mathbf{R})$.

Note that the other Baire classes $Ba_\xi(E, \mathbf{R})$ of real-valued functions on E can be naturally introduced by iterating the limit process and using the method of transfinite recursion on $\xi < \omega_1$ (in this connection, see [6], [120], [154] and Chapter 7 of the present book). Here we are mainly interested in properties of functions belonging to $Ba_1(E, \mathbf{R})$.

The following simple properties follow directly from the definition of the class $Ba_1(E, \mathbf{R})$:

(1) $Ba_1(E, \mathbf{R})$ is a linear algebra over the field $\mathbf{R}$; in other words, if $f \in Ba_1(E, \mathbf{R})$, $g \in Ba_1(E, \mathbf{R})$, $a \in \mathbf{R}$ and $b \in \mathbf{R}$, then

$$af + bg \in Ba_1(E, \mathbf{R}),$$

$$f \cdot g \in Ba_1(E, \mathbf{R});$$

(2) if $f \in Ba_1(E, \mathbf{R})$, $g \in Ba_1(E, \mathbf{R})$ and $g(x) \neq 0$ for all $x \in E$, then

$$f/g \in Ba_1(E, \mathbf{R});$$

(3) if $f \in Ba_1(E, \mathbf{R})$ and $\phi : \,]a, b[\, \to \mathbf{R}$ is a continuous function such that $ran(f) \subset \,]a, b[$, then

$$\phi \circ f \in Ba_1(E, \mathbf{R}).$$

Exercise 1. Verify the validity of the assertions (1), (2), and (3) above.

Exercise 2. Suppose that E is a separable topological space (i.e., E contains a countable everywhere dense subset). Show that

$$card(Ba_1(E, \mathbf{R})) \leq \mathbf{c}.$$

In fact, show that if E is nonempty and separable, then

$$card(Ba_1(E, \mathbf{R})) = \mathbf{c}.$$

Let us consider some other, less trivial, properties of $Ba_1(E, \mathbf{R})$.

Lemma 1. *Let E be a topological space, $\{a_n : 1 \leq n < \omega\}$ be a sequence of strictly positive real numbers, such that*

$$\sum_{1 \leq n < \omega} a_n < +\infty,$$

and let $\{f_n : 1 \leq n < \omega\} \subset Ba_1(E, \mathbf{R})$ be a sequence of functions such that

$$|f_n(x)| < a_n \qquad (x \in E, \ 1 \leq n < \omega).$$

Define a function

$$f \ : \ E \to \mathbf{R}$$

by the formula

$$f(x) = f_1(x) + f_2(x) + ... + f_n(x) + ... \qquad (x \in E).$$

Then f also belongs to $Ba_1(E, \mathbf{R})$.

Proof. First, note that the function f is well defined since the series

$$f_1(x) + f_2(x) + ... + f_n(x) + ... \qquad (x \in E)$$

converges uniformly with respect to $x \in E$ (in view of the relations $|f_n(x)| < a_n$ for all $x \in E$ and $n \geq 1$). Further, because $f_n \in Ba_1(E, \mathbf{R})$ for any natural number $n \geq 1$, we can write

$$f_n(x) = \lim_{k \to +\infty} f_{n,k}(x) \qquad (x \in E),$$

where $f_{n,k}$ $(k = 1, 2, ...)$ are some real-valued continuous functions on E. Without loss of generality, we may assume that

$$|f_{n,k}(x)| \leq a_n \qquad (x \in E, \ n = 1, 2, ..., \ k = 1, 2, ...).$$

Now, let us put

$$h_k(x) = f_{1,k}(x) + f_{2,k}(x) + ... + f_{k,k}(x) \qquad (x \in E, \ k = 1, 2, ...).$$

Clearly, all functions h_k are continuous on E and it suffices to show that

$$f(x) = \lim_{k \to +\infty} h_k(x) \qquad (x \in E).$$

For this purpose, fix a real $\varepsilon > 0$. There exists a natural number m such that

$$a_{m+1} + a_{m+2} + ... + a_i + ... < \varepsilon/3.$$

Consequently, we have

$$|f_{m+1}(x)| + |f_{m+2}(x)| + ... + |f_i(x)| + ... < \varepsilon/3 \qquad (x \in E),$$

$$|f_{m+1,k}(x)| + |f_{m+2,k}(x)| + ... + |f_{i,k}(x)| + ... \leq \varepsilon/3 \qquad (x \in E, \ k = 1, 2, ...).$$

For any $x \in E$ and $k > m$, we can write the inequalities

$$|f(x) - h_k(x)| \leq |f_1(x) - f_{1,k}(x)| + |f_2(x) - f_{2,k}(x)| + ... + |f_m(x) - f_{m,k}(x)| +$$

$$|f_{m+1}(x)| + |f_{m+2}(x)| + ... + |f_i(x)| + ... +$$

$$|f_{m+1,k}(x)| + |f_{m+2,k}(x)| + ... + |f_{k,k}(x)|$$

$$\leq |f_1(x) - f_{1,k}(x)| + |f_2(x) - f_{2,k}(x)| + ... + |f_m(x) - f_{m,k}(x)| + 2\varepsilon/3.$$

If $x \in E$ is fixed, then we can find $k_0 < \omega$ so large that for all natural numbers $k > k_0$ the relation

$$|f_1(x) - f_{1,k}(x)| + |f_2(x) - f_{2,k}(x)| + ... + |f_m(x) - f_{m,k}(x)| < \varepsilon/3$$

will be satisfied. But this relation immediately yields the inequality

$$|f(x) - h_k(x)| < \varepsilon$$

for all integers $k > k_0$. Therefore, we get

$$lim_{k \to +\infty} h_k(x) = f(x) \qquad (x \in E),$$

which completes the proof of Lemma 1.

We need this lemma in order to prove the following result due to Baire.

Theorem 1. *Let E be a topological space. Suppose that a sequence*

$$\{f_n : n < \omega\} \subset Ba_1(E, \mathbf{R})$$

is given uniformly convergent to a function

$$f : E \to \mathbf{R}.$$

Then we have $f \in Ba_1(E, \mathbf{R})$.

Proof. According to our assumption, for any natural number k, there exists a natural number n_k such that

$$|f(x) - f_{n_k}(x)| < \frac{1}{2^{k+1}} \qquad (x \in E).$$

Evidently, we may assume that

$$n_0 < n_1 < ... < n_k < ... .$$

Let us consider the series of functions

$$(f_{n_1} - f_{n_0}) + (f_{n_2} - f_{n_1}) + ... + (f_{n_{k+1}} - f_{n_k}) + ...$$

Because the inequalities

$$|f_{n_{k+1}}(x) - f_{n_k}(x)| \leq |f_{n_{k+1}}(x) - f(x)| + |f_{n_k}(x) - f(x)| < \frac{1}{2^{k+2}} + \frac{1}{2^{k+1}} < \frac{1}{2^k}$$

hold for all $x \in E$, we can apply Lemma 1 to this series. In this way, we obtain that the function

$$g = (f_{n_1} - f_{n_0}) + (f_{n_2} - f_{n_1}) + ... + (f_{n_{k+1}} - f_{n_k}) + ...$$

belongs to $Ba_1(E, \mathbf{R})$. But it is easy to see that

$$g = lim_{k \to +\infty} f_{n_{k+1}} - f_{n_0} = f - f_{n_0}.$$

According to property (1), we finally get

$$f = g + f_{n_0} \in Ba_1(E, \mathbf{R}).$$

The theorem has thus been proved.

Remark 1. Theorem 1 implies that, for $E \neq \emptyset$, the family of all bounded functions from $Ba_1(E, \mathbf{R})$ becomes a Banach space with respect to the norm of uniform convergence or, equivalently, with respect to the standard sup-norm

$$||f|| = sup_{x \in E} |f(x)|.$$

Remark 2. Theorem 1 can be directly generalized to the case of the Baire class $Ba_\xi(E, \mathbf{R})$, where ξ is an arbitrary ordinal number strictly less than ω_1. The proof essentially remains the same as above (cf. [120]).

Lemma 2. Let E be a topological space and let $g \in Ba_1(E, \mathbf{R})$. Then, for every $t \in \mathbf{R}$, the sets

$$g^{-1}(] - \infty, t[) = \{x \in E : g(x) < t\},$$

$$g^{-1}(]t, +\infty[) = \{x \in E : g(x) > t\}$$

are F_σ-subsets of E.

Proof. Take any $t \in \mathbf{R}$. Because $g \in Ba_1(E, \mathbf{R})$, there exists a sequence $\{g_n : n < \omega\} \subset Ba_0(E, \mathbf{R})$ such that

$$lim_{n \to +\infty} g_n(x) = g(x) \quad (x \in E).$$

It is not difficult to verify the following relations:

$$g(x) < t \Leftrightarrow (\exists k < \omega)(\exists n < \omega)(\forall m \in [n, \omega[)(g_m(x) \leq t - 1/k),$$

$$g(x) > t \Leftrightarrow (\exists k < \omega)(\exists n < \omega)(\forall m \in [n, \omega[)(g_m(x) \geq t + 1/k).$$

These relations yield at once that the above-mentioned sets

$$\{x \in E : g(x) < t\}, \quad \{x \in E : g(x) > t\}$$

are F_σ-subsets of E, and the lemma is proved.

Lemma 3. *Let E be a normal topological space, $g : E \to \mathbf{R}$ be a function and suppose that*

$$ran(g) = \{t_1, t_2, ..., t_k\}.$$

If, for any integer $i \in [1, k]$, the set

$$X_i = \{x \in E : g(x) = t_i\} = g^{-1}(t_i)$$

is an F_σ-subset of E, then $g \in Ba_1(E, \mathbf{R})$.

Proof. Obviously, we can write

$$E = X_1 \cup X_2 \cup ... \cup X_k,$$

$$X_i = F_{i,0} \cup F_{i,1} \cup ... \cup F_{i,n} \cup ... \quad (i = 1, 2, ..., k),$$

where all $F_{i,n}$ $(1 \leq i \leq k, \ n < \omega)$ are closed subsets of E and

$$F_{i,0} \subset F_{i,1} \subset ... \subset F_{i,n} \subset ... \ .$$

Let us put

$$F_n = F_{1,n} \cup F_{2,n} \cup ... \cup F_{k,n} \quad (n = 0, 1, 2, ...)$$

and define a function

$$g_n : F_n \to \mathbf{R} \quad (n = 0, 1, 2, ...)$$

by the formula

$$g_n(x) = t_i \ \ iff \ \ x \in F_{i,n}.$$

Because the finite family of closed sets $\{F_{1,n}, F_{2,n}, ..., F_{k,n}\}$ is disjoint, the function g_n is continuous on the closed set F_n. By the Tietze-Urysohn theorem (see, e.g., [85] or [120]), g_n admits a continuous extension

$$g_n^* : E \to \mathbf{R}.$$

Now, it is easy to check that

$$lim_{n\to+\infty} g_n^*(x) = g(x)$$

for all points $x \in E$. This finishes the proof of the lemma.

Lemma 4. *Let E be a topological space in which every open set is an F_σ-subset (or, equivalently, in which every closed set is a G_δ-subset). Let a set $X \subset E$ be representable in the form*

$$X = A_1 \cup A_2 \cup ... \cup A_k,$$

where all A_j ($j = 1, 2, ..., k$) are F_σ-subsets of E. Then X is representable in the form

$$X = B_1 \cup B_2 \cup ... \cup B_k,$$

where

$$B_j \subset A_j \quad (j = 1, 2, ..., k),$$

all B_j are also F_σ-subsets of E and, in addition, they are pairwise disjoint.

Proof. Obviously, we have the equality

$$X = F_1 \cup F_2 \cup ... \cup F_i \cup ...,$$

where all sets F_i ($1 \leq i < \omega$) are closed in E and each F_i is contained in some set $A_{j(i)}$. Let us put

$$C_1 = F_1, \ C_2 = F_2 \setminus F_1, ..., \ C_i = F_i \setminus (F_1 \cup ... \cup F_{i-1}), ... \ .$$

The family of sets $\{C_i : 1 \leq i < \omega\}$ is disjoint and, in view of our assumption on E, all C_i are F_σ-subsets of E. Moreover, we have

$$X = C_1 \cup C_2 \cup ... \cup C_i \cup ... \ .$$

Now, for any natural number $j \in [1, k]$, define the set B_j by the formula

$$B_j = \cup\{C_i : j \text{ is the smallest number for which } C_i \subset A_j\}.$$

Clearly, the family $\{B_1, B_2, ..., B_k\}$ is disjoint, all sets B_j $(j = 1, 2, ..., k)$ are F_σ-subsets of E and

$$B_j \subset A_j \quad (j = 1, 2, ..., k),$$

$$X = B_1 \cup B_2 \cup ... \cup B_k.$$

Lemma 4 has thus been proved.

Recall that a topological space is perfectly normal if E is normal and each open set in E is an F_σ-subset of E. For such spaces the following important statement due to Lebesgue is true.

Theorem 2. *Let E be a perfectly normal space and let*

$$f \; : \; E \to \mathbf{R}$$

be a function. These three assertions are equivalent:
(1) $f \in Ba_1(E, \mathbf{R})$;
(2) for any $t \in \mathbf{R}$, both sets $\{x \in E : f(x) < t\}$ and $\{x \in E : f(x) > t\}$ are F_σ-subsets of E;
(3) for any open set $U \subset \mathbf{R}$, the preimage $f^{-1}(U)$ is an F_σ-subset of E.

Proof. The equivalence (2) $\Leftrightarrow$ (3) is trivial and the implication (1) $\Rightarrow$ (2) was established by Lemma 2 (even for an arbitrary topological space E). Consequently, it remains to prove the implication (2) $\Rightarrow$ (1). Suppose that (2) is valid and suppose first that $ran(f) \subset \,]0, 1[$, i.e.,

$$0 < f(x) < 1 \quad (x \in E).$$

For any integer $n \geq 1$, consider the sequence $\{t_0, t_1, ..., t_n\}$ of points of $\mathbf{R}$ determined by conditions:

$$t_0 = 0, \quad t_{j+1} - t_j = 1/n \quad (j = 0, ..., n - 1).$$

In particular, we have $t_n = 1$. Further, introduce the sets:

$$A_0 = \{x \in E : f(x) < t_1\},$$

$$A_n = \{x \in E : f(x) > t_{n-1}\},$$

$$A_j = \{x \in E : t_{j-1} < f(x) < t_{j+1}\} \quad (j = 1, ..., n - 1).$$

Obviously, we have the equality

$$E = A_0 \cup A_1 \cup ... \cup A_n$$

and all A_j $(j = 0, 1, ..., n)$ are F_σ-subsets of E. Applying Lemma 4, we get another representation

$$E = B_0 \cup B_1 \cup ... \cup B_n,$$

where all B_j $(j = 0, 1, ..., n)$ are also F_σ-subsets of E, are pairwise disjoint and

$$B_j \subset A_j \qquad (j = 0, 1, ..., n).$$

Now, define a function

$$f_n : E \to \mathbf{R}$$

by putting

$$f_n(x) = t_j \quad iff \quad x \in B_j.$$

According to Lemma 3, the function f_n belongs to $Ba_1(E, \mathbf{R})$.

Take an arbitrary point $x \in E$. Then $x \in B_j$ for some integer $j \in [0, n]$. If $j = 0$, then

$$t_0 < f(x) < t_1, \quad f_n(x) = t_0, \quad |f(x) - f_n(x)| < 1/n.$$

If $j = n$, then we have

$$t_{n-1} < f(x) < t_n, \quad f_n(x) = t_n, \quad |f(x) - f_n(x)| < 1/n.$$

Finally, if $1 \le j \le n - 1$, then

$$t_{j-1} < f(x) < t_{j+1}, \quad f_n(x) = t_j, \quad |f(x) - f_n(x)| < 2/n.$$

These relations show that

$$lim_{n \to +\infty} f_n(x) = f(x)$$

uniformly with respect to $x \in E$. By virtue of Theorem 1, we obtain that $f \in Ba_1(E, \mathbf{R})$.

Suppose now that $f : E \to \mathbf{R}$ is an arbitrary function satisfying (2). Fix any increasing homeomorphism

$$\phi : \mathbf{R} \to \;]0, 1[$$

and consider the function $\phi \circ f$. This function also satisfies (2) and

$$ran(\phi \circ f) \subset \;]0, 1[.$$

As demonstrated above, $\phi \circ f \in Ba_1(E, \mathbf{R})$. Consequently,

$$f = \phi^{-1} \circ (\phi \circ f) \in Ba_1(E, \mathbf{R}),$$

which completes the proof of Theorem 2.

Example 2. Let E be a perfectly normal space and let X be a subset of E. Denote by f_X the characteristic function (i.e., indicator) of X. It is easy to verify that if X is closed in E, then relation (2) of Theorem 2 is satisfied for $f = f_X$. Therefore, according to this theorem, we have

$$f_X \in Ba_1(E, \mathbf{R}).$$

Now, if Y is an open subset of E, then, taking into account the equality

$$f_Y = 1 - f_{E \setminus Y},$$

we see that $f_Y \in Ba_1(E, \mathbf{R})$, too.

The above-mentioned facts follow also from Lemma 3.

Example 3. Let E be a subinterval of $\mathbf{R}$ and let

$$f : E \to \mathbf{R}$$

be a monotone function. It is easy to check that, for any $t \in \mathbf{R}$, both sets

$$\{x \in E : f(x) < t\}, \quad \{x \in E : f(x) > t\}$$

are some subintervals of E. Because each interval in E is an F_σ-subset of E, we infer (in view of Theorem 2) that $f \in Ba_1(E, \mathbf{R})$.

The same conclusion is true for those $f : E \to \mathbf{R}$ which are of finite variation on E. Indeed, such functions are representable in the form of the difference of two increasing functions on E (see, e.g., [154], [180]) and it suffices to refer to property (1) of the class $Ba_1(E, \mathbf{R})$.

Example 4. Let E be again a perfectly normal space and let

$$f : E \to \mathbf{R}$$

be an upper semicontinuous function. According to the definition of upper semicontinuous functions, for any $t \in \mathbf{R}$, the set $\{x \in E : f(x) < t\}$ is open in E and, hence, is an F_σ-subset of E. At the same time, the set

$$\{x \in E : f(x) > t\} = \cup_{n < \omega} \{x \in E : f(x) \geq t + 1/(n+1)\}$$

is the union of countably many closed sets, i.e., is also an F_σ-subset of E. Applying again Theorem 2, we deduce that

$$f \in Ba_1(E, \mathbf{R}).$$

From this fact it immediately follows that $g \in Ba_1(E, \mathbf{R})$ for any lower semicontinuous function $g : E \to \mathbf{R}$.

Actually, the characteristic function f_X of a closed set $X \subset E$ (see Example 2) is upper semicontinuous.

Exercise 3. Show that the set $\mathbf{Q} \subset \mathbf{R}$ of all rational numbers is not a G_δ-subset of $\mathbf{R}$. Deduce from this circumstance that the characteristic function $f_{\mathbf{Q}}$, the so-called Dirichlet function, does not belong to $Ba_1(\mathbf{R}, \mathbf{R})$ (actually, $f_{\mathbf{Q}}$ is of second Baire class).

More generally, prove that if E is an uncountable Polish space without isolated points and X is a countable everywhere dense subset of E, then the characteristic function f_X does not belong to $Ba_1(E, \mathbf{R})$.

Exercise 4. Let E be a topological space and let

$$f : E \to \mathbf{R}$$

be a function. For any $x \in E$, define

$$\Omega_f(x) = \inf_{V \in \mathcal{V}(x)} diam(f(V)),$$

where $\mathcal{V}(x)$ denotes the filter of all neighborhoods of x and $diam(f(V))$ stands for the diameter of the set $f(V)$. Denote also by $D(f)$ the set of all discontinuity points of f.

Verify that:

(a) $\Omega_f(x) = 0$ if and only if f is continuous at x (equivalently, $\Omega_f(x) > 0$ if and only if x is a discontinuity point of f);

(b) for any $t \in \mathbf{R}$, the set

$$\{x \in E : \Omega_f(x) \geq t\}$$

is closed in E;

(c) the set $D(f)$ is representable in the form

$$D(f) = E_1 \cup E_2 \cup ... \cup E_n \cup ...,$$

where

$$E_n = \{x \in E : \Omega_f(x) \geq 1/n\}$$

for each integer $n \geq 1$.

Conclude from (b) and (c) that $D(f)$ is an F_σ-subset of E (therefore, the set $C(f)$ of all continuity points of f is a G_δ-subset of E).

Let $\mathcal{B}$ be a base of open sets in $\mathbf{R}$ and let

$$\mathcal{F} = \{Y \subset \mathbf{R} : \mathbf{R} \setminus Y \in \mathcal{B}\}.$$

Prove the equality

$$D(f) = \cup\{cl(f^{-1}(Y)) \setminus f^{-1}(Y) \ : \ Y \in \mathcal{F}\}.$$

Generalize these results to an arbitrary function $f : E \to E'$, where E' is a metric space. Some applications of these facts will be presented below (see, for instance, Chapter 7).

Exercise 5. Let E be a topological space. We say that E is resolvable if E admits a representation in the form $E = A \cup B$, where A and B are some disjoint everywhere dense subsets of E (this notion was first introduced by Hewitt).

As a rule, topological spaces used in various questions of mathematical analysis turn out to be resolvable. In particular, prove that:

(i) every locally compact topological space without isolated points is resolvable;

(ii) every Hausdorff topological vector space over $\mathbf{R}$ (whose dimension is not equal to zero) is resolvable.

Fix a resolvable space E and let X be an F_σ-subset of E. Show that there exists a function

$$f : E \to \mathbf{R}$$

such that X coincides with the set $D(f)$ of all discontinuity points of f.

To do this, first represent X in the form

$$X = F_1 \cup F_2 \cup ... \cup F_n \cup ...,$$

where all sets F_n are closed in E and $F_n \subset F_{n+1}$ for each natural number $n \geq 1$. Further, put $F_0 = \emptyset$ and, for any integer $n \geq 1$, define a function

$$f_n : E \to \{0, 1\}$$

satisfying the relations:

(a) f_n is equal to zero at all points of the set $E \setminus F_n$;

(b) $\Omega_{f_n}(x) = 1$ if $x \in F_n$.

Now, take a sequence $\{a_n : n \geq 1\}$ of strictly positive real numbers, such that

$$a_{n+1} + a_{n+2} + ... + a_k + ... < a_n \quad (n = 1, 2, ...).$$

For example, it suffices to put

$$a_n = \frac{1}{3^n} \quad (n \geq 1).$$

Finally, consider the function

$$f = a_1 f_1 + a_2 f_2 + \ldots + a_n f_n + \ldots .$$

This function is well defined since the series on the right-hand side of the above equality converges uniformly on E.

Verify that:

(c) f is continuous at all points of the set $E \setminus X$;

(d) for any integer $n \geq 1$ and for all points $x \in F_n \setminus F_{n-1}$, we have

$$\Omega_f(x) \geq a_n - \sum_{k>n} a_k > 0.$$

Conclude from (c) and (d) that $D(f) = X$.

Exercise 6. Let E be a perfectly normal topological space and let

$$f : E \to \mathbf{R}$$

be a function whose graph is closed in the product space $E \times \mathbf{R}$. Applying the Kuratowski lemma on closed projections (i.e., Lemma 1 of the Introduction), show that f is of first Baire class.

Give an example of a function $f : \mathbf{R} \to \mathbf{R}$ whose graph is closed in $\mathbf{R} \times \mathbf{R}$ and whose discontinuity points constitute a nonempty perfect set in $\mathbf{R}$.

Recall that a topological space E is Baire if no nonempty open subset of E is of first category in E. For such an E, an important result (due to Baire) is well known, which yields an essential information about the structure of the set $D(f)$ of discontinuity points of an arbitrary function $f : E \to \mathbf{R}$ belonging to the class $Ba_1(E, \mathbf{R})$.

Theorem 3. *Let E be a Baire space and let $f \in Ba_1(E, \mathbf{R})$. Then the set $D(f)$ of all discontinuity points of f is of first category in E. In particular, we have*

$$C(f) \cap U \neq \emptyset$$

for any nonempty open set $U \subset E$.

Proof. As we know (see Exercise 4 above), the relation

$$C(f) = E \setminus D(f) = \bigcap_{1 \leq n < \omega} \{x \in E : \Omega_f(x) < 1/n\}$$

is valid, where all sets $\{x \in E : \Omega_f(x) < 1/n\}$ are open in E. So it suffices to demonstrate that all these sets are everywhere dense in E. In other words, it suffices to show that, for any $\varepsilon > 0$ and for any nonempty open set $U \subset E$, there exists a nonempty open set $W \subset U$ such that

$$(\forall x \in W)(\forall y \in W)(|f(x) - f(y)| < \varepsilon).$$

Taking into account the fact that f belongs to $Ba_1(E, \mathbf{R})$, choose a sequence $\{f_k : k < \omega\} \subset Ba_0(E, \mathbf{R})$ satisfying the relation

$$f(x) = lim_{k \to +\infty} f_k(x) \qquad (x \in E).$$

Further, for any natural number k, define the set

$$X_k = \{x \in E : (\forall i \geq k)(\forall j \geq k)(|f_i(x) - f_j(x)| \leq \varepsilon/3)\}.$$

All sets X_k $(k < \omega)$ are closed in E and

$$(\forall k < \omega)(X_k \subset X_{k+1}), \quad E = \cup\{X_k : k < \omega\}.$$

Consequently, we have

$$U = (U \cap X_0) \cup (U \cap X_1) \cup ... \cup (U \cap X_k) \cup ... \ .$$

Because E is a Baire space, there exists a natural number n such that

$$int(U \cap X_n) \neq \emptyset.$$

Let $V \subset U \cap X_n$ be a nonempty open subset of E. If x is an arbitrary point of V, then

$$(\forall i \geq n)(\forall j \geq n)(|f_i(x) - f_j(x)| \leq \varepsilon/3).$$

Putting $j = n$ and tending i to $+\infty$, we get

$$(\forall x \in V)(|f(x) - f_n(x)| \leq \varepsilon/3).$$

Therefore, we can write

$$|f(y) - f(x)| \leq |f(y) - f_n(y)| + |f_n(y) - f_n(x)| + |f_n(x) - f(x)| \leq$$

$$2\varepsilon/3 + |f_n(y) - f_n(x)|$$

for any two points x and y from V. Finally, because f_n is a continuous function, there exists a nonempty open set $W \subset V$ such that

$$(\forall x \in W)(\forall y \in W)(|f_n(y) - f_n(x)| < \varepsilon/3).$$

This gives at once the relation

$$(\forall x \in W)(\forall y \in W)(|f(y) - f(x)| < \varepsilon),$$

which ends the proof of the Baire theorem.

Remark 3. The proof presented above is based on the classical argument due to Baire (cf. also [154], [162]).

Remark 4. More general versions of the Baire theorem (with further information about it) can be found in [120].

The next exercise shows that the main part of the Baire theorem remains true for an arbitrary topological space E.

Exercise 7. Let E be a topological space and let $f \in Ba_1(E, \mathbf{R})$. Demonstrate that the set $D(f)$ is of first category in E (for this purpose, use again Exercise 4 and Lemma 2).

Another way to show this fact is based on the Banach statement (see Exercise 30 from the Introduction) which leads to a representation of E in the form

$$E = E' \cup E'',$$

where E' is an open Baire subspace of E and E'' is a first category closed subset of E. Applying Theorem 3 to the set $D(f|E')$, we get the required result.

Deduce from this result that if $f \in Ba_1(E, \mathbf{R})$ and X is an arbitrary subspace of E, then the set $D(f|X)$ is of first category in X. In particular, claim that if E is a complete metric space, $f \in Ba_1(E, \mathbf{R})$ and X is a nonempty closed subspace of E, then there exist points in X at which $f|X$ is continuous.

Exercise 8. According to the definition of Luzin, a topological space X is always of first category (or X is perfectly meager) if each nonempty dense in itself subset of X is of first category in X.

Let E be a topological space and let X be a subspace of E such that, for every perfect set $P \subset E$, the set $X \cap P$ is of first category in P. Show that the space X is always of first category (apply Exercise 11 from the Introduction).

Luzin proved that there exists an uncountable subspace X of $\mathbf{R}$ which is always of first category (see [136] or [120]). Other constructions of uncountable universally small sets can be found in [167] and [235] (cf. also Chapter 10 of this book).

Exercise 9. Let E be a hereditarily Lindelöf topological space always of first category and let $f : E \to \mathbf{R}$ be a function. Demonstrate that, for each subspace X of E, the set $D(f|X)$ is of first category in X. For this purpose, begin with establishing the fact that X admits a representation in the form

$$X = Y \cup Z,$$

where Y is dense in itself, Z is at most countable and $Y \cap Z = \emptyset$. Then verify the inclusion

$$D(f|X) \subset Y \cup (X' \cap Z),$$

where X' denotes the set of all accumulation points of X (in E). Finally, observe that both sets Y and $X' \cap Z$ are of first category in X.

Exercise 10. Let E be a subspace of $\mathbf{R}$ satisfying the following relations:
(a) $card(E) = \mathbf{c}$;
(b) E is always of first category.
The existence of such a subspace of $\mathbf{R}$ will be demonstrated in Chapter 10 under Martin's Axiom (see Theorem 7 of that chapter).
Prove that there exists a function

$$f : E \to \mathbf{R}$$

having the following properties:
(c) f is not Borel (consequently, f does not belong to $Ba_1(E, \mathbf{R})$);
(d) for any subspace Z of E, the set $D(f|Z)$ is of first category in Z.

This classical result is due to Luzin. It shows that property (d) of f does not imply the relation $f \in Ba_1(E, \mathbf{R})$.

It is useful to compare the above-mentioned result with Exercise 13 given below.

Exercise 11. Let E be a hereditarily Lindelöf topological space and let

$$F_0 \supset F_1 \supset ... \supset F_\xi \supset ... \quad (\xi < \omega_1)$$

be a decreasing (with respect to the inclusion relation) ω_1-sequence of closed subsets of E. Prove that there exists an ordinal $\alpha < \omega_1$ such that

$$(\forall \xi \in [\alpha, \omega_1[)(F_\xi = F_\alpha).$$

This result is known as the Baire stationarity principle.
In particular, take an arbitrary closed subset X of E and define by transfinite recursion an ω_1-sequence $\{X_\xi : \xi < \omega_1\}$ in the following manner:

$X_0 = X$;

$X_{\xi+1} = (X_\xi)'$ for any $\xi < \omega_1$, where $(X_\xi)'$ denotes the set of all accumulation points of X_ξ;

$X_\xi = \cap\{X_\zeta : \zeta < \xi\}$ for any limit ordinal $\xi < \omega_1$.

Applying the Baire stationarity principle to $\{X_\xi : \xi < \omega_1\}$, show that X admits a representation in the form

$$X = Y \cup Z, \quad Y \cap Z = \emptyset,$$

where Y is a perfect subset of E and Z is at most countable (the Cantor-Bendixson theorem).

Give another proof of the same result that does not use the method of transfinite induction. For this purpose, consider the set of all condensation points of X and take it as Y. Then define the set Z by the equality

$$Z = X \setminus Y.$$

Note that a certain generalization of the Baire stationarity principle was obtained by Luzin for decreasing ω_1-sequences of F_σ-subsets of $\mathbf{R}$ (in this connection, see [136]).

Exercise 12. Let E be a topological space. We say that E is scattered if E does not contain a nonempty dense in itself subset.

Demonstrate that the following two assertions are equivalent:

(a) E is scattered;

(b) E can be represented in the form of an injective α-sequence

$$E = \{e_\xi : \xi < \alpha\},$$

where α is some ordinal and, for any $\xi < \alpha$, the element e_ξ is an isolated point of the set $\{e_\zeta : \xi \leq \zeta < \alpha\}$.

Note that the implication (b) $\Rightarrow$ (a) is trivial. Supposing now that (a) is valid, use the method of transfinite recursion for obtaining the required representation of E.

Finally, demonstrate that every topological space X can be represented in the form

$$X = Y \cup Z,$$

where Y is a perfect subset of X, Z is a scattered subset of X and $Y \cap Z = \emptyset$ (this classical result is due to Cantor).

Exercise 13. Let E be a separable metric space and let $g : E \to \mathbf{R}$ be a function such that, for every nonempty closed set $F \subset E$, there exists a point $x \in F$ at which the restricted function $g|F$ is continuous. Prove that g is of first Baire class (this result is due to Baire).

The following argument enables to establish the above-mentioned result. Take any $a \in \mathbf{R}$ and $b \in \mathbf{R}$ such that $a < b$, and denote

$$A = \{x \in E : g(x) > a\}, \quad B = \{x \in E : g(x) < b\}.$$

Clearly, we have the equality

$$E = A \cup B.$$

Further, construct by transfinite recursion an ω_1-sequence

$$F_0 \supset F_1 \supset ... \supset F_\xi \supset ... \quad (\xi < \omega_1)$$

of closed subsets of E. Put $F_0 = E$. Suppose that, for a given $\xi < \omega_1$, the partial family $\{F_\zeta : \zeta < \xi\}$ has already been defined.

If ξ is a limit ordinal, then we put

$$F_\xi = \cap\{F_\zeta : \zeta < \xi\}.$$

If $\xi = \eta + 1$, consider the set F_η. Only two cases are possible.
1. $F_\eta = \emptyset$. In this case, we define

$$F_\xi = F_\eta = \emptyset.$$

2. $F_\eta \neq \emptyset$. In this case, there exists a point $x \in F_\eta$ at which the function $g|F_\eta$ is continuous. Consequently, there exists an open neighborhood $V(x)$ of x such that
$$F_\eta \cap V(x) \subset A \quad \vee \quad F_\eta \cap V(x) \subset B.$$
Then we define
$$F_\xi = F_\eta \setminus V(x).$$

Proceeding in this way, we will be able to construct the sets F_ξ ($\xi < \omega_1$). Observe now that, for each ordinal $\xi < \omega_1$, we have

$$F_\xi \setminus F_{\xi+1} \subset A \quad \vee \quad F_\xi \setminus F_{\xi+1} \subset B$$

and, according to the Baire stationarity principle, for some $\alpha < \omega_1$, the equalities
$$\emptyset = F_\alpha = F_{\alpha+1} = ... = F_\xi = ... \quad (\alpha \leq \xi < \omega_1)$$
are valid. Deduce from these facts that there exist two sets A' and B' such that

$$A' \subset A, \quad B' \subset B, \quad A' \cup B' = E, \quad A' \cap B' = \emptyset$$

and both A' and B' are F_σ-subsets of E.

Now, let $\{b_n : n < \omega\}$ be a strictly decreasing sequence of real numbers satisfying the relation $\lim_{n \to +\infty} b_n = a$. Put again

$$A = \{x \in E : g(x) > a\}$$

and, for each $n < \omega$, define

$$B_n = \{x \in E : g(x) < b_n\}.$$

As above, show the existence of F_σ-sets A'_n and B'_n such that

$$A'_n \subset A, \quad B'_n \subset B_n, \quad A'_n \cup B'_n = E, \quad A'_n \cap B'_n = \emptyset.$$

Finally, denote

$$X = \cup\{A'_n : n < \omega\}.$$

Verify that

$$X = \{x \in E : g(x) > a\} = A$$

and hence A is an F_σ-subset of E.

By using a similar argument, demonstrate that $B = \{x \in E : g(x) < b\}$ is an F_σ-subset of E, too.

Conclude, in view of the Lebesgue theorem (i.e., Theorem 2 of this chapter) that the function g is of first Baire class.

Remark 5. More general versions of the result presented above can be found in [120] where a different argument is applied. Namely, it is proved there that if E is a complete metric space and $g : E \to \mathbf{R}$ is a function, then the following two assertions are equivalent:

(1) $g \in Ba_1(E, \mathbf{R})$;

(2) for any nonempty closed set $F \subset E$, there exists a point $x \in F$ at which the restricted function $g|F$ is continuous.

Theorem 3 and the preceding exercise establish this equivalence in the case of a Polish space E. For a nonseparable complete metric space E, the proof of (1) $\Leftrightarrow$ (2) relies on properties of the so-called Montgomery operation (see, e.g., [120]). Note that this operation needs uncountable forms of the Axiom of Choice.

Remark 6. As mentioned above, for a function $g : \mathbf{R} \to \mathbf{R}$, the following two assertions are equivalent:

(1) g is of first Baire class;

(2) for any nonempty closed set $F \subset \mathbf{R}$, there are points $x \in F$ at which the function $g|F$ is continuous.

In other words, we have a certain characterization of functions acting from $\mathbf{R}$ into $\mathbf{R}$ and belonging to the first Baire class. Unfortunately, we do not have an analogous nice characterization of derivatives on $\mathbf{R}$ which form an important proper subclass of $Ba_1(\mathbf{R}, \mathbf{R})$ (see Example 1).

Exercise 14. Let E be a Polish topological space and let $f : E \to \mathbf{R}$ be a function whose set of discontinuity points is at most countable. Show, by applying Exercise 13, that $f \in Ba_1(E, \mathbf{R})$.

Infer from this fact that any function

$$g : [a, b] \to \mathbf{R}$$

of finite variation on a segment $[a, b]$ belongs to the first Baire class (another way to establish this result was pointed out in Example 3).

Exercise 15. Let C denote the classical Cantor discontinuum on $[0, 1]$. Define two functions

$$f : [0, 1] \to \mathbf{R}, \quad g : [0, 1] \to \mathbf{R}$$

by the following formulas:

$f(x) = 1$ if $x \in C$ and $f(x) = 0$ if $x \in [0, 1] \setminus C$;

$g(x) = 1$ if x is not an end-point of a removed interval for C, and $g(x) = 0$ if x is an end-point of some removed interval for C.

Verify that:

(a) f is upper semicontinuous and, hence, is of first Baire class;

(b) $g|C$ is discontinuous at all points of C, hence g is not of first Baire class;

(c) $D(f) = D(g) = C$.

The last relation shows, in particular, that the sets of continuity points of f and g are identical but f and g have essentially different descriptive structures.

For a while, let us turn our attention to real-valued functions of two variables and let us briefly discuss their descriptive structure from the viewpoint of descriptive properties of the corresponding partial functions of one variable.

We restrict our consideration to real-valued functions defined on the topological product of two metric spaces.

Let X and Y be metric spaces and let

$$h : X \times Y \to \mathbf{R}$$

be a function of two variables. In many cases, it is important to know whether h belongs to the first Baire class if we have some information about the descriptive structure of all partial functions

$$h(x, \cdot) : Y \to \mathbf{R} \quad (x \in X),$$

$$h(\cdot, y) : X \to \mathbf{R} \quad (y \in Y).$$

Here is a simple (but useful) sufficient condition which enables us to claim that $h \in Ba_1(X \times Y, \mathbf{R})$.

Theorem 4. *If the partial functions $h(x, \cdot)$ and $h(\cdot, y)$ are continuous for all $x \in X$ and $y \in Y$, then h is of first Baire class.*

Proof. Take any nonempty closed subset A of $\mathbf{R}$ and, for every natural number $n \geq 1$, denote

$$V_n(A) = \{t \in \mathbf{R} : inf\{|t - a| : a \in A\} < 1/n\}.$$

It is easy to verify that

$$h(x, y) \in A \Leftrightarrow (\forall n \geq 1)(\exists x' \in X)(d(x, x') < 1/n \, \& \, h(x', y) \in V_n(A)),$$

where d stands for the metric in X. This relation directly implies the equality

$$h^{-1}(A) = \cap_{n \geq 1}(\cup_{x' \in X}\{(x, y) \in X \times Y : d(x', x) < 1/n \, \& \, h(x', y) \in V_n(A)\}).$$

Our assumption on h yields at once that, for any $x' \in X$, the set

$$\{(x, y) \in X \times Y : d(x', x) < 1/n \, \& \, h(x', y) \in V_n(A)\}$$

is open in the product space $X \times Y$. Consequently, $h^{-1}(A)$ is a G_δ-subset of $X \times Y$. Therefore, for any open set $B \subset \mathbf{R}$, the preimage $h^{-1}(B)$ turns out to be an F_σ-subset of $X \times Y$. In view of Theorem 2, we claim that h is of first Baire class. This completes the proof.

Remark 7. Theorem 4 can be generalized to those functions

$$h : X \times Y \to \mathbf{R}$$

for which all $h(x, \cdot)$ $(x \in X)$ are continuous and all $h(\cdot, y)$ $(y \in Y)$ belong to the class $Ba_\xi(X, \mathbf{R})$ where ξ is a fixed ordinal number strictly less than ω_1. Such a generalization is presented, e.g., in [120]. It turns out that any function h with the above-mentioned property belongs to the Baire class $Ba_{\xi+1}(X \times Y, \mathbf{R})$. In this context, it is reasonable to underline that the Montgomery operation plays again an essential role for obtaining the generalized result.

From Theorem 4 we easily get the following useful statement.

Theorem 5. *If a function $h : X \times Y \to \mathbf{R}$ is continuous with respect to each of its variables $x \in X$ and $y \in Y$, then h is continuous almost everywhere on $X \times Y$ in the sense of the Baire category.*

Proof. Indeed, by virtue of Theorem 4, our function h belongs to the class $Ba_1(X \times Y, \mathbf{R})$. It remains to apply the Baire theorem on the structure of the set of discontinuity points of functions belonging to the first Baire class (see Theorem 3 and Exercise 7 of this chapter).

Remark 8. A question naturally arises whether a function

$$h : \mathbf{R} \times \mathbf{R} \to \mathbf{R}$$

continuous with respect to each of its variables is continuous almost everywhere in the sense of the two-dimensional Lebesgue measure

$$\lambda_2 = \lambda \times \lambda.$$

Tolstov [221] answered this question negatively: he constructed an example of a function

$$g : \mathbf{R} \times \mathbf{R} \to \mathbf{R}$$

satisfying the following relations:

(1) g is continuous with respect to each of its variables;

(2) the set $D(g)$ has a strictly positive λ_2-measure.

A useful discussion of the topic concerning the separate and joint continuity of functions of several variables is presented in the work by Piotrowski [168].

Remark 9. Also, it is natural to ask about descriptive properties of a function $h : \mathbf{R} \times \mathbf{R} \to \mathbf{R}$ whose all corresponding partial functions are of first Baire class. It turns out that the descriptive structure of such a function can be very complicated. For instance, we will see in Chapter 16 that there exists a function

$$g : \mathbf{R} \times \mathbf{R} \to \mathbf{R}$$

satisfying the following relations:

(1) g is lower semicontinuous with respect to each of its variables;

(2) g is not measurable in the Lebesgue sense (i.e., g is not measurable with respect to λ_2).

On the other hand, it should be mentioned that if a function

$$h : X \times Y \to \mathbf{R}$$

is continuous with respect to $y \in Y$ and is measurable (in some sense) with respect to $x \in X$, then h turns out to be measurable (in an appropriate sense) on the product space $X \times Y$. Actually, these two properties of h are fundamental for functions of two variables and play a significant role in various questions of mathematical analysis, the theory of ordinary differential equations, optimization theory, probability and random processes (cf. Chapters 17 and 18).

Let E be a topological space, $\{f_n : n < \omega\}$ be a sequence of real-valued continuous functions on E and suppose that

$$f(x) = lim_{n \to +\infty} f_n(x) \qquad (x \in E).$$

According to the Baire definition, we have $f \in Ba_1(E, \mathbf{R})$. Naturally, one can ask about necessary and sufficient conditions under which this f is also continuous (i.e., $f \in Ba_0(E, \mathbf{R})$). A convenient sufficient condition is well known from the standard course of mathematical analysis. Namely, if $\{f_n : n < \omega\}$ converges uniformly on E, then f is continuous. However, this condition is very far from being necessary.

Exercise 16. Give an example of a sequence

$$f_n : [0, 1] \to \mathbf{R} \qquad (n < \omega)$$

of continuous functions, which converges pointwise to a continuous function

$$f : [0, 1] \to \mathbf{R},$$

but this convergence is not uniform on $[0, 1]$.

We are going to give here one necessary and sufficient condition for the continuity of the limit function.

Theorem 6. *Let E be a topological space, $\{f_n : n < \omega\}$ be a sequence of real-valued continuous functions on E and let*

$$f(x) = lim_{n \to +\infty} f_n(x) \qquad (x \in E).$$

Then the following two assertions are equivalent:

(1) f is continuous on E;
(2) for any real ε > 0 and for any natural number m, there exists a
natural number n ≥ m such that the set

$$\{x \in E : |f(x) - f_n(x)| < \varepsilon\}$$

is open in E.

Proof. Let us check (1) ⇒ (2). Suppose that relation (1) is valid. Then, for any ε > 0 and for any m < ω, the set

$$\{x \in E : |f(x) - f_m(x)| < \varepsilon\}$$

is open in E. We thus see that, in this case, relation (2) holds automatically.

Let us verify (2) ⇒ (1). Suppose that (2) is valid. We must show that f is continuous at each point $x_0 \in E$. Take an arbitrary ε > 0. In view of (2), there exists a strictly increasing sequence

$$(n_1, n_2, ..., n_k, ...)$$

of natural numbers such that all sets

$$E_k = \{x \in E : |f(x) - f_{n_k}(x)| < \varepsilon/3\}$$

are open in E. Moreover, because $\{f_n : n < \omega\}$ converges to f pointwise, we have the equality

$$E = \cup\{E_k : 1 \leq k < \omega\}.$$

Therefore, there exists a natural number k_0 such that $x_0 \in E_{k_0}$. We thus claim that E_{k_0} is an open neighborhood of x_0. Now, for any point $x \in E_{k_0}$, we may write

$$|f(x) - f(x_0)| \leq |f(x) - f_{n_{k_0}}(x)| + |f_{n_{k_0}}(x) - f_{n_{k_0}}(x_0)| + |f_{n_{k_0}}(x_0) - f(x_0)|$$

$$\leq 2\varepsilon/3 + |f_{n_{k_0}}(x) - f_{n_{k_0}}(x_0)|.$$

Because the function $f_{n_{k_0}}$ is continuous, there exists a neighborhood $U(x_0)$ of x_0 such that

$$U(x_0) \subset E_{k_0}, \quad (\forall x \in U(x_0))(|f_{n_{k_0}}(x) - f_{n_{k_0}}(x_0)| < \varepsilon/3).$$

This immediately implies

$$|f(x) - f(x_0)| < \varepsilon \quad (x \in U(x_0)),$$

which shows the continuity of f at x_0.

Theorem 6 has thus been proved.

Remark 10. The above theorem is a particular case of a more general result concerning functions of the class $Ba_\xi(E, \mathbf{R})$, where ξ is an arbitrary ordinal number strictly less than ω_1. Actually, if

$$\{f_n : n < \omega\} \subset Ba_\xi(E, \mathbf{R})$$

and we have

$$lim_{n \to +\infty} f_n(x) = f(x) \qquad (x \in E),$$

then an appropriate generalization of Theorem 6 yields necessary and sufficient conditions for the validity of the relation $f \in Ba_\xi(E, \mathbf{R})$. For details, see [120].

Exercise 17. Deduce from Theorem 6 the following classical result of Arzelá on the continuity of a limit function.

Let E be a quasicompact topological space, let $\{f_n : n < \omega\} \subset Ba_0(E, \mathbf{R})$ and let

$$f(x) = lim_{n \to +\infty} f_n(x) \qquad (x \in E).$$

Then these two assertions are equivalent:

(a) f is continuous on E;

(b) $\{f_n : n < \omega\}$ converges quasi-uniformly to f; in other words, for any real $\varepsilon > 0$, there exist a finite family $\{V_1, V_2, ..., V_k\}$ of open sets in E and a finite family $\{n_1, n_2, ..., n_k\}$ of natural numbers, such that

$$E = V_1 \cup V_2 \cup ... \cup V_k,$$

$$V_i \subset \{x \in E : |f(x) - f_{n_i}(x)| < \varepsilon\} \qquad (i = 1, 2, ..., k).$$

Note that in (b) it is not required that the natural numbers $n_1, n_2, ..., n_k$ would be arbitrarily large.

Exercise 18. Deduce from the above-mentioned result of Arzelá another classical theorem due to Dini.

Namely, let E be a quasicompact topological space and let $\{f_n : n < \omega\}$ be a monotone sequence of real-valued continuous functions on E. Suppose also that

$$f(x) = lim_{n \to +\infty} f_n(x) \qquad (x \in E).$$

Then the following two assertions are equivalent:

(a) f is continuous on E;

(b) $\{f_n : n < \omega\}$ converges uniformly to f.

Exercise 19. Let E be a topological space, X be a subset of E and let $e \in E$ be an accumulation point of X. Suppose, in addition, that a sequence $\{f_n : n < \omega\}$ of real-valued functions on X is given such that:

(a) the series $\sum_{n<\omega} f_n(x)$ converges uniformly with respect to $x \in X$ and

$$f(x) = \sum_{n<\omega} f_n(x) \qquad (x \in X);$$

(b) for each $n < \omega$, there exists a $lim_{x \to e} f_n(x)$ and

$$a_n = lim_{x \to e} f_n(x).$$

Demonstrate that the series $\sum_{n<\omega} a_n$ converges to some $a \in \mathbf{R}$ and we have the equality

$$lim_{x \to e} f(x) = a.$$

Exercise 20. Let $\{f_n : n < \omega\}$ be a sequence of real-valued differentiable functions defined on a segment $T \subset \mathbf{R}$. Suppose that the series $\sum_{n<\omega} f_n$ is convergent at some point t_0 of T and the series $\sum_{n<\omega} f'_n$ is uniformly convergent on T.

By using the result of the previous exercise, demonstrate that:

(a) the series $\sum_{n<\omega} f_n$ converges uniformly on T;
(b) the function $f = \sum_{n<\omega} f_n$ is differentiable on T;
(c) for all points $t \in T$, we have

$$f'(t) = \sum_{n<\omega} f'_n(t).$$

Infer from this fact that the family of all bounded derivatives on T is a Banach subspace of $Ba_1(T, \mathbf{R})$ with respect to the norm of uniform convergence (i.e., with respect to the standard sup-norm).

3. Semicontinuous functions that are not countably continuous

Luzin's theorem on the structure of Lebesgue measurable functions acting from $\mathbf{R}$ into $\mathbf{R}$ is one of the most fundamental statements in real analysis and has numerous applications. Let us recall the formulation of this classical theorem. It is convenient for us to give the formulation in terms of partial functions (cf. the Introduction). As usual, we denote by λ the Lebesgue measure on $\mathbf{R}$.

Let $f : \mathbf{R} \to \mathbf{R}$ be a partial function. We shall say that f is measurable in the Lebesgue sense (or, simply, f is λ-measurable) if the set $f^{-1}(U)$ is Lebesgue measurable for every open subset U of $\mathbf{R}$.

It immediately follows from this definition that the domain of any Lebesgue measurable partial function f is a λ-measurable subset of $\mathbf{R}$ (because of the equality $dom(f) = f^{-1}(\mathbf{R})$).

Theorem 1. *For any partial function $g : \mathbf{R} \to \mathbf{R}$ that is measurable in the Lebesgue sense and for any real $\varepsilon > 0$, there exists a closed set $F \subset \mathbf{R}$ such that*

$$F \subset dom(g), \quad \lambda(dom(g) \setminus F) < \varepsilon$$

and the restriction $g|F$ is continuous.

We omit the standard proof of Theorem 1 (see, for example, [135], [154]). Note only that there are various generalizations and extensions of this classical result (see, e.g., Exercise 1 from Chapter 7, Exercises 5 and 6 from Chapter 12).

Let us mention a direct consequence of the above-mentioned Luzin theorem.

Theorem 2. *Let $f : \mathbf{R} \to \mathbf{R}$ be a Lebesgue measurable function. Then there exists a disjoint countable covering $\{A_n : n < \omega\}$ of $\mathbf{R}$ such that*

$$\lambda(A_0) = 0$$

and, for each integer $n \geq 1$, the set A_n is closed in $\mathbf{R}$ and the function $f|A_n$ is continuous.

Exercise 1. Using the notation of Theorem 2, show that if

$$A_0 = \emptyset,$$

then f is a function of first Baire class (see Chapter 2) and, in particular, f is a Borel function. Conclude from this fact that, for a general Lebesgue measurable function $f : \mathbf{R} \to \mathbf{R}$, the set A_0 cannot be ignored.

Taking into account Theorem 2 and Exercise 1, it is natural to pose the following question:

Let $f : \mathbf{R} \to \mathbf{R}$ be a Borel function. Does there exist a countable covering $\{A_n : 1 \leq n < \omega\}$ of $\mathbf{R}$ such that the function $f|A_n$ is continuous for each natural number $n \geq 1$?

This question was originally raised by Luzin many years ago. Adian and Novikov [1] gave a negative answer to it (see also Sierpiński [201]).

Moreover, these authors were able to construct an upper semicontinuous function

$$f : \mathbf{R} \to \mathbf{R}$$

that does not admit a decomposition into countably many continuous partial functions (in other words, f is not countably continuous).

We are going to present here the construction of Adian and Novikov. Then we will consider some related results that are due to other authors and are also motivated by the Luzin problem posed above.

First, let us introduce some preliminary definitions.

Let E be a topological space and let $\mathcal{I}$ be a σ-ideal of subsets of E.

We shall say that a function

$$f : E \to \mathbf{R}$$

is countably continuous $(mod(\mathcal{I}))$ if there exists a covering $\{A_n : n < \omega\}$ of E such that $A_0 \in \mathcal{I}$ and the restriction $f|A_n$ is continuous for each natural number $n \geq 1$.

Example 1. Let E be a topological space of second category on itself and let $\mathcal{K}(E)$ denote the σ-ideal of all first category subsets of E. We know that, for any function $f : E \to \mathbf{R}$ possessing the Baire property, there exists a first category set $X \subset E$ such that $f|(E \setminus X)$ is continuous (see Exercise 13 from the Introduction). Thus, f is countably continuous $(mod(\mathcal{K}(E)))$.

Example 2. In view of Theorem 2, every Lebesgue measurable function $f : \mathbf{R} \to \mathbf{R}$ is countably continuous $(mod(\mathcal{I}(\lambda)))$, where $\mathcal{I}(\lambda)$ denotes the σ-ideal of all Lebesgue measure zero subsets of $\mathbf{R}$.

If a σ-ideal $\mathcal{I}$ is trivial, i.e. $\mathcal{I} = \{\emptyset\}$, and a function $f : E \to \mathbf{R}$ is countably continuous $(mod(\mathcal{I}))$, then we simply say that f is countably continuous.

To give the construction of Adian and Novikov, we need several auxiliary propositions.

Lemma 1. *Let E be a topological space, let $\{f_n : n < \omega\}$ be a sequence of real-valued upper (lower) semicontinuous functions on E and suppose that*

$$f(x) = lim_{n \to +\infty} f_n(x) \qquad (x \in E)$$

uniformly with respect to $x \in E$. Then f is also upper (lower) semicontinuous on E.

Proof. Obviously, it suffices to consider the case where all f_n $(n < \omega)$ are upper semicontinuous. Take any real $\varepsilon > 0$ and fix $x_0 \in E$. For each $n < \omega$ and for each $x \in E$, we can write

$$f(x) - f(x_0) \le |f(x) - f_n(x)| + (f_n(x) - f_n(x_0)) + |f_n(x_0) - f(x_0)|$$

in view of the elementary inequality

$$a + b \le |a| + |b| \qquad (a \in \mathbf{R}, \ b \in \mathbf{R}).$$

According to our assumption, there exists $m < \omega$ such that

$$|f(x) - f_n(x)| < \varepsilon/3 \qquad (x \in E, \ n = m, m+1, ...).$$

Consequently, we have

$$f(x) - f(x_0) \le (f_m(x) - f_m(x_0)) + 2\varepsilon/3 \qquad (x \in E).$$

Because f_m is upper semicontinuous at x_0, there exists a neighborhood $V(x_0)$ of x_0 such that

$$f_m(x) - f_m(x_0) < \varepsilon/3 \qquad (x \in V(x_0)).$$

This immediately implies the relation

$$f(x) - f(x_0) < \varepsilon \qquad (x \in V(x_0)).$$

Actually, we have proved that if all functions f_n $(n < \omega)$ are upper semicontinuous at x_0 and the sequence $\{f_n : n < \omega\}$ converges uniformly to f, then f is also upper semicontinuous at x_0.

The following exercise contains a more general result.

Exercise 2. Let E be a topological space, x_0 be a point from E, and let $\{f_n : n < \omega\}$ be a sequence of real-valued functions on E. Suppose that this sequence converges quasi-uniformly to a function f (see Exercise 17 from Chapter 2) and that all f_n $(n < \omega)$ are upper (lower) semicontinuous at the point x_0. Prove that f is also upper (lower) semicontinuous at x_0.

The next two lemmas contain some easy results concerning extensions of semicontinuous partial functions.

Lemma 2. *Let E be a topological space, X be a subset of E and let*

$$f : X \to \mathbf{R}$$

be a bounded from above (respectively, from below) upper (respectively, lower) semicontinuous function. Then there exists a function

$$f^* : cl(X) \to \mathbf{R}$$

that extends f, is upper (respectively, lower) semicontinuous and satisfies the equality

$$sup_{x \in X} f(x) = sup_{x \in cl(X)} f^*(x)$$

(respectively, the equality $inf_{x \in X} f(x) = inf_{x \in cl(X)} f^*(x)$).

Proof. We shall consider only the case of a bounded from above upper semicontinuous function $f : X \to \mathbf{R}$. Let z be any point from $cl(X)$. We put

$$f^*(z) = limsup_{x \to z, x \in X} f(x).$$

Obviously, the function f^* extends f (since f is upper semicontinuous). Let us verify that f^* is also upper semicontinuous at all points of $cl(X)$.

Take again an arbitrary point $z \in cl(X)$ and fix a real number t such that $f^*(z) < t$. Then, for sufficiently small reals $\varepsilon > 0$, we have

$$f^*(z) \leq t - \varepsilon.$$

Consider any such ε. By the definition of f^*, we claim that there exists an open neighborhood $V(z)$ of z for which

$$sup_{x \in V(z) \cap X} f(x) \leq t - \varepsilon/2.$$

In view of the same definition, we get

$$sup_{x \in V(z) \cap cl(X)} f^*(x) \leq t - \varepsilon/2$$

or, equivalently,

$$f^*(x) \leq t - \varepsilon/2 < t \qquad (x \in V(z) \cap cl(X)).$$

This establishes that the set $\{x \in cl(X) : f^*(x) < t\}$ is open in $cl(X)$, i.e. f^* is upper semicontinuous on $cl(X)$. Finally, the equality

$$sup_{x \in X} f(x) = sup_{x \in cl(X)} f^*(x)$$

also follows from the definition of f^*. Lemma 2 has thus been proved (cf. Exercise 22 of the Introduction).

Lemma 3. *Let E be a normal topological space, X be a subset of E and let $[a,b]$ be a segment in $\mathbf{R}$. Suppose that*

$$f : X \to [a,b]$$

is an upper (respectively, lower) semicontinuous function. Then there exists an upper (respectively, lower) semicontinuous function

$$f^* : E \to [a,b]$$

which extends the original f.

Proof. It suffices to consider only the case of a lower semicontinuous function $f : X \to [a,b]$. By virtue of Lemma 2, we may assume without loss of generality that X is a closed subset of E. Further, because X is completely regular, we have

$$f(x) = sup_{i \in I} f_i(x) \qquad (x \in X),$$

where $\{f_i : i \in I\}$ is some family of real-valued continuous functions on X and

$$ran(f_i) \subset [a,b]$$

for all $i \in I$ (cf. Exercise 18 from the Introduction). By the Tietze-Urysohn theorem, for any index $i \in I$, there exists a continuous function

$$f_i^* : E \to [a,b]$$

extending f_i. Now, let us define

$$f^*(x) = sup_{i \in I} f^*(x) \qquad (x \in E).$$

Clearly, f^* is the required lower semicontinuous extension of f. This ends the proof of Lemma 3.

Lemma 4. *Let E be a topological space, $f : E \to \mathbf{R}$ be a function and let X be a subset of E. If $f|X$ is not countably continuous, then f is not countably continuous.*

The proof of this lemma is left to the reader.

Lemma 5. *Suppose that there exists a real-valued bounded upper semi-continuous function on the Cantor space C, which is not countably continuous. Then an analogous function exists on any uncountable Polish space E.*

Proof. It is well known that E contains a topological copy of C (see, e.g., [120]), so we may assume that $C \subset E$. Because C is compact, it can be regarded as a closed subset of E. Let

$$f : C \to \mathbf{R}$$

be a bounded upper semicontinuous function that is not countably continuous. According to Lemma 3, there exists a bounded upper semicontinuous function

$$f^* : E \to \mathbf{R}$$

extending f. By virtue of Lemma 4, f^* is not countably continuous. This completes the proof of Lemma 5.

Lemma 6. *Let E be a nonempty Polish topological space without isolated points. There exists a disjoint countable family $\{C_n : n < \omega\}$ of subsets of E such that:*
(1) each C_n is homeomorphic to the Cantor discontinuum C;
(2) each C_n is nowhere dense in E;
(3) the set $\cup\{C_n : n < \omega\}$ is everywhere dense in E.

Proof. Denote by $\{U_n : n < \omega\}$ a base of nonempty open sets in E. We are going to construct the required family by using ordinary recursion. Suppose that, for a given natural number n, the partial family $\{C_i : i < n\}$ has already been defined. Consider the set

$$U = U_n \setminus \cup\{C_i : i < n\}.$$

Clearly, U is a nonempty open set in E without isolated points. It can easily be shown that there exists a countable set $D \subset U$ dense in U and such that $U \setminus D$ is also dense in U. Now, because $U \setminus D$ is a G_δ-subset of E, we claim that $U \setminus D$ is a Polish space without isolated points. Consequently, there exists a homeomorphic copy X of C contained in $U \setminus D$. It suffices to put $C_n = X$. Proceeding in this manner, we will be able to construct the required family $\{C_n : n < \omega\}$.

The reader can easily check himself (herself) the validity of relations (1), (2), and (3) for this family.

Exercise 3. Let E be a Polish topological space and let μ be a nonzero σ-finite Borel measure on E vanishing on all singletons in E. Show that there exists a subset P of E satisfying the following relations:

(a) P is homeomorphic to the Cantor space C;

(b) P is nowhere dense in E;

(c) $\mu(P) > 0$.

In fact, this exercise can be regarded as a generalization of Exercise 9 from Chapter 1.

Lemma 7. *Let E be again a nonempty Polish space without isolated points. There exists a countable family*

$$\{C_{n_1,n_2,...,n_k} : n_1 \geq 1,\ n_2 \geq 1,\ ...\ ,\ n_k \geq 1,\ k < \omega\}$$

of subsets of E satisfying the following conditions:

(1) $C_\emptyset = E$;

(2) for all integers $k \geq 1, n_1 \geq 1, n_2 \geq 1, ..., n_k \geq 1$, the set $C_{n_1,n_2,...,n_k}$ is homeomorphic to the Cantor discontinuum C;

(3) $C_{n_1,n_2,...,n_k,n_{k+1}} \subset C_{n_1,n_2,...,n_k}$ and $C_{n_1,n_2,...,n_k,n_{k+1}}$ is nowhere dense in $C_{n_1,n_2,...,n_k}$;

(4) the set $\cup\{C_{n_1,n_2,...,n_k,n} : 1 \leq n < \omega\}$ is everywhere dense in the set $C_{n_1,n_2,...,n_k}$;

(5) if $(n_1, n_2, ..., n_k) \neq (m_1, m_2, ..., m_k)$, then

$$C_{n_1,n_2,...,n_k} \cap C_{m_1,m_2,...,m_k} = \emptyset;$$

(6) for any natural number $k \geq 1$, we have

$$diam(C_{n_1,n_2,...,n_k}) < 1/2^k.$$

Proof. This lemma can easily be deduced from Lemma 6 by using the induction method. The details are left to the reader. Let us only remark that all sets

$$C(k) = \cup\{C_{n_1,n_2,...,n_k} : n_1 \geq 1, n_2 \geq 1, ..., n_k \geq 1\} \qquad (k < \omega)$$

are everywhere dense in E. This fact will be applied below.

Lemma 8. *Under the notation of Lemma 7 and of its proof, the set*

$$R = \cap\{C(k) : k < \omega\}$$

is everywhere dense in E. Moreover, the same R is everywhere dense in each set $C_{n_1, n_2, \ldots, n_k}$.

Proof. Let us establish the density of R in E. We use the standard argument. Take a closed ball B in E. Since $\cup\{C_n : n \geq 1\}$ is dense in E, there exists C_{n_1} such that

$$int(B) \cap C_{n_1} \neq \emptyset.$$

Analogously, since $\cup\{C_{n_1,n} : n \geq 1\}$ is dense in C_{n_1}, there exists C_{n_1,n_2} such that

$$int(B) \cap C_{n_1,n_2} \neq \emptyset.$$

Proceeding in this manner, we will be able to define by recursion the sequence of sets

$$C_{n_1} \supset C_{n_1,n_2} \supset \ldots \supset C_{n_1,n_2,\ldots,n_k} \supset \ldots$$

such that

$$int(B) \cap C_{n_1,n_2,\ldots,n_k} \neq \emptyset \quad (k < \omega).$$

It follows from this circumstance that

$$B \cap C_{n_1} \cap C_{n_1,n_2} \cap \ldots \cap C_{n_1,n_2,\ldots,n_k} \cap \ldots \neq \emptyset.$$

In view of the inclusion

$$C_{n_1} \cap C_{n_1,n_2} \cap \ldots \cap C_{n_1,n_2,\ldots,n_k} \cap \ldots \subset R,$$

we immediately obtain

$$B \cap R \neq \emptyset,$$

which shows that R is everywhere dense in E. By using the same argument, one can prove the density of R in any set $C_{n_1,n_2,\ldots,n_k}$ (it suffices to replace E by the last set).

Lemma 9. *We preserve the notation of the two preceding lemmas. For each natural number $k \geq 1$, define a function*

$$f_k : E \to [0,1]$$

by the formula: $f_k(x) = 0$ if $x \notin C(k)$, and

$$f_k(x) = \frac{1}{2^{n_1 + n_2 + \ldots + n_k}}$$

if $x \in C_{n_1,n_2,...,n_k}$.

Then the following relations hold:
(1) $0 \leq f_k(x) \leq 1/2^k$ for any $x \in E$;
(2) f_k is upper semicontinuous on E.

Proof. Relation (1) is almost trivial. Indeed, in view of the inequalities

$$n_1 \geq 1, \quad n_2 \geq 1, \quad ..., \quad n_k \geq 1,$$

we have

$$0 \leq f_k(x) \leq \frac{1}{2^{n_1+n_2+...+n_k}} \leq \frac{1}{2^k}.$$

Let us establish the validity of relation (2). Take any $x \in E$ and show that f_k is upper semicontinuous at x. Assume first that $x \in C_{n_1,n_2,...,n_k}$. We must verify that

$$limsup_{i \to +\infty} f_k(x_i) \leq f_k(x)$$

whenever a sequence $\{x_i : i < \omega\} \subset E$ converges to x. Suppose to the contrary that

$$limsup_{i \to +\infty} f_k(x_i) > f_k(x)$$

for some sequence $\{x_i : i < \omega\} \subset E$ converging to x. Then among the members of $\{x_i : i < \omega\}$ there are infinitely many points $x_i \in C_{m_1,m_2,...,m_k}$ where

$$m_1 + m_2 + ... + m_k < n_1 + n_2 + ... + n_k.$$

Hence there exists a fixed $(m_1, m_2, ..., m_k)$ satisfying the above-mentioned inequality and such that infinitely many points x_i belong to $C_{m_1,m_2,...,m_k}$. By virtue of the relation $lim_{i \to +\infty} x_i = x$, we obtain

$$x \in C_{m_1,m_2,...,m_k} \cap C_{n_1,n_2,...,n_k},$$

whence it follows that

$$C_{m_1,m_2,...,m_k} \cap C_{n_1,n_2,...,n_k} \neq \emptyset,$$

which is impossible because $(m_1, m_2, ..., m_k) \neq (n_1, n_2, ..., n_k)$. The contradiction obtained shows the validity of the inequality

$$limsup_{i \to +\infty} f_k(x_i) \leq f_k(x).$$

The case when $x \notin C(k)$ can be considered analogously. This ends the proof of the lemma.

Lemma 10. *Preserve the same notation as in Lemma 9 and define a function*

$$f : E \to \mathbf{R}$$

by the formula

$$f(x) = f_1(x) + f_2(x) + ... + f_k(x) + ... \qquad (x \in E).$$

Then the following relations are satisfied:
(1) f is upper semicontinuous on E;
(2) $0 \le f(x) \le 1$ for all $x \in E$;
(3) if $x \in C_{n_1, n_2, ..., n_k} \setminus C(k+1)$, then

$$f(x) = \frac{1}{2^{n_1}} + \frac{1}{2^{n_1 + n_2}} + ... + \frac{1}{2^{n_1 + n_2 + ... + n_k}};$$

(4) if $x \in C_{n_1} \cap C_{n_1, n_2} \cap ... \cap C_{n_1, n_2, ..., n_k} \cap ...,$ then

$$f(x) = \frac{1}{2^{n_1}} + \frac{1}{2^{n_1 + n_2}} + ... + \frac{1}{2^{n_1 + n_2 + ... + n_k}} + ...$$

Proof. Note first that our f is well defined because the function series $\sum_{k \ge 1} f_k$ converges uniformly on E. Further, in view of Lemmas 1 and 9, f is upper semicontinuous on E. Relation (2) also follows from Lemma 9. Relations (3) and (4) are more or less trivial and are left to the reader.

Lemma 10 has thus been proved.

Now, we are able to present the following result (essentially due to Adian and Novikov).

Theorem 1. *Let E be a topological copy of the Cantor space C. Under the same notation, the function f defined in the previous lemma is not countably continuous.*

Proof. Take any countable covering $\{D_n : n \ge 1\}$ of the space E and denote

$$A_n = R \cap D_n \qquad (n \ge 1).$$

In this way we come to the countable covering $\{A_n : n \ge 1\}$ of the set R.

First, we shall show that there exist an integer $k \ge 1$ and a k-sequence $(n_1, n_2, ..., n_k)$ of natural numbers, such that the set A_k is not nowhere dense in the set $C_{n_1, n_2, ..., n_k}$, i.e.,

$$int(cl(A_k \cap C_{n_1, n_2, ..., n_k})) \ne \emptyset,$$

where the topological operations *int* and *cl* act in the space $C_{n_1,n_2,...,n_k}$.

Suppose otherwise and fix $n_1 \geq 1$. According to our assumption, for A_1, there exists a nonempty clopen set B_1 in C_{n_1} such that

$$A_1 \cap B_1 = \emptyset.$$

Because $\cup\{C_{n_1,n} : n \geq 1\}$ is dense in C_{n_1}, there exist C_{n_1,n_2} and a nonempty clopen set $B_2 \subset C_{n_1,n_2}$ such that

$$B_2 \subset C_{n_1,n_2} \cap B_1, \quad A_2 \cap B_2 = \emptyset.$$

Continuing this process by recursion, we obtain a sequence $n_1, n_2, ..., n_k, ...$ of natural numbers and a decreasing sequence of sets $B_1, B_2, ..., B_k, ...$ such that

$$A_k \cap B_k = \emptyset \quad (k \geq 1).$$

Because all B_k are compacts, we have

$$\cap\{B_k : k \geq 1\} \neq \emptyset.$$

Also, we can write

$$\cap\{B_k : k \geq 1\} \subset R = \cup\{A_k : k \geq 1\},$$

which yields a contradiction with the relation

$$(\cap\{B_k : k \geq 1\}) \cap (\cup\{A_k : k \geq 1\}) = \emptyset.$$

The contradiction obtained shows that, for some A_k and $C_{n_1,n_2,...,n_k}$, the set $A_k \cap C_{n_1,n_2,...,n_k}$ is not nowhere dense in $C_{n_1,n_2,...,n_k}$. Choose any nonempty clopen set $B \subset C_{n_1,n_2,...,n_k}$ in which A_k is dense. Obviously, we may write

$$B \cap A_k =$$

$$(\cup\{B \cap A_k \cap C_{n_1,n_2,...,n_k,i} : i \leq 2m\}) \cup (\cup\{B \cap A_k \cap C_{n_1,n_2,...,n_k,i} : i \geq 2m+1\})$$

for any natural number $m \geq 1$. The set

$$\cup\{B \cap A_k \cap C_{n_1,n_2,...,n_k,i} : i \leq 2m\}$$

is nowhere dense in $B \cap A_k$. Therefore, the set

$$\cup\{B \cap A_k \cap C_{n_1,n_2,...,n_k,i} : i \geq 2m+1\}$$

is everywhere dense in $B \cap A_k$.

Let us establish that $f|A_k$ is discontinuous at each point of $B \cap A_k$. For this purpose, take any $x_0 \in B \cap A_k$. Because $x_0 \in R$, we have

$$f(x_0) = \frac{1}{2^{n_1}} + \frac{1}{2^{n_1+n_2}} + ... + \frac{1}{2^{n_1+n_2+...+n_k}} + \frac{1}{2^{n_1+n_2+...+n_k+m}} + ...,$$

where $m \geq 1$ is some natural number. Take

$$\varepsilon_0 = \frac{1}{2^{n_1+n_2+...+n_k+2m}}.$$

As said above, every neighborhood of x_0 contains points x belonging to the set

$$\cup\{B \cap A_k \cap C_{n_1,n_2,...,n_k,i} : i \geq 2m + 1\}.$$

Consequently, for all such x, we may write

$$f(x) \leq \frac{1}{2^{n_1}} + ... + \frac{1}{2^{n_1+n_2+...+n_k}} + \frac{1}{2^{n_1+n_2+...+n_k+2m}} \leq f(x_0) - \varepsilon_0,$$

which yields at once that $f|A_k$ is discontinuous at x_0. In this manner, we have established that $f|A_k$ is not continuous and hence the original function f cannot be countably continuous.

The result presented above is essentially based on specific properties of the classical Cantor space C. However, by taking into account Lemma 5, one can easily conclude that the following more general statement is also true.

Theorem 2. *Let E be an arbitrary uncountable Polish space. Then there exists an upper (lower) semicontinuous bounded function*

$$f : E \to \mathbf{R}$$

that is not countably continuous.

After Theorem 2 was proved, a number of publications have appeared and extended the investigation of those real-valued functions that possess a rather good descriptive structure but are not countably continuous.

In particular, Jackson and Mauldin proved in [76] that the Lebesgue measure

$$\lambda : \mathcal{F}([0,1]) \to [0,1]$$

restricted to the family of all nonempty compact subsets of $[0,1]$, equipped with the Hausdorff metric, is not countably continuous. Note, in this context, that λ is upper semicontinuous (see Exercise 19 from Chapter 1).

Another proof of the fact that λ is not countably continuous was given by van Mill and Pol in their joint paper [148]. In the same paper they considered many other examples and constructions of semicontinuous functions that fail to be countably continuous. One of their results states that in the Banach space V of bounded functions defined on $[0, 1]$ and belonging to $Ba_1([0, 1], \mathbf{R})$ the family of all countably continuous functions forms a first category set. An analogous result holds true for the Banach space V_d of all bounded derivatives on $[0, 1]$ (cf. Exercise 20 from Chapter 2, where it is indicated that the space of bounded derivatives is Banach with respect to the norm of uniform convergence).

Exercise 4. Let C denote again the classical Cantor discontinuum on $[0, 1]$, let $[0, 1]^\omega$ denote the Hilbert cube and let

$$\phi : C \to [0, 1]^\omega$$

be a Peano type function (see, e.g., Chapter 1). Let us define two functions

$$f_1 : [0, 1]^\omega \to C,$$

$$f_2 : [0, 1]^\omega \to C$$

by the formulas

$$f_1(x) = min(\phi^{-1}(x)), \quad f_2(x) = max(\phi^{-1}(x)) \quad (x \in [0, 1]^\omega).$$

Demonstrate that:
(1) both compositions $\phi \circ f_1$ and $\phi \circ f_2$ coincide with the identical transformation of $[0, 1]^\omega$ (in particular, both f_1 and f_2 are injections);
(2) f_1 is lower semicontinuous;
(3) f_2 is upper semicontinuous.

Exercise 5. Preserve the notation of Exercise 4. By using the Hurewicz theorem stating that the Hilbert cube $[0, 1]^\omega$ is not a countable union of its zero-dimensional subspaces (see, for instance, [120] or [147]) and that any nonempty subspace of C is zero-dimensional, prove that neither f_1 nor f_2 are countably continuous.

Remark 1. The result presented in Exercise 5 is due to van Mill and Pol [148]. In Chapter 10 we shall return to real-valued semicontinuous functions which are not countably continuous. Moreover, some other constructions of such functions will be given in that chapter and interesting connections with so-called Luzin sets on $\mathbf{R}$ will be indicated.

Remark 2. Laczkovich obtained a natural generalization of Theorem 2 to functions of higher Baire classes, which act from an uncountable Polish space into $\mathbf{R}$ (unpublished manuscript). His result was reproved and generalized by Cichoń and Morayne in [36]. Their approach is essentially based on the classical techniques of universal functions (cf. Exercise 6 from Chapter 7).

Further information about this topic can be found, e.g., in works by Cichoń, Morayne, Pawlikowski, Solecki [37] and Darji [44], [45] (cf. also [38]).

4. Singular monotone functions

This chapter is devoted to some elementary properties of monotone functions acting from $\mathbf{R}$ into $\mathbf{R}$, and to some widely known examples of strange monotone functions. Let

$$f : \mathbf{R} \to \mathbf{R}$$

be a partial function. We recall that f is said to be increasing (respectively, strictly increasing) if, for any two points $x \in dom(f)$ and $y \in dom(f)$, the relation $x \le y$ (respectively, $x < y$) implies the relation $f(x) \le f(y)$ (respectively, $f(x) < f(y)$).

Analogously the notion of a decreasing (respectively, strictly decreasing) partial function can be introduced. It is easy to see that f is increasing (strictly increasing) if and only if $-f$ is decreasing (strictly decreasing).

A partial function from $\mathbf{R}$ into $\mathbf{R}$ is said to be monotone (respectively, strictly monotone) if it is either increasing or decreasing (respectively, either strictly increasing or strictly decreasing).

We shall consider below only increasing partial functions (this, of course, does not restrict the generality of our considerations).

Let $f : \mathbf{R} \to \mathbf{R}$ be an increasing bounded partial function and suppose that

$$dom(f) \ne \emptyset.$$

Fix a point t from $dom(f)$. For any $x \le t$, put

$$f^*(x) = inf\{f(z) : z \in dom(f), \ x \le z \le t\}$$

and, for any $x \ge t$, put

$$f^*(x) = sup\{f(z) : z \in dom(f), \ t \le z \le x\}.$$

It can easily be demonstrated that f^* is an increasing bounded function acting from $\mathbf{R}$ into $\mathbf{R}$ and extending f.

We thus see that every increasing bounded partial function admits an increasing bounded extension defined on the whole $\mathbf{R}$. A similar argument shows that every increasing partial function $f : \mathbf{R} \to \mathbf{R}$ can be extended to an increasing function defined on some subinterval of $\mathbf{R}$. In view of this circumstance, we primarily will be dealing with increasing functions acting from a subinterval of $\mathbf{R}$ into $\mathbf{R}$.

If $f : \mathbf{R} \to \mathbf{R}$ is an increasing function and $x \in \mathbf{R}$, then there exist limits

$$lim_{y \to x, \; y > x} f(y) = f(x+),$$

$$lim_{y \to x, \; y < x} f(y) = f(x-),$$

and we have the inequalities

$$f(x-) \leq f(x) \leq f(x+).$$

Obviously f is continuous at x if and only if

$$f(x-) = f(x) = f(x+).$$

Thus, x is a discontinuity point of f if and only if

$$f(x-) < f(x) \quad \vee \quad f(x) < f(x+).$$

More generally, let

$$g : \mathbf{R} \to \mathbf{R}$$

be an arbitrary function and let $x \in \mathbf{R}$. We say that x is a simple discontinuity point for g if there exist $g(x-)$ and $g(x+)$, but

$$g(x-) \neq g(x) \quad \vee \quad g(x+) \neq g(x).$$

Evidently, if g is a monotone function, then all discontinuity points for g are its simple discontinuity points.

We have the following useful result.

Theorem 1. *Let f be an arbitrary function acting from $\mathbf{R}$ into $\mathbf{R}$. Then the set of all simple discontinuity points for f is at most countable.*

Proof. Denote by E the set of all simple discontinuity points for f. Denote also:

$$E_1 = \{x \in E \; : \; f(x-) < f(x+)\},$$

$$E_2 = \{x \in E \; : \; f(x-) > f(x+)\},$$

$$E_3 = \{x \in E \; : \; f(x) < f(x-) = f(x+)\},$$

$$E_4 = \{x \in E \; : \; f(x) > f(x-) = f(x+)\}.$$

Then we can write

$$E = E_1 \cup E_2 \cup E_3 \cup E_4,$$

ant it suffices to demonstrate that each set E_i $(i = 1, 2, 3, 4)$ is at most countable. Because the argument is similar for all E_i, we shall only show that

$$card(E_1) \leq \omega.$$

For this purpose, let us define a mapping

$$x \to (p(x), q(x), r(x)) \qquad (x \in E_1)$$

such that

$$p(x) \in \mathbf{Q}, \quad q(x) \in \mathbf{Q}, \quad r(x) \in \mathbf{Q},$$

$$x \in \,]q(x), r(x)[,$$

$$(\forall t \in \,]q(x), x[)(f(t) < p(x)),$$

$$(\forall t \in \,]x, r(x)[)(f(t) > p(x)).$$

The existence of such a mapping is obvious. Now, it is not hard to show that this mapping is injective. Indeed, suppose to the contrary that, for some distinct points x_1 and x_2 from $\mathbf{R}$, we have the equality

$$(p(x_1), q(x_1), r(x_1)) = (p(x_2), q(x_2), r(x_2)).$$

Without loss of generality we may assume that $x_1 < x_2$. Choose any point t from $]x_1, x_2[$. Then we must have simultaneously

$$f(t) < p(x_2) = p(x_1),$$

$$f(t) > p(x_1) = p(x_2),$$

which is impossible. The contradiction obtained establishes that the mapping

$$x \to (p(x), q(x), r(x)) \qquad (x \in E_1)$$

is injective, which, obviously, implies the relation

$$card(E_1) \leq card(\mathbf{Q} \times \mathbf{Q} \times \mathbf{Q}) = \omega.$$

Theorem 1 has thus been proved.

Exercise 1. Recall that a function $f : \mathbf{R} \to \mathbf{R}$ possesses the Darboux property if, for each subinterval $[a, b]$ of $\mathbf{R}$, the range of f contains the segment with the end-points $f(a)$ and $f(b)$.

Demonstrate that any function with the Darboux property has no simple discontinuity points. In particular, infer from this fact that if f is the derivative of some function acting from $\mathbf{R}$ into $\mathbf{R}$, then f has no simple discontinuity points.

As a trivial consequence of Theorem 1, we obtain that, for any monotone function f acting from $\mathbf{R}$ into $\mathbf{R}$, the set $D(f)$ of all discontinuity points of f is at most countable.

Exercise 2. Let $E = \{x_n \; : \; n \in \mathbf{N}\}$ be an arbitrary countable subset of $\mathbf{R}$ and let $\{r_n \; : \; n \in \mathbf{N}\}$ be a countable family of strictly positive real numbers, such that

$$\sum_{n \in \mathbf{N}} r_n < +\infty.$$

For any $x \in \mathbf{R}$, let us put

$$f(x) = \sum_{n \in N(x)} r_n,$$

where

$$N(x) = \{n \in \mathbf{N} \; : \; x_n < x\}.$$

In this way a certain function f acting from $\mathbf{R}$ into $\mathbf{R}$ is defined. Show that:
  (a) f is increasing;
  (b) f is continuous at each point from $\mathbf{R} \setminus E$;
  (c) for any natural index n, we have

$$f(x_n+) - f(x_n-) = r_n;$$

in particular, f is discontinuous at each point of the given set E.

Deduce from this result that if E is everywhere dense in $\mathbf{R}$ (for example, if $E = \mathbf{Q}$), then the function f constructed above has an everywhere dense set of its discontinuity points.

We now are going to present the classical Lebesgue theorem concerning the differentiation of monotone functions. For this purpose, we need three simple lemmas (cf. [154]).

First, let us recall the notion of a derived number for functions acting from $\mathbf{R}$ into $\mathbf{R}$. Suppose that $[a, b]$ is a segment of $\mathbf{R}$ and that $f : [a, b] \to \mathbf{R}$ is a function. Let $x \in [a, b]$. We shall say that $t \in \mathbf{R} \cup \{-\infty, +\infty\}$ is a derived number (or a Dini derived number) of f at x if there exists a sequence $\{x_n \; : \; n \in \mathbf{N}\}$ of points from $[a, b]$ tending to x, such that

$$(\forall n \in \mathbf{N})(x_n \neq x), \quad lim_{n \to +\infty} \frac{f(x_n) - f(x)}{x_n - x} = t.$$

In this case, we shall write
$$t = f'_D(x).$$

One more remark. For any two real numbers t_1 and t_2, it will be convenient to denote below by the symbol $[t_1, t_2]$ the segment of $\mathbf{R}$ with the end-points t_1 and t_2. Thus, we do not assume in this notation that $t_1 < t_2$.

Lemma 1. *Let $f : [a, b] \to \mathbf{R}$ be a strictly increasing function, let q be a positive real number and let X be a subset of $[a, b]$ such that, for any point $x \in X$, there exists at least one derived number $f'_D(x) \le q$. Then we have the inequality*
$$\lambda^*(f(X)) \le q\lambda^*(X),$$

where λ is the standard Lebesgue measure on $\mathbf{R}$ and λ^* denotes the outer measure associated with λ.

Proof. Fix an arbitrary $\varepsilon > 0$ and take an open subset G of $[a, b]$ such that
$$X \subset G, \quad \lambda(G) \le \lambda^*(X) + \varepsilon.$$
Consider the family of segments

$$\mathcal{V} = \{[f(x), f(x+h)] \ : \ x \in X, \ h \ne 0, \ [x, x+h] \subset G,$$

$$\frac{f(x+h) - f(x)}{h} \le q + \varepsilon\}.$$

Clearly, this family forms a Vitali covering for the set $f(X)$. Consequently, there exists a disjoint countable family

$$\{[f(x_n), f(x_n + h_n)] \ : \ n \in \mathbf{N}\} \subset \mathcal{V}$$

such that

$$\lambda(f(X) \setminus \cup\{[f(x_n), f(x_n + h_n)] \ : \ n \in \mathbf{N}\}) = 0.$$

Note that, because our function f is strictly increasing, the countable family of segments
$$\{[x_n, x_n + h_n] \ : \ n \in \mathbf{N}\}$$
is disjoint, too, and the union of this family is contained in G. So we can write

$$\lambda^*(f(X)) \le \sum_{n \in \mathbf{N}} |f(x_n + h_n) - f(x_n)| \le (q + \varepsilon) \sum_{n \in \mathbf{N}} |h_n|$$

$$\leq (q + \varepsilon)\lambda(G) \leq (q + \varepsilon)(\lambda^*(X) + \varepsilon).$$

Taking account of the fact that $\varepsilon > 0$ was chosen arbitrarily, we conclude that

$$\lambda^*(f(X)) \leq q\lambda^*(X),$$

and the proof of Lemma 1 is complete.

Lemma 2. *Let* $f : [a, b] \to \mathbf{R}$ *be a strictly increasing function, let* q *be a positive real number and let* X *be a subset of* $[a, b]$ *such that, for any point* $x \in X$, *there exists at least one derived number* $f'_D(x) \geq q$. *Then we have the inequality*

$$\lambda^*(f(X)) \geq q\lambda^*(X).$$

Proof. As we know, the set of all discontinuity points for f is at most countable. Taking this fact into account, we may assume without loss of generality that f is continuous at each point belonging to the given set X. Now, if $q = 0$, then there is nothing to prove. So let us suppose that $q > 0$. Pick an arbitrary $\varepsilon > 0$ for which $q - \varepsilon > 0$. There exists an open set $G \subset \mathbf{R}$ such that

$$f(X) \subset G, \quad \lambda(G) \leq \lambda^*(f(X)) + \varepsilon.$$

Consider the family of segments

$$\mathcal{V} = \{[x, x + h] : x \in X, \ h \neq 0, \ [f(x), f(x + h)] \subset G,$$

$$\frac{f(x + h) - f(x)}{h} \geq q - \varepsilon\}.$$

Obviously, this family forms a Vitali covering for the set X. Consequently, there exists a disjoint countable family

$$\{[x_n, x_n + h_n] : n \in \mathbf{N}\} \subset \mathcal{V}$$

for which we have

$$\lambda(X \setminus \cup\{[x_n, x_n + h_n] : n \in \mathbf{N}\}) = 0.$$

Again, since our f is strictly increasing, the countable family of segments

$$\{[f(x_n), f(x_n + h_n)] : n \in \mathbf{N}\}$$

must be disjoint, too, and the union of this family is contained in G. Hence we may write

$$(q - \varepsilon)\lambda^*(X) \leq (q - \varepsilon)\sum_{n \in \mathbf{N}} |h_n| \leq \sum_{n \in \mathbf{N}} |f(x_n + h_n) - f(x_n)|$$

$$\leq \lambda(G) \leq \lambda^*(f(X)) + \varepsilon.$$

Taking account of the fact that $\varepsilon > 0$ is arbitrarily small, we come to the desired inequality

$$q\lambda^*(X) \leq \lambda^*(f(X)).$$

Lemma 2 has thus been proved.

Lemma 3. *Let* $f : [a, b] \to \mathbf{R}$ *be a strictly increasing function and let*

$$X = \{x \in [a, b] : \text{there exist two distinct derived numbers of } f \text{ at } x\}.$$

Then X *is a set of* λ*-measure zero.*

Proof. For any two rational numbers p and q satisfying the inequalities

$$0 \leq p < q,$$

let us denote

$$X_{p,q} = \{x \in [a, b] : \text{there exists a derived number of } f \text{ at } x$$

less than p, *and there exists a derived number of* f *at* x *greater than* $q\}$. Clearly, we have

$$X = \cup\{X_{p,q} : 0 \leq p < q, \ p \in \mathbf{Q}, \ q \in \mathbf{Q}\}.$$

So it suffices to show that each set $X_{p,q}$ is of λ-measure zero. Indeed, according to Lemma 1, we may write

$$\lambda^*(f(X_{p,q})) \leq p\lambda^*(X_{p,q}).$$

At the same time, according to Lemma 2, we have

$$\lambda^*(f(X_{p,q})) \geq q\lambda^*(X_{p,q}).$$

These two inequalities yield

$$q\lambda^*(X_{p,q}) \leq p\lambda^*(X_{p,q})$$

or, equivalently,

$$0 \leq (p - q)\lambda^*(X_{p,q}).$$

Because $p - q < 0$, we must have

$$\lambda^*(X_{p,q}) = 0.$$

This ends the proof of the lemma.

We are now ready to present the classical Lebesgue theorem on the differentiability (almost everywhere) of monotone functions.

Theorem 2. *Let* $f : [a,b] \to \mathbf{R}$ *be a monotone function. Then* f *is differentiable at almost all (with respect to* λ*) points of* $[a,b]$.

Proof. Obviously, we may suppose that f is increasing. Moreover, because the set of all differentiability points for f coincides with the set of all differentiability points for f_1, where

$$f_1(x) = f(x) + x \qquad (x \in [a,b]),$$

and f_1 is strictly increasing, we may assume without loss of generality that our original function f is also strictly increasing. Now, in view of Lemma 3, it suffices only to demonstrate that the set

$$X = \{x \in [a,b] \ : \ for\ each\ n \in \mathbf{N},\ there\ exists$$

$$a\ derived\ number\ f'_D(x) \geq n\}$$

is of λ-measure zero. But this follows directly from Lemma 2, because, in conformity with this lemma, we may write

$$n\lambda^*(X) \leq \lambda^*(f(X)) \leq f(b) - f(a),$$

$$\lambda^*(X) \leq \frac{f(b) - f(a)}{n}$$

for every natural number $n \geq 1$, which immediately yields the required equality

$$\lambda^*(X) = 0.$$

This completes the proof of Theorem 2.

It follows at once from this theorem that a nowhere differentiable real-valued function f defined on a segment $[a,b]$ is simultaneously nowhere monotone on $[a,b]$, i.e., there does not exist a nondegenerate subinterval of $[a,b]$ on which f is monotone.

Exercise 3. Let $f : \mathbf{R} \to \mathbf{R}$ be a continuous function. Demonstrate that the following two assertions are equivalent:
(a) f is injective;
(b) f is strictly monotone.

Show that there exist continuous functions

$$g : \mathbf{R} \to \mathbf{R}$$

which cannot be represented in the form

$$g = g_1 + g_2,$$

where g_1 and g_2 are monotone functions acting from $\mathbf{R}$ into $\mathbf{R}$.

On the other hand, by using the method of transfinite induction, prove that any function

$$h : \mathbf{R} \to \mathbf{R}$$

is representable in the form $h = h_1 + h_2$, where both functions

$$h_1 : \mathbf{R} \to \mathbf{R}, \quad h_2 : \mathbf{R} \to \mathbf{R}$$

are injective.

In this context, let us point out that if h is Lebesgue measurable, then both h_1 and h_2 can be chosen to be Lebesgue measurable, too (see [139]).

Exercise 4. Let $f : [a, b] \to \mathbf{R}$ be an increasing continuous function. Show that

$$\int_a^b f'(t)dt \leq f(b) - f(a).$$

Give an example where this inequality is strict (cf. Theorem 4 below). In addition, demonstrate that if

$$\int_a^b f'(t)dt = f(b) - f(a),$$

then the function f is absolutely continuous on the whole segment $[a, b]$.

Exercise 5. Let λ denote, as usual, the standard Lebesgue measure on $\mathbf{R}$ and let X be an arbitrary Lebesgue measure zero subset of $\mathbf{R}$. Then there exists a sequence $\{U_n : n \in \mathbf{N}\}$ of open subsets of $\mathbf{R}$, such that

$$X \subset U_n, \quad \lambda(U_n) < 1/2^n \quad (n \in \mathbf{N}).$$

For any $n \in \mathbf{N}$, let us define

$$f_n(x) = \lambda(U_n \cap \,] - \infty, x]) \quad (x \in \mathbf{R}).$$

Then f_n is increasing, continuous and

$$0 \leq f_n(x) \leq 1/2^n$$

for all $x \in \mathbf{R}$. Further, define

$$f_X(x) = \sum_{n \in \mathbf{N}} f_n(x) \qquad (x \in \mathbf{R}).$$

Show that the function f_X is increasing, continuous and, for any point $x \in X$, the equality

$$lim_{y \to x,\ y \neq x} \frac{f_X(y) - f_X(x)}{y - x} = +\infty$$

holds true.

For our further considerations, we need the following useful result due to Fubini.

Theorem 3. *Let $\{F_n : n \in \mathbf{N}\}$ be a sequence of positive (i.e., non-negative) increasing functions given on a segment $[a, b] \subset \mathbf{R}$, such that the series $\sum_{n \in \mathbf{N}} F_n(x)$ converges for each point $x \in [a, b]$ and*

$$F(x) = \sum_{n \in \mathbf{N}} F_n(x) \qquad (x \in [a, b]).$$

Then, for almost all (with respect to the Lebesgue measure λ restricted to $[a, b]$) points $x \in [a, b]$, the equality

$$F'(x) = \sum_{n \in \mathbf{N}} F_n'(x)$$

is satisfied.

Proof. Clearly, F is an increasing function on $[a, b]$ and we may write

$$F'(x) \geq \sum_{n \in \mathbf{N}} F_n'(x)$$

for all those points $x \in [a, b]$ at which the derivatives

$$F'(x),\ F_0'(x),\ F_1'(x),\ \ldots,\ F_n'(x),\ \ldots$$

exist. In view of Theorem 2, the series of functions $\sum_{n \in \mathbf{N}} F_n'$ is convergent almost everywhere on $[a, b]$. Now, denote

$$S_n(x) = \sum_{m \leq n} F_m(x)$$

and, for any natural number k, choose an index $n(k)$ such that

$$F(b) - S_{n(k)}(b) \leq 1/2^k.$$

Because all F_n are positive and increasing, we also have

$$0 \leq F(x) - S_{n(k)}(x) \leq 1/2^k$$

for each $x \in [a, b]$. This implies that the series of increasing functions

$$\sum_{k \in \mathbf{N}} (F(x) - S_{n(k)}(x))$$

converges uniformly on $[a, b]$ to some increasing function. Applying Theorem 2 once more, we easily infer that the series

$$\sum_{k \in \mathbf{N}} (F(x) - S_{n(k)}(x))'$$

converges at almost all points $x \in [a, b]$. From this fact we also claim that

$$lim_{k \to +\infty} (F(x) - S_{n(k)}(x))' = 0$$

for almost all $x \in [a, b]$, i.e.,

$$F'(x) = lim_{k \to +\infty} (F_0'(x) + F_1'(x) + \dots + F_{n(k)}'(x))$$

for almost all $x \in [a, b]$. But this immediately yields that

$$F' = \sum_{n \in \mathbf{N}} F_n'$$

almost everywhere on $[a, b]$. The theorem has thus been proved.

The next exercise provides an application of this theorem to the differentiation of an indefinite Lebesgue integral.

Exercise 6. Let f be a positive lower semicontinuous function given on a segment $[a, b]$. We recall (see Exercise 18 from the Introduction) that f can be represented in the form

$$f = sup_{n \in \mathbf{N}} f_n,$$

where all functions f_n are positive, too, and continuous. Derive from this fact that f can be also represented in the form

$$f = \sum_{n \in \mathbf{N}} g_n,$$

where all functions g_n are positive and continuous.

Let now f be a positive, Lebesgue integrable, lower semicontinuous function on $[a, b]$ and let

$$F(x) = \int_a^x f(t)dt \qquad (x \in [a, b]).$$

Show, by applying Theorem 3 and the fact formulated above, that

$$F'(x) = f(x)$$

for almost all (with respect to the Lebesgue measure) points $x \in [a, b]$.

Let g be a positive, Lebesgue integrable function on $[a, b]$. Show that, for each $\varepsilon > 0$, there exists a lower semicontinuous function f on $[a, b]$, such that $g \leq f$ and

$$\int_a^b g(t)dt + \varepsilon > \int_a^b f(t)dt.$$

Deduce from this fact that there exists a sequence $\{f_n : n \in \mathbf{N}\}$ of Lebesgue integrable lower semicontinuous functions, such that:

(a) $f_{n+1} \leq f_n$ for any $n \in \mathbf{N}$;
(b) $g \leq f_n$ for any $n \in \mathbf{N}$;
(c) $lim_{n\to+\infty} f_n(x) = g(x)$ for almost all points $x \in [a, b]$.

In particular, we may write

$$(f_0 - g) = \sum_{n \in \mathbf{N}} (f_n - f_{n+1})$$

almost everywhere on $[a, b]$. Observe that

$$f_0 - g \geq 0,$$

$$f_n - f_{n+1} \geq 0 \qquad (n \in \mathbf{N}).$$

Putting

$$F(x) = \int_a^x (f_0 - g)(t)dt \qquad (x \in [a, b]),$$

$$F_n(x) = \int_a^x (f_n - f_{n+1})(t)dt \qquad (x \in [a, b])$$

and applying Theorem 3 again, demonstrate that

$$(\int_a^x g(t)dt)' = g(x)$$

for almost all $x \in [a, b]$. Finally, prove the Lebesgue theorem stating that if h is an arbitrary real-valued Lebesgue integrable function on $[a, b]$, then

$$(\int_a^x h(t)dt)' = h(x)$$

for almost all $x \in [a, b]$.

The exercise presented above shows us that the classical Lebesgue theorem concerning the differentiation of a function H defined by

$$H(x) = \int_a^x h(t)dt \qquad (x \in [a, b]),$$

can be logically deduced from Theorem 3. However, this approach has a weak side because it does not yield the description of the set of those points $x \in [a, b]$ at which

$$H'(x) = h(x).$$

We now turn our attention to the construction of a strictly increasing function

$$g : \mathbf{R} \to \mathbf{R}$$

whose derivative vanishes almost everywhere. Such a construction is essentially based on Theorem 3.

Let us recall that the first step of the construction of the Cantor set on $\mathbf{R}$ is that we remove from the segment $[0, 1]$ the open interval $]1/3, 2/3[$. Let us put at this step

$$f(x) = 0 \qquad (x \leq 0),$$
$$f(x) = 1 \qquad (x \geq 1),$$
$$f(x) = 1/2 \qquad (x \in \ ]1/3, 2/3[).$$

Now, suppose that on the n-th step of the construction we have already defined the function f for all those points which belong to the union of the removed (at this and earlier steps) intervals. Obviously, we obtain a finite family $\{[a_i, b_i] \ : \ 1 \leq i \leq m\}$ of pairwise disjoint segments on $[0, 1]$. It is easy to check that $m = 2^n$, but we do not need this fact for our further purposes. Pick any segment $[a_i, b_i]$ from the above-mentioned family. Taking into account the inductive assumption, we may put

$$f(x) = \frac{f(a_i-) + f(b_i+)}{2}$$

for all points
$$x \in \ ](2a_i + b_i)/3, (2b_i + a_i)/3[.$$

So we have defined our function f for all points belonging to the union of all intervals removed at the $(n+1)$-th step. Continuing the process in this way, we will be able to construct f on the set $\mathbf{R} \setminus C$, where C denotes the Cantor set. From the definition of f immediately follows that f is increasing and continuous on its domain. Moreover, it is easily seen that f can be uniquely extended to an increasing continuous function

$$f \ : \ \mathbf{R} \to [0,1].$$

Because f is constant on each removed interval, we obviously have

$$f'(x) = 0 \qquad (x \in \mathbf{R} \setminus C),$$

i.e., the derivative of f vanishes almost everywhere on $\mathbf{R}$.

Thus, we have shown that there exists a nonconstant increasing bounded continuous function f from $\mathbf{R}$ into $\mathbf{R}$, whose derivative is zero almost everywhere. Now, let p and q be any two points of $\mathbf{R}$ such that $p < q$. Because f is not constant, there are some points x and y from $\mathbf{R}$ such that $f(x) < f(y)$. Evidently, $x < y$ and there exists a homothety (or translation) h of the plane $\mathbf{R}^2$, for which

$$h((x,0)) = (p,0), \qquad h((y,0)) = (q,0).$$

Let f^* denote the function from $\mathbf{R}$ into $\mathbf{R}$, whose graph coincides with the image of the graph of f with respect to h. Then we may assert that f^* is also an increasing bounded continuous function, whose derivative vanishes almost everywhere, and
$$f^*(p) < f^*(q).$$

In virtue of the remarks made above, we can formulate and prove the following classical result concerning the existence of strictly increasing continuous singular functions.

Theorem 4. *There exists a function*

$$g \ : \ \mathbf{R} \to \mathbf{R}$$

satisfying these three conditions:
    (1) g is continuous and strictly increasing;
    (2) $(\forall x \in \mathbf{R})(0 \le g(x) \le 1)$;
    (3) the derivative of g is zero almost everywhere on $\mathbf{R}$.

Proof. Let $\{(p_n, q_n) : n \in \mathbf{N}\}$ denote the countable family of all pairs of rational numbers, such that $p_n < q_n$. According to the argument presented above, for each natural index n, there exists a function

$$g_n : \mathbf{R} \to [0, 1]$$

such that:

(a) g_n is continuous and increasing;
(b) $0 \leq g_n(x) \leq 1/2^{n+1}$ for all $x \in \mathbf{R}$;
(c) the inequality $g_n(p_n) < g_n(q_n)$ holds true;
(d) the derivative of g_n vanishes almost everywhere.

It follows from (b) that the series $\sum_{n \in \mathbf{N}} g_n$ is uniformly convergent. So we may consider the function

$$g = \sum_{n \in \mathbf{N}} g_n$$

that is continuous and increasing because of (a). Evidently,

$$0 \leq g(x) \leq 1 \quad (x \in \mathbf{R}).$$

In accordance with (c), we also have

$$g(p_n) < g(q_n) \quad (n \in \mathbf{N}),$$

which immediately implies that g is a strictly increasing function. Finally, taking into account condition (d) and applying Theorem 3, we conclude that the derivative of g equals zero almost everywhere on $\mathbf{R}$.

Exercise 7. Let f be any continuous increasing function acting from $\mathbf{R}$ into $\mathbf{R}$, whose derivative is equal to zero almost everywhere on $\mathbf{R}$. For each half-open subinterval $[a, b[$ of $\mathbf{R}$, let us put

$$\mu([a, b[) = f(b) - f(a).$$

Show that μ can be uniquely extended to a σ-finite Borel measure on $\mathbf{R}$ (denoted by the same symbol μ) that is diffused (i.e., vanishes on all one-element subsets of $\mathbf{R}$) and is singular with respect to the standard Lebesgue measure λ. The latter means that there exists a Borel subset X of $\mathbf{R}$ for which we have

$$\lambda(X) = 0, \quad \mu(\mathbf{R} \setminus X) = 0.$$

Formulate the converse assertion and prove it by utilizing the Vitali covering theorem.

Exercise 8. Demonstrate that:

(a) an increasing function $f : \mathbf{R} \to \mathbf{R}$ is continuous from the right if and only if it is upper semicontinuous;

(b) an increasing function $f : \mathbf{R} \to \mathbf{R}$ is continuous from the left if and only if it is lower semicontinuous.

Starting with these facts, give an example of a monotone function

$$g : \mathbf{R} \to \mathbf{R}$$

that is upper (lower) semicontinuous and whose discontinuity points form an everywhere dense subset of $\mathbf{R}$ (cf. Exercise 2 of this chapter).

Exercise 9. Let $f : \mathbf{R} \to \mathbf{R}$ be an arbitrary bounded from above function. We define a function

$$f^* : \mathbf{R} \to \mathbf{R}$$

by putting

$$f^*(x) = sup_{y<x} f(y) \qquad (x \in \mathbf{R}).$$

Verify that f^* is increasing and lower semicontinuous.

In addition, suppose that the original function f is increasing. Show that f and f^* have the same set of continuity points. In general, f^* does not coincide with f. Check that f^* coincides with f if and only if f is continuous from the left.

Exercise 10. Let $\{t_n : n < \omega\}$ be an arbitrary sequence of real numbers. Prove that at least one of the following three assertions is valid:

(a) $\{t_n : n < \omega\}$ contains an infinite strictly increasing subsequence;

(b) $\{t_n : n < \omega\}$ contains an infinite strictly decreasing subsequence;

(c) $\{t_n : n < \omega\}$ contains infinitely many terms which are equal to each other.

Give a straightforward proof of this result. On the other hand, show that the same result is a direct consequence of the Ramsey combinatorial theorem for countable graphs [170].

Conclude from the said above that any partial function

$$f : \mathbf{R} \to \mathbf{R}$$

satisfying the relation

$$card(dom(f)) \geq \omega$$

either is strictly monotone on some infinite subset of $\mathbf{R}$ or is constant on some infinite subset of $\mathbf{R}$.

The last result cannot be extended to uncountable subsets of $dom(f)$ (within the theory **ZFC**). Indeed, we shall see in our further considerations that (under **CH**) there exists a function

$$g : \mathbf{R} \to \mathbf{R},$$

which is not monotone on any uncountable subset of **R**. The construction of such a function is very similar to the classical construction of a Sierpiński-Zygmund function (cf. Chapter 7) or to the classical construction of a Luzin set on **R** (cf. Chapter 10).

Exercise 11. Let $f : \mathbf{R} \to \mathbf{R}$ be an arbitrary continuous function. Prove that there exists a nonempty perfect set $P \subset \mathbf{R}$ such that the restriction $f|P$ is monotone on P.

This can be done by using the following (fairly standard) argument. Suppose that there exists no nondegenerate subinterval of **R** on which f is decreasing. In this case, construct by recursion a dyadic system

$$(T_{i_1 i_2 \ldots i_k})_{i_1 \in \{0,1\}, i_2 \in \{0,1\}, \ldots, i_k \in \{0,1\}, k \geq 0}$$

of nondegenerate compact subintervals of **R** satisfying the relations:

(a) $T_{i_1 i_2 \ldots i_k i_{k+1}} \subset T_{i_1 i_2 \ldots i_k}$;

(b) $T_{i_1 i_2 \ldots i_k 0} \cap T_{i_1 i_2 \ldots i_k 1} = \emptyset$;

(c) $diam(T_{i_1 i_2 \ldots i_k}) < 1/2^k$;

(d) if $(i_1, i_2, \ldots, i_k) \prec (j_1, j_2, \ldots, j_k)$, then $x < y$ and $f(x) < f(y)$ for all points $x \in T_{i_1 i_2 \ldots i_k}$ and $y \in T_{j_1 j_2 \ldots j_k}$ (here $\preceq$ denotes the standard lexicographical ordering on the set of all k-sequences whose terms belong to $\{0,1\}$).

Finally, put

$$P = \bigcap_{k \geq 0} (\cup \{T_{i_1 i_2 \ldots i_k} : i_1, i_2, \ldots, i_k \in \{0,1\}\})$$

and verify that f is increasing on the nonempty perfect set P.

Establish the same result for those functions

$$f : \mathbf{R} \to \mathbf{R},$$

which are Lebesgue measurable or possess the Baire property (reduce this more general situation to the case of a continuous real-valued function given on a nonempty perfect subset of **R**).

In connection with the previous exercise, see also [16].

Exercise 12. Give an example of a Peano type mapping

$$f = (f_1, f_2) : [0, 1] \to [0, 1]^2$$

such that both coordinate functions f_1 and f_2 are differentiable at almost all points of $[0, 1]$ (in the sense of λ) and the equalities

$$f_1' = f_2' = 0$$

hold almost everywhere on $[0, 1]$ (in the same sense).

It is useful to compare this exercise with Exercise 13 from Chapter 1.

5. Everywhere differentiable nowhere monotone functions

As mentioned in the preceding chapter, if a function

$$f \; : \; \mathbf{R} \to \mathbf{R}$$

is nowhere differentiable, then f is nowhere monotone, i.e., there does not exist a nondegenerate subinterval of $\mathbf{R}$ on which f is monotone.

This chapter is devoted to some constructions of functions also acting from $\mathbf{R}$ into $\mathbf{R}$, differentiable everywhere but nowhere monotone. The question of the existence of such functions is obviously typical for classical mathematical analysis. And it should be noticed that many mathematicians of the end of the 19th century and of the beginning of the 20th century tried to present various constructions of the above-mentioned functions. As a rule, their constructions were either incorrect or, at least, incomplete. As pointed out in [83], the first explicit construction of such a function was given by Köpcke in 1889. Another example was suggested by Pereno in 1897 (this example is presented in [74], pp. 412–421). In addition, Denjoy gave in his extensive work [48] a proof of the existence of an everywhere differentiable nowhere monotone function, as a consequence of his deep investigations concerning trigonometric series and their convergence.

Afterwards, a number of distinct proofs of the existence of such functions were given by several authors (see, e.g., [65], [83], [126], [229]).

We begin with the discussion of the construction presented in [83]. This construction is completely elementary and belongs to classical mathematical analysis. We need some easy auxiliary propositions.

Lemma 1. *Let r and s be two strictly positive real numbers. The following assertions hold:*
(1) if $r > s$, then

$$(r - s)/(r^2 - s^2) < 2/r;$$

(2) if $r > 1$ and $s > 1$, then

$$(r + s - 2)/(r^2 + s^2 - 2) < 2/s.$$

Proof. Indeed, we have

$$(r - s)/(r^2 - s^2) = 1/(r + s) < 1/r < 2/r.$$

Thus, (1) is valid. Further, it can easily be checked that the inequality of (2) is equivalent to the inequality

$$(r - s)^2 + (r - 1)(s - 1) + r^2 + r + 3s > 5$$

which, obviously, is true under our assumptions $r > 1$ and $s > 1$. This completes the proof of Lemma 1.

Lemma 2. *Let ϕ be the function from $\mathbf{R}$ into $\mathbf{R}$ defined by*

$$\phi(x) = (1 + |x|)^{-1/2} \qquad (x \in \mathbf{R})$$

and let a and b be any two distinct real numbers. Then we have

$$(1/(b - a)) \int_a^b \phi(x)dx < 4min(\phi(a), \phi(b)).$$

Proof. Without loss of generality we may assume that $a < b$. Only three cases are possible.

1. $0 \leq a < b$. In this case, taking into account (1) of Lemma 1, we can write

$$(1/(b - a)) \int_a^b \phi(x)dx = 2((1 + b)^{1/2} - (1 + a)^{1/2})/((1 + b) - (1 + a))$$

$$< 4/(1 + b)^{1/2} = 4min(\phi(a), \phi(b)).$$

2. $a < b \leq 0$. This case can be reduced to the previous one, because of the evenness of our function ϕ.

3. $a < 0 < b$. In this case, taking into account (2) of Lemma 1, we can write

$$(1/(b - a)) \int_a^b \phi(x)dx = 2((1 + b)^{1/2} + (1 - a)^{1/2} - 2)/((1 + b) + (1 - a) - 2)$$

$$< 4min(\phi(a), \phi(b)).$$

This ends the proof of Lemma 2.

Lemma 3. *Let $n > 0$ be a natural number and let*

$$\psi \; : \; \mathbf{R} \to \mathbf{R}$$

be any function of the form

$$\psi(x) = \sum_{1 \le k \le n} c_k \phi(d_k(x - t_k)) \qquad (x \in \mathbf{R}),$$

where ϕ is the function from Lemma 2,

$$c_1, \; c_2, \; \ldots, \; c_n,$$

$$d_1, \; d_2, \; \ldots, \; d_n$$

are strictly positive real numbers, and

$$t_1, \; t_2, \; \ldots, \; t_n$$

are real numbers. Then

$$(1/(b - a)) \int_a^b \psi(x)dx < 4min(\psi(a), \psi(b))$$

for all distinct reals a and b.

Proof. The assertion follows immediately from Lemma 2, by taking into account the fact that

$$(1/(b - a)) \int_a^b \phi(d(x - t))dx = (1/(d(b - t) - d(a - t))) \int_{d(a-t)}^{d(b-t)} \phi(x)dx$$

for any $d > 0$ and $t \in \mathbf{R}$.

Lemma 4. *Let $(\psi_n)_{n \ge 1}$ be a sequence of functions as in Lemma 3. For any point $x \in \mathbf{R}$ and for each integer $n \ge 1$, let us define*

$$\Psi_n(x) = \int_0^x \psi_n(z)dz$$

and suppose that, for some $a \in \mathbf{R}$, the series $\sum_{n \ge 1} \psi_n(a)$ is convergent. Denote

$$\sum_{n \ge 1} \psi_n(a) = s < +\infty.$$

Then we have:

(1) the series $F(x) = \sum_{n \geq 1} \Psi_n(x)$ converges uniformly on every bounded subinterval of $\mathbf{R}$;
   (2) the function F is differentiable at the point a and

$$F'(a) = s.$$

In particular, if

$$\sum_{n \geq 1} \psi_n(z) = f(z) < +\infty$$

for each $z \in \mathbf{R}$, then the function F is differentiable everywhere on $\mathbf{R}$ and the equality

$$F' = f$$

holds true.

Proof. Take any $b \in \mathbf{R}$ satisfying the relation $b \geq |a|$. In view of Lemma 3, for all points $x \in [-b, b]$ and for all integers $n \geq 1$, we may write

$$|\Psi_n(x)| \leq |\int_0^a \psi_n(z)dz| + |\int_a^x \psi_n(z)dz| \leq$$

$$4|a|\psi_n(a) + 4|x - a|\psi_n(a) \leq 12b\psi_n(a).$$

This shows the uniform convergence of

$$\sum_{n \geq 1} \Psi_n(x) \qquad (x \in [-b, b])$$

on the segment $[-b, b]$.
   Further, let $\varepsilon > 0$ be given. Pick a natural number $k > 0$ such that

$$10 \cdot \sum_{n > k} \psi_n(a) < \varepsilon.$$

Because all functions ψ_n are continuous on the whole $\mathbf{R}$ (in particular, at a), there exists some $\delta > 0$ such that

$$|(1/h)\int_a^{a+h} \psi_n(z)dz - \psi_n(a)| < \varepsilon/2k,$$

whenever

$$0 < |h| < \delta, \qquad 1 \leq n \leq k.$$

Consequently, assuming that $0 < |h| < \delta$ and applying Lemma 3 again, we get

$$|(F(a+h) - F(a))/h - s| = |\sum_{n \geq 1}((1/h) \int_a^{a+h} \psi_n(z)dz - \psi_n(a))| \leq$$

$$\sum_{1 \leq n \leq k} |(1/h) \int_a^{a+h} \psi_n(z)dz - \psi_n(a)| + \sum_{n > k}((1/h) \int_a^{a+h} \psi_n(z)dz + \psi_n(a))$$

$$< \varepsilon/2 + \sum_{n > k} 5\psi_n(a) < \varepsilon.$$

We thus conclude that

$$F'(a) = s,$$

and the lemma is proved.

Lemma 5. *Let $n > 0$ be a natural number, let I_1, ..., I_n be pairwise disjoint nondegenerate segments on $\mathbf{R}$ and let t_k denote the midpoint of I_k for each natural number $k \in [1, n]$. Fix any strictly positive real numbers*

$$\varepsilon, \ y_1, \ \cdots, \ y_n.$$

Then there exists a function ψ as in Lemma 3 such that, for all natural numbers $k \in [1, n]$, the following relations are fulfilled:
(1) $\psi(t_k) > y_k$;
(2) $(\forall x \in I_k)(\psi(x) < y_k + \varepsilon)$;
(3) $(\forall x \in \mathbf{R} \setminus (I_1 \cup ... \cup I_n))(\psi(x) < \varepsilon)$.

Proof. Let us denote

$$c_k = y_k + \varepsilon/2 \qquad (1 \leq k \leq n).$$

Then, for each integer $k \in [1, n]$, define

$$\phi_k(x) = c_k\phi(d_k(x - t_k)) \qquad (x \in \mathbf{R}),$$

where $d_k > 0$ is chosen so large that

$$(\forall x \in \mathbf{R} \setminus I_k)(\phi_k(x) < \varepsilon/2n).$$

Finally, put

$$\psi = \phi_1 + \ldots + \phi_n.$$

Then, taking into account the fact that

$$max_{x \in \mathbf{R}} \phi_k(x) = \phi_k(t_k) = c_k \qquad (1 \leq k \leq n),$$

it is easy to check that the function ψ satisfies relations (1), (2) and (3).

Lemma 6. *Let any two disjoint countable subsets*

$$\{t_k \ : \ k \in \mathbf{N}, \ k \geq 1\},$$

$$\{r_k \ : \ k \in \mathbf{N}, \ k \geq 1\}$$

of $\mathbf{R}$ *be given. Then there exists a function*

$$F \ : \ \mathbf{R} \to \mathbf{R}$$

such that:
  (1) F is differentiable everywhere on $\mathbf{R}$;
  (2) $0 < F'(x) \leq 1$ for all $x \in \mathbf{R}$;
  (3) $F'(t_k) = 1$ for each $k \in \mathbf{N} \setminus \{0\}$;
  (4) $F'(r_k) < 1$ for each $k \in \mathbf{N} \setminus \{0\}$.

Proof. We are going to construct by recursion the sequence $(\psi_n)_{n \geq 1}$ of functions as in Lemma 3 with some additional properties. Namely, denoting

$$f_n = \sum_{1 \leq k \leq n} \psi_k,$$

we wish the following conditions to be satisfied:
  (i) for any integer $n \geq 1$ and for all integers $k \in [1, n]$, we have

$$f_n(t_k) > 1 - 1/n;$$

  (ii) for any integer $n \geq 1$ and for each point $x \in \mathbf{R}$, we have

$$f_n(x) < 1 - 1/(n+1);$$

  (iii) for any integer $n \geq 1$ and for all integers $k \in [1, n]$, we have

$$\psi_n(r_k) < 1/(2n \cdot 2^n).$$

First, choose a nondegenerate segment I_1 with midpoint t_1, such that $r_1 \notin I_1$ and apply Lemma 5 with

$$\varepsilon = 1/4, \quad y_1 = 1/4.$$

Evidently, we obtain ψ_1 and $f_1 = \psi_1$, such that relations (i), (ii) and (iii) are fulfilled for $n = 1$.

Suppose now that, for a natural number $n > 1$, we have already defined the functions

$$\psi_1, \ldots, \psi_k, \ldots, \psi_{n-1}$$

satisfying the corresponding analogues of (i) - (iii) for $n - 1$. Pick disjoint nondegenerate segments $I_1, \ldots, I_n$ in such a way that:

(a) t_k is the midpoint of I_k for each integer $k \in [1, n]$;

(b) $I_k \cap \{r_1, \ldots, r_n\} = \emptyset$ for each integer $k \in [1, n]$;

(c) for any integer $k \in [1, n]$ and for any point $x \in I_k$, we have the inequality

$$f_{n-1}(x) < f_{n-1}(t_k) + \delta,$$

where

$$\delta = 1/(n(n+1)) - 1/(2n \cdot 2^n).$$

We now can apply Lemma 5 with

$$\varepsilon = 1/(2n \cdot 2^n)$$

and

$$y_k = 1 - (1/n) - f_{n-1}(t_k) \qquad (1 \le k \le n).$$

Applying the above-mentioned lemma, we get the function ψ_n. Clearly,

$$\psi_n(r_k) < \varepsilon = 1/(2n \cdot 2^n)$$

for all natural numbers $k \in [1, n]$, so relation (iii) holds true. Further, for any natural number $k \in [1, n]$, we also have

$$f_n(t_k) = f_{n-1}(t_k) + \psi_n(t_k) > f_{n-1}(t_k) + y_k = 1 - 1/n,$$

which shows that relation (i) holds true, too. Finally, in order to verify (ii), fix any point $x \in \mathbf{R}$. If, for some integer $k \in [1, n]$, the point x belongs to I_k, then we may write

$$f_n(x) = f_{n-1}(x) + \psi_n(x) < f_{n-1}(t_k) + \delta + y_k + \varepsilon =$$

$$1 - 1/n + 1/(n(n+1)) = 1 - 1/(n+1).$$

If x does not belong to $I_1 \cup \ldots \cup I_n$, then

$$f_n(x) = f_{n-1}(x) + \psi_n(x) < 1 - 1/n + \varepsilon < 1 - 1/(n+1).$$

Thus, relation (ii) holds true, too. Proceeding in this manner, we are able to construct the required sequence $(\psi_n)_{n \geq 1}$. Putting

$$f = \sum_{n \geq 1} \psi_n = lim_{n \to +\infty} f_n$$

and

$$F(x) = \int_0^x f(z)dz \qquad (x \in \mathbf{R}),$$

we obtain the function

$$F \; : \; \mathbf{R} \to \mathbf{R}.$$

In view of Lemma 4, we also get

$$F'(x) = f(x) \qquad (x \in \mathbf{R}).$$

Further, the definition of F immediately implies

$$0 < F'(x) \leq 1 \qquad (x \in \mathbf{R}),$$

$$F'(t_k) = 1 \qquad (k \in \mathbf{N}, \; k \geq 1).$$

Now, fix an integer $k \geq 1$ and let n be a natural number strictly greater than k. Then

$$F'(r_k) = f_{n-1}(r_k) + \sum_{m \geq n} \psi_m(r_k) <$$

$$1 - 1/n + \sum_{m \geq n} 1/(2m \cdot 2^m) < 1 - 1/n + 1/2n = 1 - 1/2n < 1.$$

This completes the proof of Lemma 6.

Theorem 1. *There exists a function*

$$H \; : \; \mathbf{R} \to \mathbf{R}$$

such that:
  (1) H is differentiable everywhere on $\mathbf{R}$;
  (2) H' is bounded on $\mathbf{R}$;
  (3) H is monotone on no nondegenerate subinterval of $\mathbf{R}$.

Proof. Denote by

$$\{t_n \; : \; n \in \mathbf{N}, \; n > 0\},$$

$$\{r_n \; : \; n \in \mathbf{N}, \; n > 0\}$$

any two disjoint countable everywhere dense subsets of $\mathbf{R}$. Using the result of the previous lemma, take any two everywhere differentiable functions

$$F \ : \ \mathbf{R} \to \mathbf{R},$$

$$G \ : \ \mathbf{R} \to \mathbf{R},$$

satisfying the relations:

(a) $0 < F'(x) \leq 1$ and $0 < G'(x) \leq 1$ for all $x \in \mathbf{R}$;

(b) $F'(t_n) = 1$ and $F'(r_n) < 1$ for each natural number $n \geq 1$;

(c) $G'(r_n) = 1$ and $G'(t_n) < 1$ for each natural number $n \geq 1$.

Now, define

$$H = F - G.$$

Obviously, we have

$$H'(t_n) > 0, \quad H'(r_n) < 0 \quad (n \in \mathbf{N}, \ n \geq 1).$$

Because both the sets

$$\{t_n \ : \ n \in \mathbf{N}, \ n > 0\},$$

$$\{r_n \ : \ n \in \mathbf{N}, \ n > 0\}$$

are everywhere dense in $\mathbf{R}$, we infer that H cannot be monotone on any subinterval of $\mathbf{R}$. Also, the relation

$$-1 < H'(x) < 1 \quad (x \in \mathbf{R})$$

implies that H' is bounded, and the theorem has thus been proved.

In fact, the preceding argument establishes the existence of many functions

$$f \ : \ \mathbf{R} \to \mathbf{R},$$

which are everywhere differentiable, nowhere monotone and such that f' is bounded on $\mathbf{R}$. Let us mention some other interesting and unusual properties of any such function f.

1. f has a point of a local maximum and a point of a local minimum in every nonempty open subinterval of $\mathbf{R}$. Actually, for each nondegenerate segment $[a, b] \subset \mathbf{R}$, we can find some points x_1 and x_2 satisfying the relations

$$a < x_1 < x_2 < b,$$

$$f'(x_1) > 0, \quad f'(x_2) < 0.$$

Let us denote
$$M = sup_{t \in [x_1, x_2]} \, f(t).$$
Then, for some $\tau \in [x_1, x_2]$, we must have
$$f(\tau) = M,$$
and it is clear that τ must be in the interior of $[x_1, x_2]$. In other words, τ is a point of a local maximum for our f.

Applying a similar argument, we can also find a point of a local minimum of f on the same nondegenerate segment $[a, b]$.

2. Because f' is bounded, the function f satisfies the so-called Lipschitz condition, i.e., for some constant $d \geq 0$, we have
$$|f(x) - f(y)| \leq d|x - y| \qquad (x \in \mathbf{R}, \; y \in \mathbf{R}).$$
Note that, in the latter relation, we may put
$$d = sup_{t \in \mathbf{R}}|f'(t)|.$$

In particular, f is absolutely continuous. This also implies that f' is Lebesgue integrable on each bounded subinterval of $\mathbf{R}$.

3. The function f' is not integrable in the Riemann sense on any non-degenerate segment $[a, b] \subset \mathbf{R}$. To see this, suppose otherwise, i.e., suppose that f' is Riemann integrable on $[a, b]$. Then, according to a well-known theorem of mathematical analysis, f' must be continuous at almost all (with respect to the standard Lebesgue measure) points of $[a, b]$ (see, e.g., [154]). Taking into account the fact that f' changes its sign on each nonempty open subinterval of $\mathbf{R}$, we infer that f' must be zero at almost all points of $[a, b]$. Consequently, f must be constant on $[a, b]$, which is impossible. The contradiction obtained yields the desired result.

4. Being a derivative, the function f' belongs to the first Baire class, i.e., it can be represented as a pointwise limit of a sequence of continuous functions (see Example 1 from Chapter 2). Hence, by virtue of the classical Baire theorem (see, e.g., [120], [154] or Theorem 3 from Chapter 2), the set of all those points of $\mathbf{R}$ at which f' is continuous is residual (co-meager), i.e., is the complement of a first category subset of $\mathbf{R}$.

5. Let us denote
$$X = \{t \in \mathbf{R} \; : \; f'(t) > 0\},$$
$$Y = \{t \in \mathbf{R} \; : \; f'(t) < 0\}.$$

The sets X and Y are disjoint, Lebesgue measurable and have the property that, for each nondegenerate segment $[a, b] \subset \mathbf{R}$, the relations

$$\lambda(X \cap [a, b]) > 0,$$

$$\lambda(Y \cap [a, b]) > 0$$

are fulfilled (where λ denotes, as usual, the Lebesgue measure on $\mathbf{R}$). In order to demonstrate this fact, suppose, for example, that

$$\lambda(Y \cap [a, b]) = 0.$$

Then we get $f'(t) \geq 0$ for almost all points $t \in [a, b]$. But this immediately implies that our function f, being of the form

$$f(x) = \int_0^x f'(t)dt + f(0) \quad (x \in \mathbf{R}),$$

is increasing on $[a, b]$, which contradicts the definition of f.

Exercise 1. Give a direct construction of two disjoint Lebesgue measurable subsets X and Y of $\mathbf{R}$, such that, for any nonempty open interval $I \subset \mathbf{R}$, the inequalities

$$\lambda(I \cap X) > 0,$$

$$\lambda(I \cap Y) > 0$$

hold true. More generally, show that there exists a countable partition

$$\{X_n \ : \ n < \omega\}$$

of $\mathbf{R}$ consisting of Lebesgue measurable sets and such that, for any nonempty open subinterval I of $\mathbf{R}$ and for any natural number n, the relation

$$\lambda(I \cap X_n) > 0$$

is fulfilled.

Exercise 2. Denote by $\mathcal{E}$ the family of all Lebesgue measurable subsets of the unit segment $[0, 1]$. For any two sets $X \in \mathcal{E}$ and $Y \in \mathcal{E}$, put

$$d(X, Y) = \lambda(X \triangle Y).$$

Identifying all those X and Y, for which $d(X, Y) = 0$, we come to the metric space $(\mathcal{E}, d)$. Check that $(\mathcal{E}, d)$ is complete and separable, i.e., is a Polish space. Further, let $\mathcal{E}'$ be a subspace of $\mathcal{E}$ consisting of all sets $X \in \mathcal{E}$ such that

$$\lambda(X \cap I) > 0,$$

$$\lambda(([0,1] \setminus X) \cap I) > 0$$

for each nondegenerate subinterval I of $[0,1]$. Show that $\mathcal{E}'$ is the complement of a first category subset of $\mathcal{E}$. Hence, according to the Baire theorem, we have

$$\mathcal{E}' \neq \emptyset.$$

It is useful to compare this fact with Exercise 1 above.

Now, we are going to present an essential generalization of Theorem 1 due to Weil (see [229]). Namely, Weil gave a proof of the existence of everywhere differentiable nowhere monotone functions (with a bounded derivative) by using the above-mentioned Baire theorem.

We recall that a function

$$f \;:\; \mathbf{R} \to \mathbf{R}$$

is a derivative if there exists at least one everywhere differentiable function

$$F \;:\; \mathbf{R} \to \mathbf{R}$$

satisfying the relation

$$(\forall x \in \mathbf{R})(F'(x) = f(x)).$$

Let us consider the set

$$D = \{f \;:\; f \text{ is a derivative and } f \text{ is bounded}\}.$$

Obviously, D is a vector space over $\mathbf{R}$. We may equip this set with a metric d defined by the formula

$$d(f,g) = sup_{x \in \mathbf{R}}|f(x) - g(x)|.$$

Clearly, the metric d produces the topology of uniform convergence. In view of the well-known theorem of analysis, a uniform limit of a sequence of bounded derivatives is a bounded derivative (cf. Exercise 20 from Chapter 2). This shows, in particular, that the pair (D, d) is a Banach space (it can easily be seen that it is nonseparable). Take any function $f \in D$ and consider the set $f^{-1}(0)$. We assert that this set is a G_δ-subset of $\mathbf{R}$. Indeed, we may write

$$f^{-1}(0) = \{x \in \mathbf{R} \;:\; lim_{n \to +\infty} n(F(x + 1/n) - F(x)) = 0\},$$

where a function

$$F : \mathbf{R} \to \mathbf{R}$$

is such that

$$F'(x) = f(x) \qquad (x \in \mathbf{R}).$$

Equivalently, we have

$$f^{-1}(0) = \bigcap_{1 \le n < \omega} (\bigcup_{n \le m < \omega} \{x \in \mathbf{R} \; : \; m(F(x + 1/m) - F(x)) < 1/n\}).$$

This formula yields at once the desired result.

Let us put

$$D_0 = \{f \in D \; : \; f^{-1}(0) \text{ is everywhere dense in } \mathbf{R}\}.$$

We need the following simple fact.

Lemma 7. *The set D_0 is a closed vector subspace of the space D. Consequently, D_0 is a Banach space, as well.*

Proof. First, we show that D_0 is closed in D. Let $\{f_k \; : \; k < \omega\}$ be a sequence of functions from D_0, converging (in metric d) to some function $f \in D$. We put

$$Z_k = f_k^{-1}(0) \qquad (k < \omega).$$

Then all the sets Z_k are everywhere dense G_δ-subsets of $\mathbf{R}$. Therefore, the set

$$Z = \cap\{Z_k \; : \; k < \omega\}$$

is an everywhere dense G_δ-subset of $\mathbf{R}$, too. Obviously, the inclusion

$$Z \subset f^{-1}(0)$$

is valid. Thus, we obtain that $f \in D_0$.

Now, let us demonstrate that D_0 is a vector subspace of D. Clearly, if $f \in D_0$ and $t \in \mathbf{R}$, then $tf \in D_0$. Further, take any two functions $g \in D_0$ and $h \in D_0$ and consider the sets

$$Z_g = g^{-1}(0),$$

$$Z_h = h^{-1}(0).$$

Then the set $Z_g \cap Z_h$ is an everywhere dense G_δ-subset of $\mathbf{R}$, and it is evident that

$$Z_g \cap Z_h \subset Z_{g+h},$$

where
$$Z_{g+h} = (g+h)^{-1}(0).$$
This shows that D_0 is a vector space.

Lemma 7 has thus been proved.

Notice now that the space D_0 is nontrivial, i.e., contains nonzero functions. For instance, this fact follows directly from Theorem 1. But it can also easily be proved by another argument.

Exercise 3. Give a direct proof (i.e., without the aid of Theorem 1) that D_0 contains nonzero elements.

Theorem 2. *Let us denote*

$$E = \{f \in D_0 \ : \ \text{there is a nondegenerate subinterval of } \mathbf{R}$$

$$\text{on which } f \text{ preserves its sign}\}.$$

Then the set E is of first category in the space D_0.

Proof. Let $\{I_n \ : \ n \in \mathbf{N}\}$ be the family of all subintervals of $\mathbf{R}$ with rational end-points. For each $n \in \mathbf{N}$, put

$$A_n = \{f \in D_0 \ : \ (\forall x \in I_n)(f(x) \geq 0)\},$$

$$B_n = \{f \in D_0 \ : \ (\forall x \in I_n)(f(x) \leq 0)\}.$$

Clearly, we have
$$E = \cup_{n \in \mathbf{N}}(A_n \cup B_n),$$

so it suffices to demonstrate that each of the sets A_n and B_n is closed and nowhere dense. We shall establish this fact only for A_n (for B_n, the argument is analogous). The closedness of A_n is trivial. In order to prove that A_n contains no ball in D_0, take any $f \in D_0$ and fix an arbitrary $\varepsilon > 0$. Because $f \in D_0$, there exists a point $x \in I_n$ such that

$$f(x) = 0.$$

Now, by starting with the existence of a nonzero bounded derivative belonging to D_0, it is easy to show that there is a function $h \in D_0$ for which

$$h(x) < 0, \quad sup_{y \in \mathbf{R}}|h(y)| < \varepsilon.$$

Let us define
$$g = f + h.$$

Then the function g belongs to the ball of D_0 with center f and radius ε. At the same time, g does not belong to A_n because

$$g(x) = f(x) + h(x) = h(x) < 0.$$

This establishes that A_n is nowhere dense in D_0, and the theorem has thus been proved.

In the next chapter, we shall consider one more proof of the existence of everywhere differentiable nowhere monotone functions, by applying some properties of the so-called density topology on $\mathbf{R}$.

Exercise 4. Let E be an arbitrary topological space. Recall that a family $\mathcal{N}$ of subsets of E is a net in E if each open subset of E can be represented as the union of some subfamily of $\mathcal{N}$ (recall also that the concept of a net, for topological spaces, was introduced by Archangelskii; obviously, it generalizes the concept of a base of a topological space). We denote by the symbol $nw(E)$ the smallest cardinality of a net in E.

Let now

$$f \; : \; E \to \mathbf{R}$$

be a function. We put:

$lmaxv(f)$ = the set of all those $t \in \mathbf{R}$ for which there exists a nonempty open subset U of E such that $t = sup(f(U))$ and, in addition, there is a point $e \in U$ such that $f(e) = sup(f(U))$;

$lminv(f)$ = the set of all those $t \in \mathbf{R}$ for which there exists a nonempty open subset V of E such that $t = inf(f(V))$ and, in addition, there is a point $e \in V$ such that $f(e) = inf(f(V))$.

Check that:

$$card(lmaxv(f)) \leq nw(E) + \omega,$$

$$card(lminv(f)) \leq nw(E) + \omega.$$

In particular, if E possesses a countable net, then the above-mentioned subsets of $\mathbf{R}$ are at most countable.

Denote also:

$slmax(f)$ = the set of all points $e \in E$ having the property that there exists a neighborhood $U(e)$ such that $f(e) > f(x)$ for each $x \in U(e)$;

$slmin(f)$ = the set of all points $e \in E$ having the property that there exists a neighborhood $V(e)$ such that $f(e) < f(x)$ for each $x \in V(e)$.

Verify that:

$$card(slmax(f)) \leq nw(E) + \omega,$$

$$card(slmin(f)) \leq nw(E) + \omega.$$

In particular, if E possesses a countable net, then the sets $slmax(f)$ and $slmin(f)$ are at most countable.

Finally, for $E = \mathbf{R}$, give an example of a continuous function f for which the latter two sets are everywhere dense in E.

6. Nowhere approximately differentiable functions

The first example of a nowhere approximately differentiable function from the Banach space $C[0,1]$ is due to Jarník (see [77]). Moreover, he showed that such functions are typical, i.e., they constitute a set whose complement is of first category in $C[0,1]$.

In this chapter we present one precise construction of a function acting from $\mathbf{R}$ into $\mathbf{R}$, which is nowhere approximately differentiable. This construction is due to Malý [140] (cf. also [39] and [40]). It is not difficult and, at the same time, is rather vivid from the geometrical point of view.

We begin with some preliminary notions and facts.

Let λ denote the standard Lebesgue measure on $\mathbf{R}$ and let X be a λ-measurable subset of $\mathbf{R}$. We recall that a point $x \in \mathbf{R}$ is said to be a density point for (of) X if

$$lim_{h \to 0,\ h>0}\ \lambda(X \cap [x-h, x+h])/2h = 1.$$

According to the classical Lebesgue theorem (see, e.g., the Introduction), almost all points of X are its density points.

Exercise 1. Let $(t_n)_{n \in \mathbf{N}}$ be a sequence of strictly positive real numbers, such that

$$lim_{n \to +\infty} t_n = 0, \quad lim_{n \to +\infty} t_n/t_{n+1} = 1.$$

Let X be a Lebesgue measurable subset of $\mathbf{R}$ and let $x \in \mathbf{R}$. Prove that the following two assertions are equivalent:

(1) x is a density point of X;
(2) $lim_{n \to +\infty} \lambda(X \cap [x - t_n, x + t_n])/2t_n = 1$.

Exercise 2. Let X be a Lebesgue measurable subset of $\mathbf{R}$ and let $x \in \mathbf{R}$. Show that the following two assertions are equivalent:

(1) x is a density point of X;
(2) $lim_{h \to 0+,\ k \to 0+} \lambda(X \cap [x - h, x + k])/(h + k) = 1$.

The notion of a density point turned out to be rather deep and fruitful not only for real analysis but also for general topology, probability theory and some other domains of mathematics. For example, by use of this notion the important concept of the density topology on $\mathbf{R}$ was introduced and investigated by several authors (Pauc, Goffman, Waterman, Nishiura, Neugebauer, Tall, and others). This topology was studied, with its further generalizations, from different points of view (see, e.g., [66], [162], [171] and [219]). We shall deal with the density topology (and with some of its natural analogues) in our considerations below.

Now, let $f : \mathbf{R} \to \mathbf{R}$ be a function and let $x \in \mathbf{R}$. We recall that f is said to be approximately continuous at x if there exists a λ-measurable set X such that:

(1) x is a density point of X;

(2) the function $f|(X \cup \{x\})$ is continuous at x.

The next two exercises show that all Lebesgue measurable functions can be described in terms of approximate continuity.

Exercise 3. Let $g : \mathbf{R} \to \mathbf{R}$ be a function, let ε be a fixed strictly positive real number and suppose that, for any λ-measurable set X with $\lambda(X) > 0$, there exists a λ-measurable set $Y \subset X$ with $\lambda(Y) > 0$, such that

$$(\forall x \in Y)(\forall y \in Y)(|g(x) - g(y)| < \varepsilon).$$

Demonstrate that there exists a λ-measurable function $h : \mathbf{R} \to \mathbf{R}$ for which we have

$$(\forall x \in \mathbf{R})(|g(x) - h(x)| < \varepsilon).$$

Infer from this fact that if the given function g satisfies the above condition for any $\varepsilon > 0$, then g is measurable in the Lebesgue sense.

Exercise 4. Let $f : \mathbf{R} \to \mathbf{R}$ be a function. By applying the result of Exercise 3 and utilizing the classical Luzin theorem on the structure of Lebesgue measurable functions (see, e.g., [154] or Chapter 3 of this book), show that the following two assertions are equivalent:

(a) the function f is measurable in the Lebesgue sense;

(b) for almost all (with respect to λ) points $x \in \mathbf{R}$, the function f is approximately continuous at x.

Exercise 5. Let $f : \mathbf{R} \to \mathbf{R}$ be a locally bounded Lebesgue measurable function and let

$$F(x) = \int_0^x f(t)dt \qquad (x \in \mathbf{R}).$$

Prove that, for any point $x \in \mathbf{R}$ at which the function f is approximately continuous, we have $F'(x) = f(x)$.

Check also that the local boundedness of f is essential here.

Let now $f : \mathbf{R} \to \mathbf{R}$ be a function and let $x \in \mathbf{R}$. We say that f is approximately differentiable at x if there exist a Lebesgue measurable set $Y \subset \mathbf{R}$, for which x is a density point, and a limit

$$lim_{y \to x, \ y \in Y, \ y \neq x} \frac{f(y) - f(x)}{y - x}.$$

This limit is denoted by $f'_{ap}(x)$ and is called an approximate derivative of f at the point x.

Exercise 6. Demonstrate that if a function $f : \mathbf{R} \to \mathbf{R}$ is approximately differentiable at $x \in \mathbf{R}$, then f is also approximately continuous at x.

Exercise 7. Check that an approximate derivative of a function

$$f : \mathbf{R} \to \mathbf{R}$$

at a point $x \in \mathbf{R}$ is uniquely determined, i.e., it does not depend on the choice of a Lebesgue measurable set Y for which x is a density point and for which the corresponding limit exists (apply Exercise 29 from the Introduction).

Check also that the family of all functions acting from $\mathbf{R}$ into $\mathbf{R}$ and approximately differentiable at x forms a vector space over $\mathbf{R}$.

Exercise 8. Show that if a function $f : \mathbf{R} \to \mathbf{R}$ is differentiable (in the usual sense) at a point $x \in \mathbf{R}$, then f is approximately differentiable at x and $f'_{ap}(x) = f'(x)$. Give an example showing that the converse assertion is not true.

For our purposes below, we need two simple auxiliary propositions.

Lemma 1. *Let $f : \mathbf{R} \to \mathbf{R}$ be a function, let x be a point of $\mathbf{R}$ and suppose that f is approximately differentiable at x. Then, for any real number $M_1 > f'_{ap}(x)$, we have*

$$lim_{h \to 0+}\lambda(\{y \in [x-h, x+h] \setminus \{x\} \ : \ (f(y) - f(x))/(y-x) \geq M_1\})/2h = 0.$$

Similarly, for any real number $M_2 < f'_{ap}(x)$, we have

$$lim_{h \to 0+}\lambda(\{y \in [x-h, x+h] \setminus \{x\} \ : \ (f(y) - f(x))/(y-x) \leq M_2\})/2h = 0.$$

Proof. Because the argument in both cases is comletely analogous, we shall consider only the case of M_1. There exists a λ-measurable set X such that x is a density point of X and

$$lim_{y \to x, y \neq x, y \in X}(f(y) - f(x))/(y-x) = f'_{ap}(x).$$

Fix $\varepsilon > 0$ for which

$$f'_{ap}(x) + \varepsilon < M_1.$$

Then there exists a real number $\delta > 0$ such that, for any strictly positive real $h < \delta$, we have

$$(\forall y \in X \cap [x - h, x + h] \setminus \{x\})((f(y) - f(x))/(y - x) \leq f'_{ap}(x) + \varepsilon).$$

But, if $\delta > 0$ is sufficiently small, then

$$\lambda(X \cap [x - h, x + h])/2h \geq 1 - \varepsilon$$

for all strictly positive reals $h < \delta$. So we obtain the relation

$$\lambda(\{y \in [x - h, x + h] \setminus \{x\} \ : \ (f(y) - f(x))/(y - x) \geq M_1\})/2h \leq \varepsilon,$$

and the lemma is proved.

Actually, in our further considerations we need only the following auxiliary assertion, which is an immediate consequence of Lemma 1.

Lemma 2. *Let* $f \ : \ \mathbf{R} \to \mathbf{R}$ *be a function, let* x *be a point of* $\mathbf{R}$ *and suppose that, for every strictly positive real number* M, *the relation*

$$liminf_{h \to 0+} \frac{\lambda(\{y \in [x - h, x + h] \setminus \{x\} \ : \ |f(y) - f(x)|/|y - x| \geq M\})}{2h} > 0$$

holds true. Then f *is not approximately differentiable at* x.

In particular, suppose that two sequences

$$\{h_k \ : \ k \in \mathbf{N}\}, \quad \{M_k \ : \ k \in \mathbf{N}\}$$

of real numbers are given, satisfying the following conditions:
(1) $h_k > 0$ and $M_k > 0$ for all integers $k \geq 0$;
(2) $lim_{k \to +\infty} h_k = 0$ and $lim_{k \to +\infty} M_k = +\infty$;
(3) the lower limit

$$liminf_{k \to +\infty} \lambda(\{y \in [x - h_k, x + h_k] \setminus \{x\} \ :$$

$$|f(y) - f(x)|/|y - x| \geq M_k\})/2h_k$$

is strictly positive.

Then we can assert that our function f is not approximately differentiable at the point x.

After these simple preliminary remarks, we are able to begin the construction of a nowhere approximately differentiable function.

First of all, let us put

$$f_1(0/9) = 0, \quad f_1(1/9) = 1/3, \quad f_1(2/9) = 0, \quad f_1(3/9) = 1/3,$$

$$f_1(4/9) = 2/3, \quad f_1(5/9) = 1/3, \quad f_1(6/9) = 2/3, \quad f_1(7/9) = 3/3,$$

$$f_1(8/9) = 2/3, \quad f_1(9/9) = 3/3$$

and extend (uniquely) this partial function to a continuous function

$$f_1 \; : \; [0, 1] \to [0, 1]$$

in such a way that f_1 becomes affine on each segment $[k/9, (k+1)/9]$ where $k = 0, 1, ..., 8$. We shall start with this function f_1. In our further construction, we also need an analogous function g acting from the segment $[0, 9]$ into the segment $[0, 3]$. Namely, we put

$$g(x) = 3f_1(x/9) \qquad (x \in [0, 9]).$$

Obviously, g is continuous and affine on each segment $[k, k+1]$ where $k = 0, 1, ..., 8$. Also, another function similar to g will be useful in our construction. Namely, we denote by g^* the function from $[0, 9]$ into $[0, 3]$, whose graph is symmetric with the graph of g, with respect to the straight line

$$\{(x, y) \in \mathbf{R} \times \mathbf{R} \; : \; y = 3/2\}.$$

In other words, we put

$$g^*(x) = 3 - g(x)$$

for all $x \in [0, 9]$. Suppose now that, for a natural number $n \geq 1$, the function

$$f_n \; : \; [0, 1] \to [0, 1]$$

has already been defined, such that:

(a) f_n is continuous;

(b) for each segment of the form $[k/9^n, (k+1)/9^n]$, where

$$k \in \{0, 1, ..., 9^n - 1\},$$

the function f_n is affine on it and the image of this segment with respect to f_n is some segment of the form $[j/3^n, (j+1)/3^n]$, where

$$j \in \{0, 1, ..., 3^n - 1\}.$$

Let us construct a function

$$f_{n+1} : [0,1] \to [0,1].$$

For this purpose, it suffices to define f_{n+1} for any segment $[k/9^n, (k+1)/9^n]$ where $k \in \{0, 1, ..., 9^n - 1\}$. Here only two cases are possible.

1. f_n is increasing on $[k/9^n, (k+1)/9^n]$. In this case, let us consider the following two sets of points of the plane:

$$\{(0,0), \ (0,3), \ (9,3), \ (9,0)\},$$

$$\{(k/9^n, f_n(k/9^n)), \ (k/9^n, f_n((k+1)/9^n)),$$

$$((k+1)/9^n, f_n((k+1)/9^n)), \ ((k+1)/9^n, f_n(k/9^n))\}.$$

Because we have the vertices of two rectangles, there exists a unique affine transformation

$$h : \mathbf{R}^2 \to \mathbf{R}^2$$

satisfying the conditions

$$h(0,0) = (k/9^n, f_n(k/9^n)), \quad h(0,3) = (k/9^n, f_n((k+1)/9^n)),$$

$$h(9,3) = ((k+1)/9^n, f_n((k+1)/9^n)), \quad h(9,0) = ((k+1)/9^n, f_n(k/9^n)).$$

Let the graph of the restriction of f_{n+1} to the segment $[k/9^n, (k+1)/9^n]$ coincide with the image of the graph of g with respect to h.

2. f_n is decreasing on $[k/9^n, (k+1)/9^n]$. In this case, let us consider the following two sets of points of the plane:

$$\{(0,0), \ (0,3), \ (9,3), \ (9,0)\},$$

$$\{(k/9^n, f_n((k+1)/9^n)), \ (k/9^n, f_n(k/9^n)),$$

$$((k+1)/9^n, f_n(k/9^n)), \ ((k+1)/9^n, f_n((k+1)/9^n))\}.$$

Here we also have the vertices of two rectangles, so there exists a unique affine transformation

$$h^* : \mathbf{R}^2 \to \mathbf{R}^2$$

satisfying the relations

$$h^*(0,0) = (k/9^n, f_n((k+1)/9^n)), \quad h^*(0,3) = (k/9^n, f_n(k/9^n)),$$

$$h^*(9,3) = ((k+1)/9^n, f_n(k/9^n)), \quad h^*(9,0) = ((k+1)/9^n, f_n((k+1)/9^n)).$$

Let the graph of the restriction of f_{n+1} to the segment $[k/9^n, (k+1)/9^n]$ coincide with the image of the graph of g^* with respect to h^*.

The function f_{n+1} has thus been determined. From the above construction immediately follows that the corresponding analogues of the conditions (a) and (b) hold true for f_{n+1}, too. In other words, f_{n+1} is continuous and, for each segment of the form $[k/9^{n+1}, (k+1)/9^{n+1}]$, where

$$k \in \{0, 1, ..., 9^{n+1} - 1\},$$

the function f_{n+1} is affine on it and the image of this segment with respect to f_{n+1} is some segment of the form $[j/3^{n+1}, (j+1)/3^{n+1}]$, where

$$j \in \{0, 1, ..., 3^{n+1} - 1\}.$$

Moreover, our construction shows that

$$(\forall x \in [0, 1])(|f_{n+1}(x) - f_n(x)| \leq 1/3^n).$$

In addition, let

$$[u, v] = [k/9^n, (k+1)/9^n]$$

be an arbitrary segment on which f_n is affine. Then it is not hard to check that

$$f_{n+1}([u, (2u+v)/3]) = f_n([u, (2u+v)/3]),$$
$$f_{n+1}([(2u+v)/3, (2v+u)/3]) = f_n([(2u+v)/3, (2v+u)/3]),$$
$$f_{n+1}([(u+2v)/3, v]) = f_n([(u+2v)/3, v]).$$

Proceeding in this way, we come to the sequence of functions

$$(f_1, \quad f_2, \quad \ldots, \quad f_n, \quad \ldots)$$

uniformly convergent to some continuous function f also acting from $[0, 1]$ into $[0, 1]$. We assert that f is nowhere approximately differentiable on the segment $[0, 1]$. In order to demonstrate this fact, let us take an arbitrary point $x \in [0, 1]$ and fix a natural number $n \geq 1$.

Clearly, there exists a number $k \in \{0, 1, ..., 9^n - 1\}$ such that

$$x \in [k/9^n, (k+1)/9^n].$$

Therefore, we have

$$f_n(x) \in [j/3^n, (j+1)/3^n]$$

for some number $j \in \{0, 1, ..., 3^n - 1\}$. For the sake of simplicity, denote

$$[u, v] = [k/9^n, (k+1)/9^n], \quad [p, q] = [j/3^n, (j+1)/3^n].$$

From the remarks made above it immediately follows that, for all natural numbers $m > n$, we have

$$f_m(x) \in [p, q]$$

and, consequently, $f(x) \in [p, q]$, too. Further, we may assume without loss of generality that f_n is increasing on $[u, v]$ (the case when f_n is decreasing on $[u, v]$ can be considered completely analogously).

Suppose first that $f(x) \leq (p + q)/2$ and put

$$D_1 = [(2v + u)/3, v].$$

Then, for each point $y \in D_1$, we may write

$$f(y) \in [(2q + p)/3, q].$$

Hence, we get

$$(f(y) - f(x))/(y - x) \geq ((2q + p)/3 - (p + q)/2)/(v - u) = (1/6)(3^n).$$

Suppose now that $f(x) \geq (p + q)/2$ and denote

$$D_2 = [u, (2u + v)/3].$$

In this case, for any point $y \in D_2$, we may write

$$f(y) \in [p, (2p + q)/3].$$

Hence, we get

$$(f(x) - f(y))/(x - y) \geq ((p + q)/2 - (2p + q)/3)/(v - u) = (1/6)(3^n).$$

Thus, in the both cases, we have

$$\lambda(\{y \in [x - 1/9^n, x + 1/9^n] \setminus \{x\} \ :$$

$$|f(y) - f(x)|/|y - x| \geq (1/6)(3^n)\}) \geq (1/3)(1/9^n)$$

or, equivalently,

$$\lambda(\{y \in [x - 1/9^n, x + 1/9^n] \setminus \{x\} \ :$$

$$|f(y) - f(x)|/|y - x| \geq (1/6)(3^n)\}) \geq (1/6)\lambda([x - 1/9^n, x + 1/9^n]).$$

The latter relation immediately yields that our function f is not approximately differentiable at x (see Lemma 2 and the comments after this lemma).

Remark 1. The function f constructed above has a number of other interesting properties (for more information concerning f, see [140] and [40]).

Now, starting with an arbitrary continuous nowhere approximately differentiable function acting from $[0,1]$ into $[0,1]$, we can easily obtain an analogous function for $\mathbf{R}$. We thus come to the following classical result (first obtained by Jarník in 1934).

Theorem 1. *There exist continuous bounded functions acting from $\mathbf{R}$ into $\mathbf{R}$, which are nowhere approximately differentiable.*

Remark 2. Actually, Jarník proved that almost all (in the sense of the Baire category) functions from the Banach space $C[0,1]$ are nowhere approximately differentiable. Clearly, this result generalizes the corresponding result of Banach and Mazurkiewicz for the usual differentiability (see Theorem 2 from the Introduction). Further investigations showed that analogous statements hold true for many kinds of generalized derivatives. The main tool for obtaining such statements is the notion of porosity of a subset X of $\mathbf{R}$ at a given point $x \subset \mathbf{R}$. However, this interesting topic is out of the scope of our book. So we only refer the reader to the fundamental paper [27] where several category analogues of Theorem 1 for generalized derivatives are discussed from this point of view.

In Chapter 15 of our book we give an application of a nowhere approximately differentiable function to the question concerning some relationships between the sup-measurability and weak sup-measurability of functions acting from $\mathbf{R} \times \mathbf{R}$ into $\mathbf{R}$.

Because the concept of an approximate derivative relies essentially on the notion of a density point, it is reasonable to introduce here the so-called density topology on $\mathbf{R}$ and to consider briefly some elementary properties of this topology.

Exercise 9. For any Lebesgue measurable subset X of $\mathbf{R}$, let us denote

$$d(X) = \{x \in \mathbf{R} \; : \; x \text{ is a density point for } X\}.$$

Further, denote by $\mathcal{T}_d$ the family of all those Lebesgue measurable sets $Y \subset \mathbf{R}$, for which $Y \subset d(Y)$. Show that:

(a) $\mathcal{T}_d$ is a topology on $\mathbf{R}$ strictly extending the standard Euclidean topology of $\mathbf{R}$;

(b) the topological space $(\mathbf{R}, \mathcal{T}_d)$ is a Baire space and satisfies the Suslin condition (i.e., no nonempty open set in $(\mathbf{R}, \mathcal{T}_d)$ is of first category and each disjoint family of nonempty open sets in $(\mathbf{R}, \mathcal{T}_d)$ is at most countable);

(c) every first category set in $(\mathbf{R}, \mathcal{T}_d)$ is nowhere dense and closed (in particular, the family of all subsets of $(\mathbf{R}, \mathcal{T}_d)$ having the Baire property coincides with the Borel σ-algebra of $(\mathbf{R}, \mathcal{T}_d)$);

(d) a set $X \subset \mathbf{R}$ is Lebesgue measurable if and only if X has the Baire property in $(\mathbf{R}, \mathcal{T}_d)$;

(e) a set $X \subset \mathbf{R}$ is of Lebesgue measure zero if and only if X is a first category subset of $(\mathbf{R}, \mathcal{T}_d)$;

(f) the space $(\mathbf{R}, \mathcal{T}_d)$ is not separable.

The above-mentioned topology $\mathcal{T}_d$ is usually called the density topology on $\mathbf{R}$. In a similar way, the density topology can be introduced for the Euclidean space $\mathbf{R}^n$ $(n \geq 2)$ equipped with the n-dimensional Lebesgue measure λ_n.

Exercise 10. Let $f : \mathbf{R} \to \mathbf{R}$ be a function and let $x \in \mathbf{R}$. Prove that the following two assertions are equivalent:

(a) f is approximately continuous at x;

(b) f regarded as a mapping from $(\mathbf{R}, \mathcal{T}_d)$ into $\mathbf{R}$ is continuous at x.

Exercise 11. By starting with the result of the previous exercise, show that the topological space $(\mathbf{R}, \mathcal{T}_d)$ is connected. For this purpose, suppose to the contrary that there exists a partition $\{A, B\}$ of $\mathbf{R}$ into two nonempty sets $A \in \mathcal{T}_d$ and $B \in \mathcal{T}_d$. Then define a function

$$f : \mathbf{R} \to \mathbf{R}$$

by putting $f(x) = 1$ for all $x \in A$, and $f(x) = -1$ for all $x \in B$. Obviously, f is a bounded continuous mapping from $(\mathbf{R}, \mathcal{T}_d)$ into $\mathbf{R}$ and hence, according to Exercise 10, f is approximately continuous at each point of $\mathbf{R}$. Further, define

$$F(x) = \int_0^x f(t)dt \qquad (x \in \mathbf{R}).$$

By applying Exercise 5 of this chapter, demonstrate that the function F is differentiable everywhere on $\mathbf{R}$ and

$$F'(x) = 1 \quad \vee \quad F'(x) = -1$$

for each $x \in \mathbf{R}$. This yields a contradiction with the Darboux property of any derivative.

One of the most interesting facts concerning the density topology states that $(\mathbf{R}, \mathcal{T}_d)$ is a completely regular topological space (see, for instance, [162] and [219]). This property of $\mathcal{T}_d$ implies some nontrivial consequences in real analysis. To illustrate, we shall sketch here a proof of the existence of everywhere differentiable nowhere monotone functions by applying the above-mentioned fact (note that this approach is due to Goffman [65]).

Exercise 12. Consider any two disjoint countable sets

$$A = \{a_n \ : \ n \in \mathbf{N}\} \subset \mathbf{R},$$

$$B = \{b_n \ : \ n \in \mathbf{N}\} \subset \mathbf{R},$$

each of which is everywhere dense in $\mathbf{R}$. Taking into account Exercise 10 and the fact that $(\mathbf{R}, \mathcal{T}_d)$ is completely regular, we can find, for any $n \in \mathbf{N}$, an approximately continuous function $f_n \ : \ \mathbf{R} \to [0,1]$ satisfying the relations

$$0 \leq f_n(x) \leq 1 \qquad (x \in \mathbf{R}),$$

$$f_n(a_n) = 1, \quad (\forall x \in B)(f_n(x) = 0).$$

Analogously, for any $n \in \mathbf{N}$, there exists an approximately continuous function $g_n \ : \ \mathbf{R} \to [0,1]$ such that

$$0 \leq g_n(x) \leq 1 \qquad (x \in \mathbf{R}),$$

$$g_n(b_n) = 1, \quad (\forall x \in A)(g_n(x) = 0).$$

Now, define a function $h \ : \ \mathbf{R} \to \mathbf{R}$ by the formula

$$h = \sum_{n \in \mathbf{N}} (1/2^n)(f_n - g_n).$$

Check that:
(a) h is bounded and approximately continuous;
(b) $h(a) > 0$ for all $a \in A$;
(c) $h(b) < 0$ for all $b \in B$.
Let H denote an indefinite integral of h. Show that:
(i) H is everywhere differentiable on $\mathbf{R}$ and $H'(x) = h(x)$ for each $x \in \mathbf{R}$;
(ii) H is nowhere monotone.
We thus see that with the aid of the density topology on $\mathbf{R}$ it is possible to give another proof of the existence of everywhere differentiable nowhere monotone functions acting from $\mathbf{R}$ into $\mathbf{R}$ (cf. the proof presented in Chapter 5).

Remark 3. The density topology on $\mathbf{R}$ can be regarded as a very particular case of the so-called von Neumann topology. Let $(E, \mathcal{S}, \mu)$ be a space with a complete probability measure (or, more generally, with a complete nonzero σ-finite measure). Then, in conformity with a deep theorem of von

Neumann and Maharam (see, e.g., [138], [162], [171], [222]), there exists a topology $T = T(\mu)$ on E such that:

(1) (E, T) is a Baire space satisfying the Suslin condition;

(2) the family of all subsets of (E, T) having the Baire property coincides with the σ-algebra $\mathcal{S}$;

(3) a set $X \subset E$ is of μ-measure zero if and only if X is of first category in (E, T).

We say that $T = T(\mu)$ is a von Neumann topology associated with the given measure space $(E, \mathcal{S}, \mu)$. Note that T, in general, is not unique. This fact is not so surprising, because the proof of the existence of T is essentially based on the Axiom of Choice. There are many nontrivial applications of a von Neumann topology in various branches of contemporary mathematics (for instance, some applications to the general theory of stochastic processes can be found in [171]).

Remark 4. For the real line $\mathbf{R}$, an interesting analogue of the density topology, formulated in terms of category and the Baire property, was introduced and considered by Wilczyński in [231]. Wilczyński's topology was then investigated by several authors. An extensive survey devoted to properties of this topology and to functions continuous with respect to it is contained in [40] (see also the list of references presented in the same work).

Remark 5. There are some invariant extensions of the Lebesgue measure λ, for which an analogue of the classical Lebesgue theorem on density points does not hold. For example, there exist a measure μ on $\mathbf{R}$ and a μ-measurable set $X \subset \mathbf{R}$, such that:

(1) μ is an extension of λ;

(2) μ is invariant under the group of all isometric transformations of $\mathbf{R}$;

(3) there is only one μ-density point for X, i.e., there exists a unique point $x \in \mathbf{R}$ for which we have

$$lim_{h \to 0+} \frac{\mu(X \cap [x - h, x + h])}{2h} = 1.$$

A more detailed account of the measure μ and its other extraordinary properties can be found in [97].

7. Blumberg's theorem and Sierpiński-Zygmund functions

In various questions of analysis, we often need to consider some nontrivial restrictions of a given function (e.g., acting from $\mathbf{R}$ into $\mathbf{R}$), which have nice additional descriptive properties. In general, these properties do not hold for the original function. In order to illustrate this, let us recall two widely known statements from the theory of real functions. The first of them is the classical theorem of Luzin concerning the structure of an arbitrary Lebesgue measurable function acting from $\mathbf{R}$ into $\mathbf{R}$. Undoubtedly, this theorem plays the most fundamental role in real analysis and measure theory.

Let λ denote the standard Lebesgue measure on the real line. Let

$$f \; : \; \mathbf{R} \to \mathbf{R}$$

be a function measurable in the Lebesgue sense. Then, according to the Luzin theorem (see, e.g., [154] or Theorem 2 from Chapter 3), there exists a sequence $\{D_n \; : \; n \in \mathbf{N}\}$ of closed subsets of $\mathbf{R}$, such that

$$\lambda(\mathbf{R} \setminus \cup\{D_n \; : \; n \in \mathbf{N}\}) = 0$$

and, for each $n \in \mathbf{N}$, the restricted function

$$f|D_n \; : \; D_n \to \mathbf{R}$$

is continuous.

It immediately follows from this important statement that, for any Lebesgue measurable function $f \; : \; \mathbf{R} \to \mathbf{R}$, there exists a continuous function

$$g \; : \; \mathbf{R} \to \mathbf{R}$$

such that

$$\lambda(\{x \in \mathbf{R} \; : \; f(x) = g(x)\}) > 0.$$

Indeed, it suffices to take a set D_n with $\lambda(D_n) > 0$ and then to extend the function $f|D_n$ to a continuous function g acting from $\mathbf{R}$ into $\mathbf{R}$ (obviously, we are dealing here with a very special case of the classical Tietze-Urysohn theorem on the existence of a continuous extension of a continuous real-valued function defined on a closed subset of a normal topological space). In particular, we have the equality

$$card(\{x \in \mathbf{R} \ : \ f(x) = g(x)\}) = \mathbf{c}$$

where $\mathbf{c}$ denotes, as usual, the cardinality of the continuum.

Also, we may formulate the corresponding analogue of the Luzin theorem for real-valued functions possessing the Baire property. This analogue is essentially due to Baire (see, for instance, [120], [162] or Exercise 13 from the Introduction).

Let

$$f \ : \ \mathbf{R} \to \mathbf{R}$$

be a function having the Baire property. Then there exists a subset D of $\mathbf{R}$ such that:

(1) the set $\mathbf{R} \setminus D$ is of first category;

(2) the function $f|D$ is continuous.

In particular, because $card(D) = \mathbf{c}$ and $cl(D) = \mathbf{R}$, we conclude that the restriction of f to some everywhere dense subset of $\mathbf{R}$ having the cardinality of the continuum turns out to be continuous.

It can easily be observed that the Luzin theorem and its analogue for the Baire property hold true in much more general situations. The following two exercises show this.

Exercise 1. Let E be a Hausdorff topological space, let μ be a finite Radon measure on E and let μ' denote the usual completion of μ. Prove that, for any μ'-measurable function

$$f \ : \ E \to \mathbf{R}$$

and for each $\varepsilon > 0$, there exists a compact set $K \subset E$ for which these two relations are fulfilled:

(a) $\mu(E \setminus K) < \varepsilon$;

(b) the restriction of f to K is continuous.

Deduce from this result that there exists an F_σ-subset Y of $\mathbf{R}$ such that

$$Y \subset f(E),$$

$$\mu'(E \setminus f^{-1}(Y)) = 0.$$

In other words, every Radon measure is perfect (the notion of a perfect measure was introduced by Gnedenko and Kolmogorov).

Let us underline that perfect probability measures play an essential role in the theory of random (stochastic) processes (cf. [171]).

Exercise 2. Let E be a Baire topological space and let E' be a topological space with a countable base. Demonstrate that, for any mapping

$$f \; : \; E \to E'$$

having the Baire property, there exists a set $D \subset E$ satisfying the following relations:

(a) if U is an arbitrary nonempty open subset of E, then the set $D \cap U$ is co-meager in U;

(b) the function $f|D$ is continuous.

Infer from (a) and (b) that the restriction of f to some everywhere dense subset of E is continuous.

The preceding results point out the nice behavior of Lebesgue measurable functions from $\mathbf{R}$ into $\mathbf{R}$ (respectively, of functions from $\mathbf{R}$ into $\mathbf{R}$ with the Baire property) on some subsets of $\mathbf{R}$ that are not small (in a certain sense). Namely, as mentioned above, an immediate consequence of the Luzin theorem is that, for any Lebesgue measurable function f acting from $\mathbf{R}$ into $\mathbf{R}$, there exists a closed subset A of $\mathbf{R}$ with strictly positive measure, such that the restriction $f|A$ is continuous. The corresponding analogue for the Baire property states that, for any function g acting from $\mathbf{R}$ into $\mathbf{R}$ and possessing the Baire property, there exists a co-meager subset B of $\mathbf{R}$ such that the restriction $g|B$ is continuous, too.

In this connection, a natural question arises: what can be said about an arbitrary function acting from $\mathbf{R}$ into $\mathbf{R}$? In other words, if an arbitrary function

$$f \; : \; \mathbf{R} \to \mathbf{R}$$

is given, is it true that $f|D$ is continuous on some set $D \subset \mathbf{R}$, which is not small (in a certain sense)?

The first topological property of D that may be considered in this respect is the density of D in $\mathbf{R}$. It turns out that this property is completely sufficient. Namely, Blumberg showed in [18] that there always exists an everywhere dense subset D of $\mathbf{R}$ for which the restriction $f|D$ is continuous.

This chapter of our book is devoted to the Blumberg theorem and to some strange functions that naturally appear when one tries to generalize this theorem in various directions. Such strange functions were first constructed by Sierpiński and Zygmund (see [203]). They are extremely discontinuous. More precisely, the restrictions of such functions to any subset of **R** having the cardinality of the continuum are always discontinuous.

Note that the Blumberg theorem is a result of the theory **ZF** & **DC** (cf. the proof of this theorem given below).

At the same time, the construction of a Sierpiński-Zygmund function cannot be carried out within **ZF** & **DC** and essentially relies on an uncountable form of the Axiom of Choice.

We begin with the following auxiliary notion.

Let f be a function acting from **R** into **R** and let x be a point of **R**.

We shall say that x is a pleasant point with respect to f (or, briefly, f-pleasant point) if, for each real $\varepsilon > 0$, there exists a neighborhood

$$V(x) = V(x, \varepsilon)$$

of x such that the set

$$\{y \in \mathbf{R} \ : \ |f(y) - f(x)| < \varepsilon\}$$

is categorically dense in $V(x)$ (i.e., the intersection of this set with any nonempty open interval contained in $V(x)$ is of second category in that interval).

In accordance with the definition above, we shall say that a point $x \in \mathbf{R}$ is unpleasant with respect to f (or, briefly, f-unpleasant point) if x is not f-pleasant.

The following key lemma shows that, for any function

$$f \ : \ \mathbf{R} \to \mathbf{R},$$

the set of all f-unpleasant points is small in the sense of the Baire category.

Lemma 1. *Let f be a function acting from **R** into **R**. Then the set of all f-unpleasant points is of first category in **R**. Consequently, the set of all f-pleasant points is co-meager on every nonempty open subinterval of **R**.*

Proof. Suppose otherwise, i.e., suppose that the set

$$A = \{x \in \mathbf{R} \ : \ x \ is \ unpleasant \ with \ respect \ to \ f\}$$

is not of first category. For each point $x \in A$, there exists a strictly positive number $\varepsilon(x)$ such that, for any neighborhood V of x, the set

$$\{y \in \mathbf{R} \ : \ |f(y) - f(x)| < \varepsilon(x)\}$$

is not categorically dense in V. Let us pick two rational numbers $r(x)$ and $s(x)$ satisfying the inequalities

$$f(x) - \varepsilon(x)/2 < r(x) < f(x) < s(x) < f(x) + \varepsilon(x)/2.$$

Further, for any pair (r, s) of rational numbers, let us put

$$A_{r,s} = \{x \in A : r(x) = r,\ s(x) = s\}.$$

Evidently, we have the equality

$$A = \cup_{(r,s) \in \mathbf{Q} \times \mathbf{Q}}\, A_{r,s}.$$

Because, by our assumption, A is not of first category, there exists a pair

$$(r_0, s_0) \in \mathbf{Q} \times \mathbf{Q}$$

such that the set A_{r_0,s_0} is not of first category, either. Consequently, there exists a nonempty open interval $]a, b[\subset \mathbf{R}$ such that the set A_{r_0,s_0} is categorically dense in $]a, b[$. Choose any point

$$x_0 \in\ ]a, b[\ \cap\ A_{r_0,s_0}.$$

For this point, we may write

$$f(x_0) - \varepsilon(x_0)/2 < r_0 < f(x_0) < s_0 < f(x_0) + \varepsilon(x_0)/2.$$

Analogously, for each point $y \in A_{r_0,s_0}$, we have

$$f(y) - \varepsilon(y)/2 < r_0 < f(y) < s_0 < f(y) + \varepsilon(y)/2.$$

Therefore,

$$|f(y) - f(x_0)| < s_0 - r_0 < \varepsilon(x_0).$$

In other words, we get the inclusion

$$A_{r_0,s_0} \subset \{y \in \mathbf{R} : |f(y) - f(x_0)| < \varepsilon(x_0)\},$$

which immediately implies that the set

$$\{y \in \mathbf{R} : |f(y) - f(x_0)| < \varepsilon(x_0)\}$$

is categorically dense in $]a, b[$ contradicting the definition of $\varepsilon(x_0)$. The contradiction obtained ends the proof of Lemma 1.

We are now ready to prove the classical Blumberg theorem.

Theorem 1. *Let f be an arbitrary function acting from $\mathbf{R}$ into $\mathbf{R}$. Then there exists an everywhere dense subset X of $\mathbf{R}$ such that the function $f|X$ is continuous.*

Proof. Because $\mathbf{R}$ is homeomorphic to the open unit interval $]0,1[$, it suffices to establish the assertion of Theorem 1 for any function

$$f :]0,1[\to]0,1[.$$

Let $\Gamma(f)$ denote the graph of f. Taking Lemma 1 into account, we can recursively construct two sequences

$$\{Z_n : n \in \mathbf{N}\}, \quad \{D_n : n \in \mathbf{N}\},$$

satisfying the following conditions:

(1) for each natural number n, the set Z_n can be represented in the form

$$Z_n = \cup\{]a_i, b_i[\times]c_i, d_i[\ : \ i \in I(n)\}$$

where $I(n)$ is a countable set, the intervals of the family $\{]a_i, b_i[\ : \ i \in I(n)\}$ are contained in $]0,1[$ and are pairwise disjoint, the set

$$pr_1(Z_n \cap \Gamma(f))$$

is categorically dense in $]0,1[$ and, for any $i \in I(n)$, the length of $]c_i, d_i[$ is strictly less than $1/(n+1)$;

(2) for each natural number n, the set D_n is a finite $(1/(n+1))$-net of $]0,1[$, i.e., for any point $t \in \ ]0,1[$, there exists a point $d \in D_n$ such that

$$|t - d| < 1/(n+1);$$

(3) the sequence of sets $\{Z_n : n \in \mathbf{N}\}$ is decreasing by inclusion;
(4) the sequence of sets $\{D_n : n \in \mathbf{N}\}$ is increasing by inclusion;
(5) for each natural number n, we have

$$D_n \subset pr_1(Z_n \cap \Gamma(f))$$

and any point from D_n is f-pleasant.

We leave the details of this construction to the reader, because they are not difficult.

As soon as the above-mentioned sequences are defined, we put

$$D = \cup\{D_n : n \in \mathbf{N}\}.$$

Then condition (2) implies that the set D is everywhere dense in $]0, 1[$, and it can easily be verified, by using conditions (1), (3), (4), (5), that the restriction of f to D is continuous.

The Blumberg theorem has thus been proved.

Remark 1. Some stronger versions of the Blumberg theorem may be obtained by using additional set-theoretical hypotheses. For instance, in [10] the situation is discussed when Martin's Axiom (or certain of its consequences) is assumed. Actually, Blumberg's theorem was analyzed and generalized in many directions. Moreover, the concept of a Blumberg topological space was introduced and investigated. Here we only formulate the corresponding definition. Namely, we say that a topological space E is a Blumberg space if, for any function

$$g \; : \; E \to \mathbf{R},$$

there exists an everywhere dense subset X of E such that the restriction $g|X$ is continuous. The class of all Blumberg spaces turns out to be rather wide and possesses a number of interesting properties. Let us point out that any Blumberg space must be a Baire space (the reader can easily verify this simple fact). A useful survey of results concerning the Blumberg theorem and its generalizations is presented in paper [23] (see also [22], [78], [184], [230]).

We thus see that any function (acting from $\mathbf{R}$ into $\mathbf{R}$) restricted to an appropriate countable everywhere dense subset of $\mathbf{R}$ becomes continuous. In this connection, the question arises whether that subset can be chosen to be uncountable. A partial negative answer to this question yields the classical Sierpiński-Zygmund function constructed by them in [203], with the aid of an uncountable form of the Axiom of Choice. This function has the property that its restriction to each subset of $\mathbf{R}$ of cardinality continuum is discontinuous. Consequently, if the Continuum Hypothesis holds, then the restriction of the same function to any uncountable subset of $\mathbf{R}$ is discontinuous, too. We shall present here the construction of the Sierpiński-Zygmund function. Moreover, we shall give a slightly more general construction of a Sierpiński-Zygmund type function possessing some additional topological properties.

For this purpose, we need several auxiliary notions and statements.

Let E be a Hausdorff topological space. We recall that E is normal if, for any two disjoint closed sets $X \subset E$ and $Y \subset E$, there exist two open sets $U \subset E$ and $V \subset E$, such that

$$X \subset U, \quad Y \subset V, \quad U \cap V = \emptyset.$$

In other words, E is normal if and only if any two disjoint closed subsets of E can be separated by disjoint open subsets of E.

We also recall the well-known Tietze-Urysohn theorem (see, e.g., [120]) stating that if E is a normal space, X is a closed subset of E and

$$f : X \to \mathbf{R}$$

is a continuous function, then there always exists a continuous function

$$f^* : E \to \mathbf{R}$$

extending f. Furthermore, if, for the original function f, we have the relation

$$ran(f) \subset [a, b] \subset \mathbf{R},$$

then the extended function f^* may be chosen to satisfy an analogous relation

$$ran(f^*) \subset [a, b].$$

In fact, the property of the extendability of continuous real-valued functions defined on closed subsets of E is equivalent to the normality of E.

Recall that a normal topological space E is perfectly normal if each closed set in E is a G_δ-subset of E (or, equivalently, if each open set in E is an F_σ-subset of E). The following exercise yields another definition of perfectly normal spaces.

Exercise 3. Let E be a normal space. Show that these two assertions are equivalent:

(a) E is perfectly normal;

(b) for any closed subset X of E, there exists a continuous function

$$f : E \to [0, 1]$$

such that $f^{-1}(\{0\}) = X$.

The next exercise indicates another important property of perfectly normal spaces, concerning the structure of the σ-algebras of their Borel subsets.

Exercise 4. Let E be a topological space such that every closed subset of E is a G_δ-set in E. Let $\mathcal{M}$ be some class of subsets of E, satisfying the following conditions:

(a) the family of all closed subsets of E is contained in $\mathcal{M}$;

(b) if $\{X_n : n \in \mathbf{N}\}$ is an increasing (by inclusion) sequence of sets belonging to $\mathcal{M}$, then the set $\cup\{X_n : n \in \mathbf{N}\}$ belongs to $\mathcal{M}$, too;

(c) if $\{Y_n \,:\, n \in \mathbf{N}\}$ is a decreasing (by inclusion) sequence of sets belonging to $\mathcal{M}$, then the set $\cap\{Y_n \,:\, n \in \mathbf{N}\}$ belongs to $\mathcal{M}$, too.

Prove, by applying the method of transfinite induction, that $\mathcal{B}(E) \subset \mathcal{M}$, i.e., each Borel subset of E belongs to $\mathcal{M}$.

The result presented in Exercise 4 enables us to define Borel subsets of a perfectly normal space E in the following manner.

First, we put:

$\mathcal{B}_0^*(E) =$ the class of all closed subsets of E.

Suppose now that, for a nonzero ordinal $\xi < \omega_1$, the classes of sets

$$\mathcal{B}_\zeta^*(E) \qquad (\zeta < \xi)$$

have already been defined.

If ξ is an odd ordinal, then we put:

$\mathcal{B}_\xi^*(E) =$ the class of all those sets that can be represented in the form $\cup\{X_n \,:\, n \in \mathbf{N}\}$ where $\{X_n \,:\, n \in \mathbf{N}\}$ is some increasing (with respect to inclusion) sequence of sets belonging to the class $\cup\{\mathcal{B}_\zeta^*(E) \,:\, \zeta < \xi\}$.

If ξ is an even ordinal, then we put:

$\mathcal{B}_\xi^*(E) =$ the class of all those sets that can be represented in the form $\cap\{Y_n \,:\, n \in \mathbf{N}\}$ where $\{Y_n \,:\, n \in \mathbf{N}\}$ is some decreasing (with respect to inclusion) sequence of sets belonging to the class $\cup\{\mathcal{B}_\zeta^*(E) \,:\, \zeta < \xi\}$.

Finally, we define

$$\mathcal{B}^*(E) = \cup\{\mathcal{B}_\xi^*(E) \,:\, \xi < \omega_1\}.$$

Then, in virtue of the result of Exercise 4, we may assert that

$$\mathcal{B}^*(E) = \mathcal{B}(E).$$

For any set $X \in \mathcal{B}(E)$, we say that X is of order ξ if

$$X \in \mathcal{B}_\xi^*(E) \setminus \cup\{\mathcal{B}_\zeta^*(E) \,:\, \zeta < \xi\}.$$

Let now E be an arbitrary topological space. We denote by the symbol $B(E, \mathbf{R})$ the family of all Borel mappings acting from E into $\mathbf{R}$. Furthermore, for any ordinal number $\xi < \omega_1$, we define by transfinite recursion the class $Ba_\xi(E, \mathbf{R})$ of functions also acting from E into $\mathbf{R}$.

First of all, we put:

$$Ba_0(E, \mathbf{R}) = C(E, \mathbf{R}),$$

where $C(E, \mathbf{R})$ is the family of all continuous real-valued functions on E (see Chapter 2). Suppose that, for a nonzero ordinal $\xi < \omega_1$, the classes of functions

$$Ba_\zeta(E, \mathbf{R}) \qquad (\zeta < \xi)$$

have already been defined. Let us denote by $Ba_\xi(E, \mathbf{R})$ the class of all those functions

$$f \ : \ E \to \mathbf{R}$$

that satisfy the following condition: there exists a sequence

$$\{f_n \ : \ n \in \mathbf{N}\} \subset \cup\{Ba_\zeta(E, \mathbf{R}) \ : \ \zeta < \xi\}$$

(certainly, depending on f) for which we have

$$f(x) = lim_{n \to +\infty} f_n(x) \qquad (x \in E).$$

In other words, $Ba_\xi(E, \mathbf{R})$ consists of all pointwise limits of sequences of functions belonging to $\cup\{Ba_\zeta(E, \mathbf{R}) \ : \ \zeta < \xi\}$.

Continuing in this manner, we are able to define all the classes

$$Ba_\xi(E, \mathbf{R}) \qquad (\xi < \omega_1).$$

Finally, we put

$$Ba(E, \mathbf{R}) = \cup\{Ba_\xi(E, \mathbf{R}) \ : \ \xi < \omega_1\}.$$

For each ordinal $\xi < \omega_1$, the family of functions

$$Ba_\xi(E, \mathbf{R}) \setminus \cup\{Ba_\zeta(E, \mathbf{R}) \ : \ \zeta < \xi\}$$

is usually called the Baire class of functions having order ξ. A real-valued function given on E is called measurable in the Baire sense if it belongs to $Ba(E, \mathbf{R})$.

Exercise 5. Let E be a topological space. Prove, by using the method of transfinite induction, that:

(a) for any $\xi < \zeta < \omega_1$, the inclusion

$$Ba_\xi(E, \mathbf{R}) \subset Ba_\zeta(E, \mathbf{R})$$

is fulfilled;

(b) $Ba(E, \mathbf{R})$ is a vector space over $\mathbf{R}$;

(c) if $f \in Ba(E, \mathbf{R})$ and $g \in Ba(E, \mathbf{R})$, then $f \cdot g \in Ba(E, \mathbf{R})$;

(d) if $f \in Ba(E, \mathbf{R})$, then $|f| \in Ba(E, \mathbf{R})$;

(e) if f and g belong to $Ba(E, \mathbf{R})$, then

$$sup(f, g) \in Ba(E, \mathbf{R}), \quad inf(f, g) \in Ba(E, \mathbf{R});$$

(f) if f and g belong to $Ba(E, \mathbf{R})$ and $g(x) \neq 0$ for all $x \in E$, then f/g also belongs to $Ba(E, \mathbf{R})$;

(g) if a sequence $\{f_n : n \in \mathbf{N}\} \subset Ba(E, \mathbf{R})$ is given, such that there exists a pointwise limit

$$f(x) = lim_{n \to +\infty} f_n(x) \quad (x \in E),$$

then f belongs to $Ba(E, \mathbf{R})$;

(h) if $g \in Ba(E, \mathbf{R})$ is such that

$$ran(g) \subset \,]a, b[$$

for some open interval $]a, b[$, then, for any continuous function

$$\phi \, : \,]a, b[\, \to \mathbf{R},$$

the function $\phi \circ g$ belongs to $Ba(E, \mathbf{R})$;

(i) each function belonging to $Ba(E, \mathbf{R})$ is a Borel mapping from E into $\mathbf{R}$; in other words, we have the inclusion

$$Ba(E, \mathbf{R}) \subset B(E, \mathbf{R}).$$

In some cases, this inclusion can be proper. For example, let us equip the ordinal number ω_1 with its order topology. Check that

$$Ba(\omega_1, \mathbf{R}) \neq B(\omega_1, \mathbf{R}).$$

Exercise 6. Let Φ be a class of functions acting from $\mathbf{R}$ into $\mathbf{R}$. We say that a function

$$h \, : \, \mathbf{R} \times \mathbf{R} \to \mathbf{R}$$

is universal for Φ if, for any function $\phi \in \Phi$, there exists a point $y = y(\phi)$ of $\mathbf{R}$ such that

$$\phi(x) = h(x, y) \quad (x \in \mathbf{R}).$$

By starting with the existence of continuous mappings of Peano type (see Chapter 1) and applying the method of transfinite induction, show that, for each ordinal $\xi < \omega_1$, there exists a function

$$h_\xi \, : \, \mathbf{R} \times \mathbf{R} \to \mathbf{R}$$

satisfying the following conditions:

(a) h_ξ is a Borel mapping from $\mathbf{R} \times \mathbf{R}$ into $\mathbf{R}$;

(b) h_ξ is universal for the class $Ba_\xi(\mathbf{R}, \mathbf{R})$.

Deduce from this fact that, for any ordinal $\xi < \omega_1$, the set

$$Ba_\xi(\mathbf{R}, \mathbf{R}) \setminus \cup \{Ba_\zeta(\mathbf{R}, \mathbf{R}) \ : \ \zeta < \xi\}$$

is not empty (i.e., there are Baire functions of order ξ). In particular, the family of the Baire classes

$$\{Ba_\xi(\mathbf{R}, \mathbf{R}) \ : \ \xi < \omega_1\}$$

is strictly increasing by inclusion.

This remarkable result was first obtained by Lebesgue (see, e.g., [120] or [154]).

Lemma 2. *For any perfectly normal space E, the equality*

$$Ba(E, \mathbf{R}) = B(E, \mathbf{R})$$

is fulfilled. Consequently, this equality holds true for an arbitrary metric space E (in particular, for $E = \mathbf{R}$).

Proof. Obviously, it suffices to demonstrate that every bounded real-valued Borel function on E belongs to the class $Ba(E, \mathbf{R})$. Because any such function is uniformly approximable by linear combinations of characteristic functions of Borel subsets of E, it is enough to show that the characteristic function of each Borel subset of E belongs to $Ba(E, \mathbf{R})$.

Let X be an arbitrary closed subset of E. Taking account of the perfect normality of E and applying the Tietze-Urysohn extension theorem, we can easily verify that the characteristic function f_X is a pointwise limit of a sequence of continuous functions on E whose ranges are contained in $[0,1]$ (cf. Exercise 3 of this chapter).

Let now ξ be an ordinal from the interval $]0, \omega_1[$ and suppose that the assertion is true for the characteristic functions of all Borel subsets of E belonging to the Borel classes of order strictly less than ξ. Take any Borel set $X \subset E$ of order ξ. According to the result of Exercise 4, we may write $X = \cup \{X_n \ : \ n \in \mathbf{N}\}$ (or, respectively, $X = \cap \{X_n \ : \ n \in \mathbf{N}\}$) where $\{X_n \ : \ n \in \mathbf{N}\}$ is an increasing (respectively, a decreasing) sequence of Borel sets in E belonging to some Borel classes of strictly lower orders. But, in the both cases, we have

$$f_X = lim_{n \to +\infty} f_{X_n}.$$

In conformity with our inductive assumption, all characteristic functions f_{X_n} belong to the Baire class $Ba(E, \mathbf{R})$. Hence the characteristic function f_X belongs to this class, too (see Exercise 5 of the present chapter). This completes the proof of Lemma 2.

Let X be a topological space, Y be a metric space and let Z be a subset of X. Suppose that a function

$$f : Z \to Y$$

is given, and let x be an arbitrary point from X. We denote

$$\Omega_f(x) = \inf_{U \in \mathcal{B}(x)} diam(f(U)),$$

where $\mathcal{B}(x)$ denotes a local base of X at x (i.e., $\mathcal{B}(x)$ is a fundamental system of open neighborhoods of x). The real number $\Omega_f(x) \geq 0$ is usually called the oscillation of a function f at a point x (cf. Exercise 4 from Chapter 2).

It can easily be observed that if x does not belong to the closure of Z, then

$$\Omega_f(x) = 0.$$

Also, for a given function

$$f : X \to Y,$$

the following two assertions are equivalent:

(1) f is continuous on X;
(2) for each point $x \in X$, we have $\Omega_f(x) = 0$.

The next auxiliary statement is due to Lavrentieff (see [127]). It has many important applications in general topology and descriptive set theory (cf., for instance, [120], [84]).

Lemma 3. *Let X be a metric space, let Y be a complete metric space and let Z be a subset of X. Suppose that a continuous function*

$$f : Z \to Y$$

is given. Then there exist a set $Z^* \subset X$ and a function

$$f^* : Z^* \to Y,$$

satisfying these three relations:

(1) $Z \subset Z^*$;
(2) Z^* is a G_δ-subset of X;
(3) f^* is a continuous extension of f.

Proof. Let $cl(Z)$ denote the closure of Z in X. We put

$$Z^* = \{z \in cl(Z) \; : \; \Omega_f(z) = 0\}.$$

Because our original function f is continuous, we have

$$(\forall z \in Z)(\Omega_f(z) = 0).$$

Consequently, the inclusion

$$Z \subset Z^*$$

is valid. Now, let z be an arbitrary point of Z^*. Then there exists a sequence
of points

$$\{z_n \; : \; n \in \mathbf{N}\} \subset Z$$

such that

$$lim_{n \to +\infty} z_n = z.$$

Taking account of the equality $\Omega_f(z) = 0$, we see that

$$\{f(z_n) \; : \; n \in \mathbf{N}\}$$

is a Cauchy sequence in Y. But Y is complete, so the above-mentioned
sequence converges to some point $y \in Y$. In addition, it can easily be
shown that y does not depend on the choice of $\{z_n \; : \; n \in \mathbf{N}\}$. So we may
put

$$f^*(z) = y.$$

In this way we get the mapping

$$f^* \; : \; Z^* \to Y$$

that is continuous because, according to our definition,

$$(\forall z \in Z^*)(\Omega_{f^*}(z) = 0).$$

Thus, it remains to demonstrate that Z^* is a G_δ-subset of X. Obviously,
the equality

$$Z^* = \cap\{V_n \; : \; n \in \mathbf{N}\} \cap cl(Z)$$

holds, where, for each $n \in \mathbf{N}$, the set V_n is defined as follows:

$$V_n = \{x \in X \; : \; \Omega_f(x) < 1/(n+1)\}.$$

Because all sets V_n ($n \in \mathbf{N}$) are open in X (cf. Exercise 4 from Chapter 2)
and the closed set $cl(Z)$ is a G_δ-subset of X, we infer that Z^* is a G_δ-subset
of X, too. This finishes the proof of Lemma 3.

Exercise 7. Let X and Y be two complete metric spaces, let A be a subset of X and let B be a subset of Y. Suppose also that

$$f \ : \ A \to B$$

is a homeomorphism between A and B. By starting with the result of Lemma 3, show that there exist two sets $A^* \subset X$ and $B^* \subset Y$ and a mapping

$$f^* \ : \ A^* \to B^*,$$

satisfying these four relations:
  (a) $A \subset A^*$ and $B \subset B^*$;
  (b) A^* is a G_δ-subset of X and B^* is a G_δ-subset of Y;
  (c) f^* is a homeomorphism between A^* and B^*;
  (d) f^* is an extension of f.
This classical result was obtained by Lavrentieff [127] and is known as the Lavrentieff theorem on extensions of homeomorphisms. It found numerous applications in topology and, especially, in descriptive set theory (see, e.g., [120] where this theorem is applied in order to prove the topological invariance of Borel classes in complete metric spaces).

Lemma 4. *Let X be a metric space, Z be a subset of X and let*

$$f \ : \ Z \to \mathbf{R}$$

be a Borel mapping. Then there exist a set $Z^* \subset X$ and a mapping

$$f^* \ : \ Z^* \to \mathbf{R},$$

satisfying the following relations:
  (1) $Z \subset Z^*$;
  (2) Z^* is a Borel subset of X;
  (3) f^* is a Borel mapping;
  (4) f^* is an extension of f.

Proof. Taking account of the equalities

$$B(Z, \mathbf{R}) = Ba(Z, \mathbf{R}) = \cup \{ Ba_\xi(Z, \mathbf{R}) \ : \ \xi < \omega_1 \},$$

we may try to apply here the method of transfinite induction.

If our f belongs to the class $Ba_0(Z, \mathbf{R})$, then we use the result contained in Lemma 3.

Now, let ξ be a nonzero ordinal number strictly less than ω_1, and suppose that the assertion is true for all functions from the family

$$\cup \{ Ba_\zeta(Z, \mathbf{R}) \ : \ \zeta < \xi \}.$$

Let f be an arbitrary function belonging to the class $Ba_\xi(Z, \mathbf{R})$. According to the definition of $Ba_\xi(Z, \mathbf{R})$, there exist two sequences

$$\{\xi_n \ : \ n \in \mathbf{N}\}, \quad \{f_n \ : \ n \in \mathbf{N}\}$$

of ordinals and functions, respectively, such that:
    (a) for each $n \in \mathbf{N}$, we have

$$\xi_n < \xi, \quad f_n \in Ba_{\xi_n}(Z, \mathbf{R});$$

(b) for any point $z \in Z$, we have

$$f(z) = lim_{n \to +\infty} \ f_n(z).$$

By the inductive assumption, for each $n \in \mathbf{N}$, there exist a Borel set $Z_n^* \subset X$ and a Borel function

$$f_n^* \ : \ Z_n^* \to \mathbf{R}$$

extending f_n (in particular, $Z \subset Z_n^*$). Let us denote

$$Z' = \cap\{Z_n^* \ : \ n \in \mathbf{N}\},$$

$$Z^* = \{z \in Z' \ : \ there \ exists \ a \ lim_{n \to +\infty} \ f_n^*(z)\}.$$

Evidently, the set Z^* is Borel in X and $Z \subset Z^*$. Further, for any point $z \in Z^*$, we may put

$$f^*(z) = lim_{n \to +\infty} \ f_n^*(z).$$

Then the mapping

$$f^* \ : \ Z^* \to \mathbf{R}$$

defined in this manner is Borel, too, and extends the original mapping f. This finishes the proof of Lemma 4.

Remark 2. A slightly more precise formulation of Lemma 4 can be found, e.g., in monograph [120] (the proof remains almost the same). But, for our further purposes, Lemma 4 is completely sufficient.

Exercise 8. Let $[0,1]^\omega$ denote, as usual, the Hilbert cube (or the Tychonoff cube of weight ω). By applying the result of Lemma 4, show that if X is a metric space, Z is a subset of X and a mapping

$$g \ : \ Z \to [0,1]^\omega$$

is Borel, then there exist a set $Z^* \subset X$ and a mapping

$$g^* \; : \; Z^* \to [0,1]^\omega,$$

satisfying the following conditions:
    (a) $Z \subset Z^*$;
    (b) Z^* is a Borel subset of X;
    (c) g^* is a Borel mapping;
    (d) g^* is an extension of g.

Now, we are ready to establish the result of Sierpiński and Zygmund (in a more general form, not only for continuous restrictions of functions but also for Borel restrictions).

Theorem 2. *There exists a function*

$$f \; : \; \mathbf{R} \to \mathbf{R}$$

such that, for any set $Z \subset \mathbf{R}$ of cardinality of the continuum, the restriction of f to Z is not a Borel mapping (from Z into $\mathbf{R}$). In particular, this restriction is not continuous.

Proof. Let Φ denote the family of all partial Borel mappings (from $\mathbf{R}$ into $\mathbf{R}$) defined on uncountable Borel subsets of $\mathbf{R}$. Clearly, we have the equality

$$card(\Phi) = \mathbf{c},$$

where $\mathbf{c}$ denotes, as usual, the cardinality of the continuum. Consequently, we may enumerate this family in the following manner:

$$\Phi = \{\phi_\beta \; : \; \beta < \alpha\},$$

where α is the smallest ordinal number with

$$card(\alpha) = \mathbf{c}.$$

Also, we can analogously enumerate the set of all points of $\mathbf{R}$, i.e., represent $\mathbf{R}$ in the form

$$\mathbf{R} = \{x_\beta \; : \; \beta < \alpha\}.$$

Now, for each $\beta < \alpha$, consider the family

$$\{\Gamma(\phi_\xi) \; : \; \xi \leq \beta\}$$

of graphs of functions from $\{\phi_\xi \; : \; \xi \leq \beta\}$ and take the set

$$(\{x_\beta\} \times \mathbf{R}) \setminus \cup \{\Gamma(\phi_\xi) \; : \; \xi \leq \beta\}.$$

Clearly, the latter set is not empty (moreover, it is of cardinality of the continuum). So we may pick a point (x_β, y_β) from this set. Let us put

$$f(x_\beta) = y_\beta \quad (\beta < \alpha).$$

Evidently, we obtain some mapping

$$f : \mathbf{R} \to \mathbf{R}.$$

Let us demonstrate that f is the desired function.

Indeed, it immediately follows from the definition of f that, for any partial Borel function

$$\phi : \mathbf{R} \to \mathbf{R}$$

defined on an uncountable Borel subset of $\mathbf{R}$, the inequality

$$card(\{x \in dom(\phi) \; : \; \phi(x) = f(x)\}) < \mathbf{c}$$

is fulfilled. Suppose for a while that there exists a set $Z \subset \mathbf{R}$ of cardinality of the continuum, such that the restriction $f|Z$ is Borel. Then, in conformity with Lemma 4, there exists a partial Borel mapping f^* extending $f|Z$ and defined on some uncountable Borel subset of $\mathbf{R}$. Consequently, f^* belongs to the family Φ, and

$$Z \subset \{x \in dom(f^*) \; : \; f^*(x) = f(x)\},$$

$$card(\{x \in dom(f^*) \; : \; f^*(x) = f(x)\}) = \mathbf{c},$$

which yields an obvious contradiction.

Theorem 2 has thus been proved.

Actually, the proof of Theorem 2 presented above shows that there are many functions of Sierpiński-Zygmund type (for instance, the cardinality of the family of all such functions is equal to $2^{\mathbf{c}}$).

Some interesting generalizations of this classical theorem are possible (see, e.g., the next exercise and Exercise 15 from Chapter 11).

Exercise 9. Give a generalization of Theorem 2 for uncountable Polish spaces. In other words, show that if E is an arbitrary uncountable Polish topological space, then there exists a function

$$f : E \to \mathbf{R}$$

such that, for each set $Z \subset E$ of cardinality of the continuum, the restriction of f to Z is not a Borel mapping.

Remark 3. In a certain sense, we may say that any Sierpiński-Zygmund function (defined on $\mathbf{R}$) is totally discontinuous with respect to the family of all subsets of $\mathbf{R}$ having the cardinality of the continuum. In other words, such a function is very bad from the point of view of continuity of its restrictions to large subsets of $\mathbf{R}$ (here "large" means that the cardinality of those subsets must be equal to $\mathbf{c}$). We shall see in the next chapter that a Sierpiński-Zygmund function is also bad from the points of view of the Lebesgue measurability and the Baire property, i.e., such a function is not measurable in the Lebesgue sense and does not possess the Baire property. Actually, this fact follows directly from the Luzin theorem on the structure of Lebesgue measurable functions and from its corresponding analogue for the functions having the Baire property. Thus, we may conclude that the Sierpiński-Zygmund construction yields an example of a function which simultaneously is not measurable in the Lebesgue sense and does not possess the Baire property. Of course, there are many other constructions of such functions (all of them are based on uncountable forms of the Axiom of Choice). The best known constructions are due to Vitali [226] and Bernstein [15]. We shall examine their constructions in our further considerations (see Chapter 8).

Remark 4. In connection with the Sierpiński-Zygmund result presented above, the following question arises naturally: does there exist a function

$$f \; : \; \mathbf{R} \to \mathbf{R}$$

such that, for any uncountable subset Z of $\mathbf{R}$, the restriction $f|Z$ is not continuous? As mentioned earlier, if the Continuum Hypothesis holds, then a Sierpiński-Zygmund function yields a positive answer to this question. Nevertheless, the question cannot be resolved within the theory $\mathbf{ZFC}$. Moreover, Shinoda demonstrated in [184] that if Martin's Axiom and the negation of the Continuum Hypothesis hold and g is an arbitrary function acting from $\mathbf{R}$ into $\mathbf{R}$, then, for each uncountable set $X \subset \mathbf{R}$, there always exists an uncountable set $Y \subset X$ such that the restriction $g|Y$ is continuous (for further details, see [184]).

Several related questions concerning the existence of a good restriction of a given function to some nonsmall subset of its domain are discussed in [38] where the corresponding references can also be found.

Exercise 10. Demonstrate that any Sierpiński-Zygmund function

$$f : \mathbf{R} \to \mathbf{R}$$

possesses the following property: for each set $X \subset \mathbf{R}$ with $card(X) = \mathbf{c}$, the restriction $f|X$ is not monotone on X (use the fact that if g is an arbitrary monotone partial function acting from $\mathbf{R}$ into itself, then the set of all discontinuity points of g is at most countable).

Deduce from this result that, under the Continuum Hypothesis, the restriction of f to each uncountable set $Y \subset \mathbf{R}$ is not monotone on Y (cf. Exercise 10 from Chapter 4).

8. Lebesgue nonmeasurable functions and functions without the Baire property

This chapter is devoted to some well-known constructions of a function acting from **R** into **R** and nonmeasurable in the Lebesgue sense (respectively, of a function acting from **R** into **R** and lacking the Baire property). Obviously, the existence of such a function is equivalent to the existence of a subset of the real line nonmeasurable in the Lebesgue sense (respectively, of a subset of this line without the Baire property).

Since the fundamental concept of the Lebesgue measure on **R** (respectively, the concept of the Baire property) was introduced, it has been extremely useful in various problems of mathematical analysis. A natural question arose whether all subsets of **R** are measurable in the Lebesgue sense (and an analogous question was posed for the Baire property). Very soon, two essentially different constructions of extraordinary subsets of **R** were discovered which gave simultaneously negative answers to these questions.

The first construction is due to Vitali [226] and the second one to Bernstein [15]. Both of them were heavily based on an uncountable form of the Axiom of Choice, so it was reasonable to ask whether it is possible to construct a Lebesgue nonmeasurable subset of **R** (or a subset of **R** without the Baire property) by using some weak forms of the Axiom of Choice which are enough for most domains of the classical mathematical analysis (for instance, the Axiom of Dependent Choices). Almost all outstanding mathematicians working that time in analysis and particularly in the theory of real functions (Borel, Lebesgue, Hausdorff, Luzin, Sierpiński, etc.) believed that there is no construction of a Lebesgue nonmeasurable subset of **R** within the theory **ZF** & **DC**. However, only after the corresponding developments in mathematical logic and axiomatic set theory and, especially, after the creation (in 1963) of the forcing method by Cohen, did it become possible to establish the needed result. We shall return to this question in our further considerations and touch briefly upon some related problems that are also interesting from the logical point of view. But first, we discuss more thoroughly analytic aspects of the problem of the existence of Lebesgue nonmeasurable sets (respectively, of sets without the Baire property).

Our starting point is one important common feature of Lebesgue measurable sets and of sets with the Baire property. We shall demonstrate that not all subsets of the real line have this feature. Obviously, this will give us the existence of required bad subsets of $\mathbf{R}$. In order to carry out our program, we first introduce the following definition.

Let X be a subset of $\mathbf{R}$. We say that this set has the Steinhaus property if there exists a real $\varepsilon > 0$ such that

$$(\forall h \in \mathbf{R})(|h| < \varepsilon \Rightarrow (X + h) \cap X \neq \emptyset).$$

In other words, a set $X \subset \mathbf{R}$ has the Steinhaus property if the corresponding difference set

$$X - X = \{x' - x'' \ : \ x' \in X, \ x'' \in X\}$$

is a neighborhood of point 0.

It turns out that, as a rule, all good subsets of $\mathbf{R}$ are either of Lebesgue measure zero, or of first category, or have the Steinhaus property. In this connection, it is reasonable to mention here that Steinhaus himself observed that all Lebesgue measurable sets on $\mathbf{R}$ with strictly positive measure have this property (see [214]). Some years later, it was also established that an analogous result is true for second category subsets of $\mathbf{R}$ having the Baire property.

Let λ denote, as usual, the standard Lebesgue measure on $\mathbf{R}$. We now formulate and prove the following classical result.

Theorem 1. *Let X be a subset of $\mathbf{R}$ satisfying at least one of these two assumptions:*
(1) $X \in dom(\lambda)$ and $\lambda(X) > 0$;
(2) $X \in \mathcal{B}a(\mathbf{R}) \setminus \mathcal{K}(\mathbf{R})$.
Then X possesses the Steinhaus property.

Proof. Suppose first that assumption (1) holds. Let x be a density point of X (see Theorem 4 from the Introduction) and let $]a, b[$ be an open interval containing x for which we have

$$\lambda(X \cap \ ]a, b[) > 2(b - a)/3.$$

Obviously, there exists a real $\varepsilon > 0$ such that

$$(\forall h \in \mathbf{R})(|h| < \varepsilon \Rightarrow \lambda(]a + h, b + h[\ \cup \ ]a, b[) \leq 4(b - a)/3).$$

Take an arbitrary $h \in \mathbf{R}$ with $|h| < \varepsilon$. We assert that

$$(X + h) \cap X \neq \emptyset.$$

Indeed, assuming for a while that $(X + h) \cap X = \emptyset$, we must have

$$\lambda(]a + h, b + h[\ \cup \ ]a, b[) \geq \lambda(((X \cap \ ]a, b[) + h) \cup (X \cap \ ]a, b[)) =$$

$$2\lambda(X \cap \ ]a, b[) > 4(b - a)/3,$$

which is impossible. Thus X possesses the Steinhaus property.

Suppose now that assumption (2) holds. Then X can be represented in the form

$$X = U \triangle X_1,$$

where U is a nonempty open set in $\mathbf{R}$ and X_1 is a first category subset of $\mathbf{R}$. Evidently, there exists a real $\varepsilon > 0$ such that

$$(\forall h \in \mathbf{R})(|h| < \varepsilon \Rightarrow (U + h) \cap U \neq \emptyset).$$

Let us fix any $h \in \mathbf{R}$ with $|h| < \varepsilon$. It is easy to check the inclusion

$$((U + h) \cap U) \setminus ((X_1 + h) \cup X_1) \subset (X + h) \cap X.$$

Taking account of the fact that $(U + h) \cap U$ is a nonempty open subset of $\mathbf{R}$ and $(X_1 + h) \cup X_1$ is a first category subset of $\mathbf{R}$, we infer that

$$((U + h) \cap U) \setminus ((X_1 + h) \cup X_1) \neq \emptyset.$$

Consequently,

$$(X + h) \cap X \neq \emptyset,$$

and this finishes the proof of Theorem 1.

The following statement is an easy consequence of Theorem 1 but, sometimes, is much more useful in practice.

Theorem 2. *Let X and Y be subsets of $\mathbf{R}$ such that at least one of these two conditions holds:*
(1) $\{X, Y\} \subset dom(\lambda)$, $\lambda(X) > 0$, $\lambda(Y) > 0$;
(2) $\{X, Y\} \subset \mathcal{B}a(\mathbf{R}) \setminus \mathcal{K}(\mathbf{R})$.
Then the vector sum

$$X + Y = \{x + y \ : \ x \in X, \ y \in Y\}$$

has a nonempty interior.

Proof. Clearly, under assumption (1), there exists an element $t \in \mathbf{R}$ such that

$$\lambda((X + t) \cap Y) > 0.$$

Actually, this relation follows from the metrical transitivity of the measure λ (also, from the Lebesgue theorem on density points).

Similarly, under assumption (2), there exists an element $r \in \mathbf{R}$ such that

$$(X + r) \cap Y \in \mathcal{B}a(\mathbf{R}) \setminus \mathcal{K}(\mathbf{R}).$$

In fact, here we have an analogue of the metrical transitivity for the Baire property.

Let us put

$$Z = (X + t) \cap Y$$

in the first case, and

$$Z = (X + r) \cap Y$$

in the second one. It suffices to show that the set $Z + Z$ has a nonempty interior. If Z is symmetric with respect to zero, then we may directly apply Theorem 1. Otherwise, we can find an element $z \in \mathbf{R}$ such that

$$\lambda((Z + z) \cap (-Z)) > 0$$

in the first case, and

$$(Z + z) \cap (-Z) \in \mathcal{B}a(\mathbf{R}) \setminus \mathcal{K}(\mathbf{R})$$

in the second one. Finally, define

$$Z' = (Z + z/2) \cap (-Z - z/2).$$

The set Z' is symmetric with respect to zero and

$$Z' \subset Z + z/2.$$

Moreover, we see that

$$\lambda(Z') > 0$$

in the first case, and

$$Z' \in \mathcal{B}a(\mathbf{R}) \setminus \mathcal{K}(\mathbf{R})$$

in the second one. Applying Theorem 1 to Z' and taking account of the relation

$$Z' - Z' = Z' + Z' \subset Z + Z + z,$$

we come to the required result.

The following exercises show that Theorems 1 and 2 have analogues in much more general situations.

Exercise 1. Let $(G, \cdot)$ be an arbitrary topological group and let X be a subset of G such that

$$X \in \mathcal{B}a(G) \setminus \mathcal{K}(G).$$

Using the Banach result on first category open sets (see Exercise 30 from the Introduction), show that the set

$$X \cdot X^{-1} = \{x \cdot y^{-1} \ : \ x \in X, \ y \in X\}$$

is a neighborhood of the neutral element of G. Sometimes, this statement is called the Banach-Kuratowski-Pettis theorem (see, e.g., [85]).

Deduce from this result that if A and B are any two subsets of G satisfying the relation

$$\{A, B\} \subset \mathcal{B}a(G) \setminus \mathcal{K}(G),$$

then the set

$$A \cdot B = \{a \cdot b \ : \ a \in A, \ b \in B\}$$

has a nonempty interior.

Exercise 2. Let $(G, \cdot)$ be a σ-compact locally compact topological group with the neutral element e and let μ be the left invariant Haar measure on G. We denote by μ' the usual completion of μ. Let X be an arbitrary μ'-measurable subset of G. Starting with the fact that μ is a Radon measure, prove that

$$lim_{g \to e} \ \mu'((g \cdot X) \cap X) = \mu'(X).$$

In particular, if $\mu'(X) > 0$, then there exists a neighborhood $U(e)$ of e such that

$$(\forall g \in U(e))(\mu'((g \cdot X) \cap X) > 0)$$

and, consequently,

$$(\forall g \in U(e))((g \cdot X) \cap X \neq \emptyset).$$

Conclude from this fact that if A and B are any two μ'-measurable subsets of G with $\mu'(A) > 0$ and $\mu'(B) > 0$, then the set $A \cdot B$ has a nonempty interior.

Now, we are ready to present the first classical construction of a subset of the real line, nonmeasurable in the Lebesgue sense and without the Baire property. As mentioned earlier, this construction is due to Vitali (see [226]).

First, let us consider a binary relation $V(x, y)$ on $\mathbf{R}$ defined by the formula

$$x \in \mathbf{R} \ \& \ y \in \mathbf{R} \ \& \ x - y \in \mathbf{Q}$$

where $\mathbf{Q}$ denotes, as usual, the set of all rational numbers. Because $\mathbf{Q}$ is a subgroup of the additive group of $\mathbf{R}$, we infer that $V(x, y)$ is an equivalence relation on $\mathbf{R}$. Consequently, we obtain the partition of $\mathbf{R}$ canonically associated with $V(x, y)$. This partition will be called the Vitali partition of $\mathbf{R}$ and will be denoted by $\mathbf{R/Q}$. Any selector of the Vitali partition will be called a Vitali subset of $\mathbf{R}$.

Theorem 3. *There exist Vitali subsets of $\mathbf{R}$. If X is an arbitrary Vitali subset of $\mathbf{R}$, then X is Lebesgue nonmeasurable and does not possess the Baire property.*

Proof. The existence of Vitali sets follows directly from the Axiom of Choice applied to the Vitali partition. Now, let X be a Vitali set and suppose for a moment that X is either Lebesgue measurable or possesses the Baire property. Then, taking account of the relation

$$\mathbf{R} = \cup\{X + q \ : \ q \in \mathbf{Q}\},$$

we infer that X must be of strictly positive measure (respectively, of second category). But this immediately yields a contradiction. Indeed, for each rational number $q \neq 0$, we have

$$(X + q) \cap X = \emptyset,$$

because X is a selector of $\mathbf{R/Q}$. Here q may be arbitrarily small. In other words, we see that our X does not have the Steinhaus property. This contradicts Theorem 1 and finishes the proof of Theorem 3.

We thus obtained that Vitali sets are very bad from the points of view of the Lebesgue measure and the Baire property. However, these sets may be rather good for other nonzero σ-finite invariant measures given on $\mathbf{R}$.

Exercise 3. Prove that there exists a measure μ on the real line, satisfying the following conditions:

(a) μ is a nonzero nonatomic complete σ–finite measure invariant under the group of all isometric transformations of $\mathbf{R}$;

(b) $dom(\lambda) \subset dom(\mu)$ where λ denotes the standard Lebesgue measure on $\mathbf{R}$;

(c) $(\forall Y \in dom(\lambda))(\lambda(Y) = 0 \Rightarrow \mu(Y) = 0)$;

(d) $(\forall Y \in dom(\lambda))(\lambda(Y) > 0 \Rightarrow \mu(Y) = +\infty)$;

(e) there exists a Vitali set X such that $X \in dom(\mu)$.

Moreover, since μ is complete and σ-finite, we can consider a von Neumann topology $\mathcal{T}(\mu)$ associated with μ. Let $\mathbf{R}^*$ denote the set of all real numbers, equipped with $\mathcal{T}(\mu)$. Then the σ-ideal $\mathcal{K}(\mathbf{R}^*)$ and the σ-algebra $\mathcal{B}a(\mathbf{R}^*)$ are invariant under the group of all translations of $\mathbf{R}^*$ and the Vitali set X mentioned in (e) belongs to $\mathcal{B}a(\mathbf{R}^*)$, i.e., possesses the Baire property with respect to $\mathcal{T}(\mu)$.

The next exercise shows that any Vitali set remains nonmeasurable with respect to each invariant extension of the Lebesgue measure λ.

Exercise 4. Prove that, for any measure ν on $\mathbf{R}$ invariant under the group $\mathbf{Q}$ and extending the measure λ, no Vitali subset of $\mathbf{R}$ is ν–measurable.

It is not hard to see that the argument used in the Vitali construction heavily relies on the assumption of the invariance of the Lebesgue measure with respect to translations of $\mathbf{R}$. This argument does not work for those nonzero σ-finite complete measures μ on $\mathbf{R}$ which are only quasiinvariant (i.e., μ is defined on a σ-algebra of subsets of $\mathbf{R}$, invariant under translations, and the σ-ideal of all μ-measure zero sets is preserved by translations, too). So the following question arises: how to prove the existence of nonmeasurable sets with respect to such a measure μ. We shall consider this question in the next chapter of the book. Namely, we shall show there that a more general algebraic construction is possible yielding the existence of nonmeasurable sets with respect to μ. The main role in that construction will be played by so-called Hamel bases of $\mathbf{R}$.

Now, we want to turn our attention to another classical construction of a Lebesgue nonmeasurable set (of a set without the Baire property). As pointed out earlier, this construction is due to Bernstein (see [15]). First, let us introduce one useful notion closely related to the Bernstein construction.

Let E be a topological space and let X be a subset of E.

We say that X is totally imperfect in E if X contains no nonempty perfect subset of E.

We say that X is a Bernstein subset of E if X and $E \setminus X$ are totally imperfect in E. Equivalently, X is a Bernstein subset of E if, for each nonempty perfect set $P \subset E$, we have

$$P \cap X \neq \emptyset, \quad P \cap (E \setminus X) \neq \emptyset.$$

It immediately follows from this definition that $X \subset E$ is a Bernstein set if and only if $E \setminus X$ is a Bernstein set.

Clearly, each subset of the real line, having cardinality strictly less than the cardinality of the continuum, is totally imperfect. The question concerning the existence of totally imperfect subsets of the real line, having the cardinality of the continuum, turns out to be rather nontrivial. For its solution, we need uncountable forms of the Axiom of Choice (cf. the next exercise).

Exercise 5. Prove, in the theory **ZF** & **DC**, that if there exists a totally imperfect subset of **R** of cardinality **c**, then there exists a Lebesgue nonmeasurable subset of **R**.

Prove also an analogous fact for the Baire property.

Exercise 6. Let n be a natural number greater than or equal to 2, and let X be a totally imperfect subset of the n–dimensional Euclidean space $\mathbf{R}^n$. Show that the set $\mathbf{R}^n \setminus X$ is connected (in the usual topological sense). Infer from this fact that any Bernstein subset of $\mathbf{R}^n$ is connected.

There are many examples of totally imperfect subsets of the Euclidean space $\mathbf{R}^n$. A wide class of such sets was introduced and investigated by Marczewski (see [216]).

Let E be a Polish topological space and let X be a subset of E. We say that X is a Marczewski subset of E if, for each nonempty perfect set $P \subset E$, there exists a nonempty perfect set $P' \subset E$ such that

$$P' \subset P, \qquad P' \cap X = \emptyset.$$

It immediately follows from this definition that every Marczewski set is totally imperfect in E, and that any subset of a Marczewski set is a Marczewski set, too. Also, it can easily be observed that any set $Y \subset E$ with $card(Y) < \mathbf{c}$ is a Marczewski set. Indeed, let us take an arbitrary nonempty perfect set $P \subset E$. Then, as we know (cf. Exercise 12 from Chapter 1), there exists a disjoint family $\{P_i \ : \ i \in I\}$ consisting of nonempty perfect sets in E and satisfying the relations

$$card(I) = \mathbf{c}, \qquad (\forall i \in I)(P_i \subset P).$$

Now, because $card(Y) < card(I)$, it is clear that there exists at least one index $i_0 \in I$ such that $P_{i_0} \cap Y = \emptyset$, and thus Y is a Marczewski set.

Let us recall the classical result of Alexandrov and Hausdorff stating that every uncountable Borel set in a Polish topological space contains a subset homeomorphic to the Cantor discontinuum (hence contains a nonempty perfect subset). Taking this result into account, we can give another, equivalent, definition of Marczewski sets. Namely, we may say that a set X lying in a Polish space E is a Marczewski set if, for each uncountable Borel subset B of E, there exists an uncountable Borel set $B' \subset E$ satisfying the relations

$$B' \subset B, \qquad B' \cap X = \emptyset.$$

In some situations, the second definition is more convenient. For instance, let E_1 and E_2 be two Polish spaces and let

$$f : E_1 \to E_2$$

be a Borel isomorphism between them. Then, for a set $X \subset E_1$, the following two assertions are equivalent:
  (1) X is a Marczewski set in E_1;
  (2) $f(X)$ is a Marczewski set in E_2.
  In other words, the Borel isomorphism f yields a one-to-one correspondence between Marczewski sets in the spaces E_1 and E_2. This fact is rather useful. For instance, suppose that we need to construct a Marczewski subset of E_1 having some additional properties that are invariant under Borel isomorphisms. Sometimes, it turns out that such a set can much more easier be constructed in E_2. Let us denote it by X'. Then we apply the Borel isomorphism f^{-1} to X' and obtain the required Marczewski set $f^{-1}(X')$ in the space E_1.
  Later, we shall demonstrate the usefulness of this idea. Namely, we shall show that there exist Marczewski subsets of $\mathbf{R}$ nonmeasurable in the Lebesgue sense (respectively, without the Baire property).
  One simple (but important) fact concerning Marczewski sets is presented in the next statement.

  Lemma 1. *Let $\{X_k : k < \omega\}$ be a countable family of Marczewski subsets of a Polish space E. Then $\cup\{X_k : k < \omega\}$ is a Marczewski set, too. In particular, if the space E is uncountable, then the family of all Marczewski subsets of E forms a σ-ideal in the Boolean algebra of all subsets of E.*

  Proof. Fix a nonempty perfect set $P \subset E$. Because X_0 is a Marczewski set, there exists a nonempty perfect set $P_0 \subset E$ such that

$$P_0 \subset P, \qquad P_0 \cap X_0 = \emptyset, \qquad diam(P_0) < 1.$$

Further, because X_1 is also a Marczewski set, there exist nonempty perfect sets $P_{00} \subset E$ and $P_{01} \subset E$ such that

$$P_{00} \subset P_0, \qquad P_{01} \subset P_0,$$

$$diam(P_{00}) < 1/2, \qquad diam(P_{01}) < 1/2,$$

$$P_{00} \cap X_1 = \emptyset, \qquad P_{01} \cap X_1 = \emptyset, \qquad P_{00} \cap P_{01} = \emptyset.$$

Proceeding in this manner, we will be able to define a dyadic system

$$\{P_{j_1 \ldots j_k} \; : \; j_1 = 0, \; j_2 \in \{0,1\}, \ldots, j_k \in \{0,1\}, \; 1 \leq k < \omega\}$$

of nonempty perfect sets in E whose diameters converge to zero, and

$$P_{j_1 \ldots j_k} \cap X_{k-1} = \emptyset$$

for each natural number $k \geq 1$. Now, putting

$$D_k = \cup\{P_{j_1 \ldots j_k} \; : \; j_1 = 0, j_2 \in \{0,1\}, \ldots, j_k \in \{0,1\}\},$$

$$D = \cap\{D_k \; : \; 1 \leq k < \omega\},$$

we obtain a nonempty perfect set $D \subset P$ satisfying the relation

$$D \cap (\cup\{X_k \; : \; k < \omega\}) = \emptyset.$$

This shows that $\cup\{X_k \; : \; k < \omega\}$ is a Marczewski subset of E.

We thus see that, in an uncountable Polish topological space E, the family of all Marczewski subsets of E forms a σ-ideal. It is usually called the Marczewski σ-ideal in E and plays an essential role in classical point set theory (cf. [24]). As mentioned above, Marczewski subsets of E can be regarded as a certain type of small sets in E. In our further considerations, we shall also deal with some other types of small sets which generate proper σ-ideals in the basic set E. For instance, we shall deal with the σ-ideal generated by all Luzin subsets of $\mathbf{R}$ (respectively, by all Sierpiński subsets of $\mathbf{R}$). In addition, we shall consider the σ-ideal of so-called universal measure zero subsets of $\mathbf{R}$ and the σ-ideal of strong measure zero subsets of $\mathbf{R}$. Various properties of these subsets will be discussed in subsequent chapters of the book (note that valuable information about different kinds of small sets can be found in [24], [120], [149], [167], and [235]).

Let us return to Bernstein sets. We now formulate and prove the classical Bernstein result on the existence of such sets.

Theorem 4. *There exists a Bernstein subset of the real line. All such subsets are Lebesgue nonmeasurable and do not possess the Baire property.*

Proof. Let α denote the least ordinal number for which

$$card(\alpha) = \mathbf{c}.$$

We know that the family of all nonempty perfect subsets of $\mathbf{R}$ is of cardinality $\mathbf{c}$. So we may denote this family by $\{P_\xi \ : \ \xi < \alpha\}$. Moreover, we may assume without loss of generality that each of the partial families

$$\{P_\xi \ : \ \xi < \alpha, \ \xi \ is \ an \ even \ ordinal\},$$

$$\{P_\xi \ : \ \xi < \alpha, \ \xi \ is \ an \ odd \ ordinal\}$$

also consists of all nonempty perfect subsets of $\mathbf{R}$. Now, applying the method of transfinite recursion, we define an α-sequence of points

$$\{x_\xi : \xi < \alpha\} \subset \mathbf{R}$$

satisfying the following two conditions:
(1) if $\xi < \zeta < \alpha$, then $x_\xi \neq x_\zeta$;
(2) for each $\xi < \alpha$, we have $x_\xi \in P_\xi$.
Suppose that, for $\beta < \alpha$, the partial β-sequence $\{x_\xi \ : \ \xi < \beta\}$ has already been defined. Take the set P_β. Obviously,

$$card(P_\beta) = \mathbf{c},$$

$$card(\{x_\xi \ : \ \xi < \beta\}) \leq card(\beta) < \mathbf{c}.$$

Hence we can write

$$P_\beta \setminus \{x_\xi \ : \ \xi < \beta\} \neq \emptyset.$$

Choose an arbitrary element x from the above-mentioned nonempty set and put $x_\beta = x$. Continuing in this manner, we will be able to construct the family $\{x_\xi \ : \ \xi < \alpha\}$ of points of $\mathbf{R}$, satisfying conditions (1) and (2). Further, we put

$$X = \{x_\xi \ : \ \xi < \alpha, \ \xi \ is \ an \ even \ ordinal\}.$$

It immediately follows from our construction that X is a Bernstein subset of $\mathbf{R}$ (because both X and $\mathbf{R} \setminus X$ are totally imperfect in $\mathbf{R}$).

It remains to demonstrate that X is not Lebesgue measurable and does not possess the Baire property.

Suppose first that X is measurable in the Lebesgue sense. Then the set $\mathbf{R} \setminus X$ is Lebesgue measurable, too, and at least one of these two sets must be of strictly positive measure. We may assume without loss of generality that $\lambda(X) > 0$. Then a well-known property of λ implies that there exists a closed set $F \subset \mathbf{R}$ contained in X and having a strictly positive measure. Because λ is a diffused (continuous) measure, we must have

$$card(F) > \omega$$

and hence $card(F) = \mathbf{c}$. Denote by F_0 the set of all condensation points of F. Obviously, F_0 is a nonempty perfect subset of $\mathbf{R}$ contained in X. But this contradicts the fact that X is a Bernstein set in $\mathbf{R}$.

Suppose now that X possesses the Baire property. Then the set $\mathbf{R} \setminus X$ possesses the Baire property, too, and at least one of these two sets must be of second category. We may assume without loss of generality that X is of second category. Consequently, we have a representation of X in the form

$$X = V \triangle Y = (V \setminus Y) \cup (Y \setminus V)$$

where V is a nonempty open subset of $\mathbf{R}$ and Y is a first category subset of $\mathbf{R}$. Applying the classical Baire theorem, we see that the set $V \setminus Y$ contains an uncountable G_δ-subset of $\mathbf{R}$. This immediately implies that X contains also a nonempty perfect subset of $\mathbf{R}$ (homeomorphic to the Cantor space), which contradicts the fact that X is a Bernstein set in $\mathbf{R}$.

A result much more general than Theorem 4 is presented in the following two exercises.

Exercise 7. Let E be an infinite set and let $\{X_j : j \in J\}$ be a family of subsets of E, such that:

(a) $card(J) \leq card(E)$;

(b) $(\forall j \in J)(card(X_j) = card(E))$.

Prove, by applying the method of transfinite induction, that there exists a family $\{Y_j : j \in J\}$ of subsets of E, satisfying the relations:

(c) $(\forall j \in J)(\forall j' \in J)(j \neq j' \Rightarrow Y_j \cap Y_{j'} = \emptyset)$;

(d) $(\forall j \in J)(\forall j' \in J)(card(X_j \cap Y_{j'}) = card(E))$.

In particular, conclude that there exists a disjoint family $\{Z_j : j \in J\}$ of subsets of E such that

$$Z_j \subset X_j, \quad card(Z_j) = card(X_j) \quad (j \in J).$$

Exercise 8. By starting with the result of the previous exercise, show that in every complete metric space E of cardinality of the continuum (hence, in every uncountable Polish topological space) there exists a Bernstein set. Moreover, demonstrate that there exists a partition $\{Y_j : j \in J\}$ of E such that:

(a) $card(J) = \mathbf{c}$;

(b) for each $j \in J$, the set Y_j is a Bernstein subset of E.

Finally, show that if the space E has no isolated points, then all Bernstein subsets of E do not have the Baire property.

The next exercise yields a characterization of Bernstein subsets of Polish spaces in terms of topological measure theory.

Exercise 9. Let E be an uncountable Polish space and let X be a subset of E. Demonstrate that the following two assertions are equivalent:

(a) X is a Bernstein subset of E;

(b) for each nonzero σ–finite diffused Borel measure μ given on E, the set X is nonmeasurable with respect to the completion of μ.

Show also that these two assertions are not, in general, equivalent for a nonseparable complete metric space E.

Exercise 10. Let us consider the first uncountable ordinal ω_1 equipped with its order topology, and let

$$\mathcal{I} = \{X \subset \omega_1 \ : \ (\exists F \subset \omega_1)(F \text{ is closed, } card(F) = \omega_1, \ F \cap X = \emptyset)\}.$$

Show that $\mathcal{I}$ is a σ–ideal of subsets of ω_1. The elements of $\mathcal{I}$ are usually called nonstationary subsets of ω_1. A set $Z \subset \omega_1$ is called a stationary subset of ω_1 if Z is not nonstationary. Let us put

$$\mathcal{S} = \mathcal{I} \cup \mathcal{I}'$$

where $\mathcal{I}'$ is the δ-filter dual to $\mathcal{I}$. Observe that $\mathcal{S}$ is the σ–algebra generated by $\mathcal{I}$. Finally, demonstrate that, for any set $X \subset \omega_1$, the following two properties are equivalent:

(a) the sets X and $\omega_1 \setminus X$ are stationary in ω_1;

(b) for every nonzero σ–finite diffused measure μ defined on $\mathcal{S}$, the set X is not measurable with respect to the completion of μ.

Any set X with the above-mentioned properties can be considered as an analogue (for the topological space ω_1) of a Bernstein subset of $\mathbf{R}$.

For more information about the σ-ideal $\mathcal{I}$ and stationary subsets of ω_1, see [71], [72], [110], and [119].

The next exercise assumes that the reader is familiar with the notion of a complete Boolean algebra (for the definition, see, e.g., [42] or [124]).

Exercise 11. Let $\mathcal{P}(\mathbf{R})$ denote the complete Boolean algebra of all subsets of the real line $\mathbf{R}$. Let $\mathcal{I}_0$ be the σ-ideal of all Lebesgue measure zero subsets of $\mathbf{R}$ and let $\mathcal{I}_1$ be the σ-ideal of all first category subsets of **R**. Consider the corresponding factor algebras $\mathcal{P}(\mathbf{R})/\mathcal{I}_0$ and $\mathcal{P}(\mathbf{R})/\mathcal{I}_1$. Show that these Boolean algebras are not complete (apply Exercise 8 of this chapter).

Exercise 12. Let E be an uncountable Polish space and let $\mathcal{I}$ be some σ-ideal of subsets of E. We say that $\mathcal{I}$ has a Borel base if, for each set $X \in \mathcal{I}$, there exists a Borel subset Y of E such that $X \subset Y$ and $Y \in \mathcal{I}$.

Suppose that all one-element subsets of E belong to $\mathcal{I}$ and that $\mathcal{I}$ possesses a Borel base. Let us put:

$\mathcal{S}(\mathcal{I})$ = the σ-algebra of subsets of E, generated by $\mathcal{B}(E) \cup \mathcal{I}$.

Here $\mathcal{B}(E)$ denotes, as usual, the Borel σ-algebra of the space E.

Let X be a Bernstein subset of E (the existence of Bernstein sets in E follows, for instance, from Exercise 8). Show that $X \notin \mathcal{S}(\mathcal{I})$.

Conclude, in particular, that if μ is a nonzero σ-finite diffused Borel measure on E, then $X \notin dom(\mu')$ where μ' stands for the completion of μ.

Formulate the corresponding consequence for the Baire property.

Exercise 13. Prove that there exists a subset X of **R** which is simultaneously a Vitali set and a Bernstein set.

All the constructions presented above were concerned with certain sets either nonmeasurable in the Lebesgue sense or without the Baire property. The existence of such sets evidently implies the existence of functions either nonmeasurable in the Lebesgue sense or without the Baire property. We now wish to consider a direct construction of a Lebesgue nonmeasurable function acting from **R** into **R**. An analogous construction can be carried out for the Baire property.

In our further considerations, we denote by the symbol λ_2 the standard two-dimensional Lebesgue measure on the Euclidean plane $\mathbf{R}^2 = \mathbf{R} \times \mathbf{R}$. Clearly, λ_2 is the completion of the product measure $\lambda \times \lambda$ where λ denotes, as usual, the standard Lebesgue measure on the real line.

We recall that a subset X of $\mathbf{R}^2$ is λ_2-thick (or λ_2-massive) in $\mathbf{R}^2$ if, for each λ_2-measurable set $Z \subset \mathbf{R}^2$ with $\lambda_2(Z) > 0$, we have

$$X \cap Z \neq \emptyset.$$

In other words, X is λ_2-thick in $\mathbf{R}^2$ if and only if the equality

$$(\lambda_2)_*(\mathbf{R}^2 \setminus X) = 0$$

is satisfied, where the symbol $(\lambda_2)_*$ denotes the inner measure associated with λ_2.

Let us point out that if a subset X of $\mathbf{R}^2$ is λ_2-measurable and λ_2-massive simultaneously, then it is of full λ_2-measure, i.e.,

$$\lambda_2(\mathbf{R}^2 \setminus X) = 0.$$

Thus, if we *a priori* know that a set $X \subset \mathbf{R}^2$ is not of full λ_2-measure but is λ_2-thick, then we can immediately conclude that X is not λ_2-measurable.

The next statement (essentially due to Sierpiński) shows that there are functions acting from $\mathbf{R}$ into $\mathbf{R}$ whose graphs are λ_2-thick subsets of the plane $\mathbf{R}^2$.

Theorem 5. *There exists a function $f : \mathbf{R} \to \mathbf{R}$ whose graph is a λ_2-thick subset of $\mathbf{R}^2$. Consequently, the following two assertions are true:*
(1) the graph of f is not a λ_2-measurable subset of $\mathbf{R}^2$;
(2) f is not a λ-measurable function.

Proof. Let α be the least ordinal number of cardinality continuum. Consider the family $\{B_\xi \ : \ \xi < \alpha\}$ consisting of all Borel subsets of $\mathbf{R}^2$ having strictly positive λ_2-measure. We are going to construct, by transfinite recursion, a family of points $\{(x_\xi, y_\xi) \ : \ \xi < \alpha\} \subset \mathbf{R}^2$ satisfying these two conditions:
(a) if $\xi < \zeta < \alpha$, then $x_\xi \neq x_\zeta$;
(b) for each $\xi < \alpha$, the point (x_ξ, y_ξ) belongs to B_ξ.
Suppose that, for an ordinal $\beta < \alpha$, the partial family

$$\{(x_\xi, y_\xi) \ : \ \xi < \beta\} \subset \mathbf{R}^2$$

has already been defined. Let us take the set B_β. Applying the classical Fubini theorem to B_β, we see that the set

$$\{x \in \mathbf{R} \ : \ \lambda(B_\beta(x)) > 0\}$$

is λ-measurable and has a strictly positive measure. Consequently, this set is of cardinality of the continuum, and there exists a point x belonging to it and distinct from all the points x_ξ ($\xi < \beta$). We put $x_\beta = x$. Then we choose an arbitrary point y from the set $B_\beta(x_\beta)$ and put $y_\beta = y$. Proceeding in this manner, we will be able to construct the required family $\{(x_\xi, y_\xi) \ : \ \xi < \alpha\}$. Now, it easily follows from condition (a) that the set

$$F = \{(x_\xi, y_\xi) \ : \ \xi < \alpha\}$$

can be regarded as the graph of a partial function acting from $\mathbf{R}$ into $\mathbf{R}$. We extend arbitrarily this partial function to a function acting from $\mathbf{R}$ into

R and denote the latter function by f. Then condition (b) implies that the graph of f is λ_2-thick in $\mathbf{R}^2$. Because there are continuumly many pairwise disjoint translates of this graph, we conclude that the graph is not of full λ_2-measure and hence it is not a λ_2-measurable subset of $\mathbf{R}^2$.

Finally, the function f is not λ-measurable. Indeed, otherwise the graph of f would be a λ_2-measure zero subset of the plane, which is impossible.

This ends the proof of the theorem.

Some generalizations of this result are presented in Chapter 11.

Exercise 14. By applying the Kuratowski-Ulam theorem, which is a topological analogue of the classical Fubini theorem (see, e.g., [120] or [162]), prove a statement for the Baire property, analogous to Theorem 5. Namely, show that there exists a function $g : \mathbf{R} \to \mathbf{R}$ such that its graph is thick in the sense of the Baire property, i.e., the graph of g intersects each subset of $\mathbf{R}^2$ having the Baire property and not belonging to the σ-ideal of all first category subsets of $\mathbf{R}^2$.

Deduce from this fact that the graph of g does not have the Baire property in $\mathbf{R}^2$ and g does not have the Baire property as a function acting from **R** into **R**.

Exercise 15. Theorem 5 with the previous exercise show us that there exist functions from **R** into **R** whose graphs are thick subsets of the plane (in particular, those graphs are nonmeasurable in the Lebesgue sense or do not have the Baire property). On the other hand, prove that there exists a measure μ on $\mathbf{R}^2$ satisfying the following conditions:

(a) μ is an extension of the Lebesgue measure λ_2;

(b) μ is invariant under the group of all translations of $\mathbf{R}^2$ and under the central symmetry of $\mathbf{R}^2$ with respect to the origin $(0,0)$;

(c) the graph of any function acting from **R** into **R** belongs to $dom(\mu)$ and, for any such graph Γ, we have $\mu(\Gamma) = 0$.

We thus see that the graphs of all functions acting from **R** into **R** are small with respect to the above-mentioned measure μ. This is a common property of all functions, which act from **R** into **R**. Another interesting common feature of all functions from **R** into **R** was described by Blumberg's theorem (see Chapter 7 of this book).

If we deal with some class of subsets of **R** that are small in a certain sense, then, as a rule, it is not easy to establish the existence of a set belonging to this class and nonmeasurable in the Lebesgue sense (or lacking the Baire property). Theorem 5 yields us that there exist functions acting from **R** into **R**, whose graphs are nonmeasurable with respect to λ_2. At the same time, as mentioned above, the graphs of such functions may be regarded as small subsets of $\mathbf{R}^2$ with respect to the measure μ of Exercise 15.

More generally, suppose that a σ-ideal $\mathcal{I}$ of subsets of $\mathbf{R}$ is given. Then the following natural question can be posed: does there exist a set $X \in \mathcal{I}$ nonmeasurable in the Lebesgue sense or without the Baire property? Obviously, the answer to this question depends on the structure of $\mathcal{I}$ and simple examples show that the answer can be negative.

However, let us consider the two classical σ-ideals:

$\mathcal{I}(\lambda)$ = the σ-ideal of all λ-measure zero subsets of $\mathbf{R}$;

$\mathcal{K}(\mathbf{R})$ = the σ-ideal of all first category subsets of $\mathbf{R}$.

These two σ-ideals are orthogonal, i.e., there exists a partition $\{A, B\}$ of $\mathbf{R}$ such that

$$A \in \mathcal{I}(\lambda), \quad B \in \mathcal{K}(\mathbf{R}).$$

The reader can easily check this simple fact, which immediately implies the existence of a Lebesgue nonmeasurable set belonging to $\mathcal{K}(\mathbf{R})$ and the existence of a Lebesgue measure zero set without the Baire property. Indeed, let X be an arbitrary Bernstein subset of $\mathbf{R}$. We put

$$X_0 = A \cap X, \quad X_1 = B \cap X.$$

Then it is easy to check that:

(1) $X_0 \in \mathcal{I}(\lambda)$ and X_0 does not possess the Baire property;

(2) $X_1 \in \mathcal{K}(\mathbf{R})$ and X_1 is not measurable in the Lebesgue sense.

A more general result is presented in the next exercise.

Exercise 16. Let X be a Bernstein subset of $\mathbf{R}$. Let Y be a λ-measurable set with $\lambda(Y) > 0$ and let Z be a subset of $\mathbf{R}$ having the Baire property but not belonging to $\mathcal{K}(\mathbf{R})$. Show that the set $X \cap Y$ is not measurable in the Lebesgue sense and that the set $X \cap Z$ does not possess the Baire property.

Returning to the question formulated above, we wish to consider it more thoroughly for the Marczewski σ-ideal in $\mathbf{R}$. In other words, it is natural to ask whether there exist Marczewski subsets of $\mathbf{R}$ nonmeasurable in the Lebesgue sense (or without the Baire property). This problem was originally raised by Marczewski (see [216]). The solution to it was independently obtained by Corazza [43] and Walsh [228]. Here we want to present their result.

First of all, we need one auxiliary proposition useful in many situations. The proof of this proposition is very similar to the argument used in the proof of Theorem 5.

Lemma 2. *Let $\{Z_j \ : \ j \in J\}$ be a family of subsets of the plane $\mathbf{R}^2$, such that:*

(1) $card(J) \leq \mathbf{c}$;

(2) for each index $j \in J$, the set of all $x \in pr_1(Z_j)$ satisfying the relation

$$card(Z_j(x)) = \mathbf{c}$$

is of cardinality of the continuum.

Then there exist a set-valued mapping $F \ : \ J \to \mathcal{P}(\mathbf{R})$ and an injective family $\{x_j \ : \ j \in J\} \subset \mathbf{R}$, such that, for any index $j \in J$, we have the equalities

$$F(j) = Z_j(x_j), \quad card(F(j)) = \mathbf{c}.$$

Proof. Obviously, we may assume without loss of generality that $card(J)$ is equal to $\mathbf{c}$. Also, we can identify the set J with the least ordinal number α for which $card(\alpha) = \mathbf{c}$. Now, we are going to define a set-valued mapping F and a family $\{x_\xi \ : \ \xi < \alpha\}$ by the method of transfinite recursion. Suppose that, for an ordinal $\beta < \alpha$, the partial families

$$\{F(\xi) \ : \ \xi < \beta\}, \quad \{x_\xi \ : \ \xi < \beta\}$$

have already been constructed. Consider the set Z_β. According to our assumption, the set of all those $x \in pr_1(Z_\beta)$ for which

$$card(Z_\beta(x)) = \mathbf{c}$$

is of cardinality of the continuum. Because $card(\{x_\xi \ : \ \xi < \beta\}) < \mathbf{c}$, there exists a point $x \in \mathbf{R}$ such that

$$(\forall \xi < \beta)(x \neq x_\xi), \quad card(Z_\beta(x)) = \mathbf{c}.$$

Therefore, we can put

$$F(\beta) = Z_\beta(x), \quad x_\beta = x.$$

In this way, we will be able to define F and $\{x_\xi \ : \ \xi < \alpha\}$ with the required properties.

By utilizing the previous lemma, it is not difficult to establish the following statement.

Theorem 6. *There exists a Marczewski subset of $\mathbf{R}^2$ nonmeasurable in the Lebesgue sense and lacking the Baire property.*

Proof. Let α be again the least ordinal number with $card(\alpha) = \mathbf{c}$ and let $\{Z_\xi \ : \ \xi < \alpha\}$ denote the family of all Borel subsets of $\mathbf{R}^2$ having a strictly positive λ_2-measure. Applying Lemma 2 to this family, we can find a set-valued mapping F and an injective family $\{x_\xi \ : \ \xi < \alpha\}$ of points of $\mathbf{R}$ with the corresponding properties. Let now $\{P_\xi \ : \ \xi < \alpha\}$ be the family of all nonempty perfect subsets of $\mathbf{R}^2$. For each $\beta < \alpha$, we put

$$y_\beta \in F(\beta) \setminus \cup \{P_\xi(x_\beta) \ : \ \xi \le \beta, \ card(P_\xi(x_\beta)) \le \omega\}.$$

Note that y_β can be defined because of the equality $card(F(\beta)) = \mathbf{c}$.

Let us check that

$$D_0 = \{(x_\xi, y_\xi) \ : \ \xi < \alpha\}$$

is a Marczewski set nonmeasurable with respect to λ_2. Indeed, D_0 can be regarded as the graph of a partial function acting from $\mathbf{R}$ into $\mathbf{R}$. From the construction of D_0 we have that D_0 is also a λ_2-thick subset of $\mathbf{R}^2$. Consequently (cf. the proof of Theorem 5), we may assert that D_0 is nonmeasurable in the Lebesgue sense. It remains to show that D_0 is a Marczewski set. In order to do this, take an arbitrary nonempty perfect subset P of $\mathbf{R}^2$. We must verify that P contains a nonempty perfect set whose intersection with D_0 is empty. If there exists at least one point $x \in \mathbf{R}$ for which

$$card(P(x)) = \mathbf{c},$$

then there is nothing to prove. Suppose now that

$$(\forall x \in \mathbf{R})(card(P(x)) \le \omega).$$

For some $\beta < \alpha$, we have $P_\beta = P$. Then, taking account of the definition of y_ξ ($\xi < \alpha$), we get

$$card(D_0 \cap P_\beta) \le card(\beta) + \omega < \mathbf{c}.$$

The latter relation easily implies that there exists a nonempty perfect subset of P having no common points with D_0.

In a similar way, starting with the family of all those Borel subsets of the plane $\mathbf{R}^2$ that do not belong to the σ-ideal $\mathcal{K}(\mathbf{R}^2)$, we can construct a Marczewski set $D_1 \subset \mathbf{R}^2$ thick in the category sense and coinciding with the graph of some partial function from $\mathbf{R}$ into $\mathbf{R}$. Then it is not hard to see that

$$D = D_0 \cup D_1$$

is a Marczewski set in $\mathbf{R}^2$ nonmeasurable in the Lebesgue sense and without the Baire property.

Exercise 17. Show that there exists a Marczewski subset D of $\mathbf{R}^2$ such that:

(a) D does not possess the Baire property;

(b) D is not measurable with respect to the completion of the product measure $\mu \times \nu$ where μ and ν are any Borel nonzero σ-finite measures on $\mathbf{R}$ vanishing on all one-element subsets of $\mathbf{R}$.

We can easily infer from Theorem 6 the existence of Marczewski subsets of $\mathbf{R}$ nonmeasurable in the Lebesgue sense (respectively, without the Baire property). Indeed, consider a Borel isomorphism

$$\phi \, : \, \mathbf{R} \to \mathbf{R}^2.$$

It is a well-known fact that ϕ can be chosen in such a way that, simultaneously, ϕ will be an isomorphism between the measures λ and λ_2. Now, if X is a λ_2-nonmeasurable Marczewski subset of $\mathbf{R}^2$, then $\phi^{-1}(X)$ is a λ-nonmeasurable Marczewski subset of $\mathbf{R}$.

Analogously, a Borel isomorphism ϕ can be chosen in such a way that it will preserve the Baire category of sets (consequently, the Baire property of sets). Now, if X is a Marczewski subset of $\mathbf{R}^2$ without the Baire property, then $\phi^{-1}(X)$ is a Marczewski subset of $\mathbf{R}$ without the Baire property.

A stronger result is contained in the next exercise.

Exercise 18. Prove (within the theory **ZF** & **DC**) that there exists a Borel isomorphism

$$\psi \, : \, \mathbf{R} \to \mathbf{R}^2$$

such that:

(a) ψ preserves the category of sets (in particular, ψ preserves the Baire property);

(b) ψ is an isomorphism between the measures λ and λ_2.

By starting with this fact and applying Theorem 6, show that there exists a Marczewski subset of the real line, nonmeasurable in the Lebesgue sense and without the Baire property.

One more example of a Lebesgue nonmeasurable function acting from $\mathbf{R}$ into $\mathbf{R}$, which also does not have the Baire property, can be obtained by using the corresponding results of Chapter 7.

Let us recall the theorem of Sierpiński and Zygmund proved in Chapter 7 and stating that there exists a function

$$f \, : \, \mathbf{R} \to \mathbf{R}$$

satisfying the following condition: for each subset X of $\mathbf{R}$ with $card(X) = \mathbf{c}$, the restriction $f|X$ is not continuous on X.

By starting with this condition, it is not hard to show that f is not Lebesgue measurable and does not possess the Baire property.

Indeed, suppose first that f has the Baire property. Then, according to a well-known theorem (see, e.g., [120], [162] or Exercise 13 from the Introduction), we can find an everywhere dense G_δ-subset A of $\mathbf{R}$ such that $f|A$ is continuous. Obviously,

$$card(A) = \mathbf{c},$$

and we obtain a contradiction with the fact that f is a Sierpiński-Zygmund function.

Now, let us demonstrate that f is not measurable in the Lebesgue sense. We will prove a much more general result asserting that f is not measurable with respect to the completion of any nonzero σ-finite diffused Borel measure on $\mathbf{R}$. Let μ be such a measure and let μ' denote the completion of μ. Because an analogue of the classical Luzin theorem holds true for μ', we can find a closed subset B of $\mathbf{R}$ with $\mu'(B) > 0$ such that the restricted function $f|B$ is continuous. Taking account of the diffusedness of μ' and of the inequality $\mu'(B) > 0$, we infer that B is uncountable and hence $card(B) = \mathbf{c}$. This yields again a contradiction with the fact that f is a Sierpiński-Zygmund function.

Exercise 19. Let K be a compact subset of $\mathbf{R}^2$. Obviously, the set $pr_1(K)$ is compact in $\mathbf{R}$. Show that there exists a lower semicontinuous (hence, Borel) mapping

$$\phi \ : \ pr_1(K) \to \mathbf{R}$$

such that the graph of ϕ is contained in K.

In addition, give an example of a compact connected subset P of $\mathbf{R}^2$ for which there exists no continuous mapping $\psi \ : \ pr_1(P) \to \mathbf{R}$ such that the graph of ψ is contained in P.

This simple result is a very particular case of much more general statements about the existence of measurable selectors for set-valued mappings measurable in various senses. For example, suppose that a Borel subset B of the Euclidean plane $\mathbf{R}^2$ is given satisfying the relation $pr_1(B) = \mathbf{R}$. Can one assert that there exists a Borel function

$$h \ : \ \mathbf{R} \to \mathbf{R}$$

such that its graph is contained in B? Such a function h is usually called a Borel uniformization of B. Luzin and Novikov (see, e.g., [133]) showed that, in general, the answer to this question is negative, i.e., there exists a Borel subset B of $\mathbf{R}^2$ with $pr_1(B) = \mathbf{R}$ for which there is no Borel uniformization.

On the other hand, suppose that we have an analytic subset A of $\mathbf{R}^2$ and consider its first projection $pr_1(A)$ that is an analytic subset of $\mathbf{R}$. Then, according to the classical theorem of Luzin, Jankov and von Neumann (see, for instance, [84]), there exists a function

$$g \; : \; pr_1(A) \to \mathbf{R}$$

satisfying the following relations:

(1) the graph of g is contained in A;

(2) g is measurable with respect to the σ-algebra generated by the family of all analytic subsets of $\mathbf{R}$.

In particular, one may assert that the above-mentioned function g has the Baire property in the restricted sense and is measurable with respect to the completion of any σ-finite Borel measure given on $\mathbf{R}$.

This important theorem has numerous applications in modern analysis and probability theory (some interesting applications are presented in [84]; cf. also [47], [125]).

Exercise 20. Let K be a compact subset of $\mathbf{R}^2$ whose all vertical sections are at most countable. Prove that K can be covered by countably many graphs of partial Borel functions acting from $\mathbf{R}$ into $\mathbf{R}$ (for a more general statement, see [133] or [136]). Using this result, show that the graph of any function $f : \mathbf{R} \to \mathbf{R}$ of Sierpiński-Zygmund type is a Marczewski subset of the Euclidean plane.

Exercise 21. Construct, by using the method of transfinite recursion, a Sierpiński-Zygmund function whose graph is a λ_2-thick subset of the Euclidean plane. Applying a similar method, construct a Sierpiński-Zygmund function whose graph is a thick subset of the plane in the category sense.

Deduce from these results that there are Marczewski subsets of the plane, nonmeasurable in the Lebesgue sense (respectively, without the Baire property).

On the other hand, show that there exists a Sierpiński-Zygmund function whose graph is a λ_2-measure zero subset of the plane. Analogously, show that there exists a Sierpiński-Zygmund function whose graph is a first category subset of the plane.

Concluding this chapter, we wish to make some remarks about logical aspects of the question concerning the existence of a Lebesgue nonmeasurable subset of the real line (or of a subset of the same line without the Baire property). Namely, in 1970, Solovay published his famous article [211] where he pointed out a model of **ZF** & **DC** in which all subsets of the real line are Lebesgue measurable and possess the Baire property. However, the existence of such a model was based on the assumption of the existence of an uncountable strongly inaccessible cardinal number and this seemed to be a weak side of the above-mentioned result. But, later, Shelah showed in his remarkable work [182] that large cardinals appeared here not accidentally. More precisely, he established that:

(1) there are models of **ZF** & **DC** in which all subsets of **R** possess the Baire property;

(2) the existence of a model of **ZF** & **DC** in which all subsets of **R** are Lebesgue measurable implies the existence of some large cardinal.

Solovay constructed also another model of set theory in which all projective subsets of **R** are Lebesgue measurable and possess the Baire property (see [211]). In this connection, it is reasonable to recall that in the Constructible Universe of Gödel there are projective subsets of **R** (even belonging to the class $\mathcal{P}r_3(\mathbf{R})$) that are not Lebesgue measurable and do not have the Baire property (for more details, see, e.g., [71], [72], [158]).

From among many other results connected with the existence of sets nonmeasurable in the Lebesgue sense (respectively, of sets without the Baire property), we want to point out the following ones:

1. Kolmogorov showed in [112] that the existence in the theory **ZF** & **DC** of a universal operation of integration for all Lebesgue measurable functions acting from $[0, 1]$ into **R** implies the existence of a Lebesgue nonmeasurable function acting from **R** into **R**. A similar result is true (in the same theory) for a universal operation of differentiation.

We thus conclude that the two fundamental operations of mathematical analysis (i.e., integration and differentiation) lead directly to real-valued functions on **R** that are nonmeasurable in the Lebesgue sense. This result seems to be interesting and important from the point of view of foundations of real analysis. In Chapter 18, we shall present some statements concerning generalized derivatives, which are closely related to the above-mentioned Kolmogorov result.

2. Sierpiński proved that the existence of a nontrivial ultrafilter in the Boolean algebra of all subsets of ω implies (within the theory **ZF** & **DC**) the existence of a subset of **R** nonmeasurable in the Lebesgue sense and without the Baire property.

The proof of this result can be found, e.g., in [193] (see also [34]).

3. Shelah and Raisonnier (see [169], [182]) established that the implication

$$\omega_1 \leq c \ \Rightarrow \ \text{there exists a Lebesgue nonmeasurable set on } \mathbf{R}$$

holds in the theory **ZF** & **DC**.

The proof of this fact presented in [169] is essentially based on some deep combinatorial properties of so-called rapid filters in ω (the notion of a rapid filter was first introduced by Mokobodzki [150]).

4. Pawlikowski showed in [164] that the Hahn-Banach theorem on extensions of partial linear continuous functionals implies (in the theory **ZF** & **DC**) the famous Banach-Tarski paradox and, hence, implies the existence of a Lebesgue nonmeasurable subset of **R**.

Various topics related to the Banach-Tarski paradox are discussed in monograph [227] by Wagon. In particular, certain sections of this monograph are specially devoted to deep relationships between equidecomposability theory and the problem of the existence of Lebesgue nonmeasurable subsets of **R**.

5. Roslanowski and Shelah have proved in their recent work [176] that there is a model of **ZFC** in which, for every function $f : \mathbf{R} \to \mathbf{R}$, there exists a set $X \subset \mathbf{R}$ nonmeasurable in the Lebesgue sense and such that the restriction $f|X$ is continuous.

It is useful to compare this result with the existence of a Sierpiński-Zygmund function (see Chapter 7).

Some other results concerning the existence of nonmeasurable sets and of sets without the Baire property are presented in [7], [33], [38], [54], [68], [73], [87], [88], [93], [100], [105], [107], [123], [153], [179], [194], [199], [209], [210], [224].

9. Hamel basis and Cauchy functional equation

In this chapter we discuss some properties of Hamel bases of the real line $\mathbf{R}$ and show their remarkable role in various constructions of strange additive functions acting from $\mathbf{R}$ into $\mathbf{R}$.

The existence of such a basis was first established by Hamel in 1905 (see [70]). Later, it was also shown that the existence of a Hamel basis cannot be proved without the aid of uncountable forms of the Axiom of Choice (in particular, it is impossible to establish the existence of such a basis within the theory $\mathbf{ZF}$ & $\mathbf{DC}$).

The construction of a Hamel basis can be done by starting with one general theorem of the theory of vector spaces (over arbitrary fields). Let us recall that, according to this general theorem, for every vector space E, there exists a basis of E, i.e., a maximal (with respect to inclusion) linearly independent subset of E. This assertion follows almost immediately from the Zorn Lemma or, equivalently, from the Axiom of Choice. So, each element of E can be represented as a linear combination of some elements of the basis and such a representation is unique.

Let us now consider the real line as a vector space over the field $\mathbf{Q}$ of all rational numbers. Then, applying the above-mentioned general theorem, we get that there are bases of $\mathbf{R}$ over $\mathbf{Q}$. Any such basis is called a Hamel basis of $\mathbf{R}$.

Exercise 1. Let E be an arbitrary vector space. Show that any two bases of E have the same cardinality. Their common cardinality is called the algebraic (or linear) dimension of the space E.

Notice that the result presented in Exercise 1 is a very particular case of a general theorem about the cardinality of any system of free generators of a free universal algebra (in this connection, see, e.g., [42]).

Let $\{e_i \ : \ i \in I\}$ be an arbitrary Hamel basis of $\mathbf{R}$. It is clear that

$$card(I) = 2^\omega = \mathbf{c}.$$

As we know, every element x of $\mathbf{R}$ may be uniquely represented in the form

$$x = \sum_{i \in I} q_i(x) \cdot e_i$$

where $(q_i(x))_{i \in I}$ is some indexed family of rational numbers, such that

$$card(\{i \in I \ : \ q_i(x) \neq 0\}) < \omega.$$

For each index $i \in I$, the rational number $q_i(x)$ is called the coordinate of x corresponding to this index.

From the purely group-theoretical point of view, this representation simply means that the additive group of the real line can be regarded as the direct sum of a family of groups $\{Q_i \ : \ i \in I\}$, satisfying the following conditions:

(1) $card(I) = \mathbf{c}$;

(2) for each index $i \in I$, the group Q_i is isomorphic to the additive group $\mathbf{Q}$ of all rationals.

In other words, we may write

$$\mathbf{R} = \sum_{i \in I} Q_i.$$

Clearly, an analogous representation will be true for the n-dimensional Euclidean space $\mathbf{R}^n$, where $n \geq 1$, for an infinite-dimensional separable Hilbert space and, more generally, for an arbitrary vector space over $\mathbf{Q}$ having the cardinality of the continuum. Thus, from the group-theoretical point of view, all these spaces are isomorphic (i.e., all of them are isomorphic to the group $\sum_{i \in I} Q_i$).

The above observation seems to be very trivial. However, by adding some argument, it yields rather nontrivial consequences. We wish to consider here one of such consequences.

Recall that the existence of Lebesgue nonmeasurable subsets of the real line was thoroughly discussed in the previous chapter of this book. More precisely, we discussed there the two classical constructions of nonmeasurable sets, due to Vitali and Bernstein, respectively. The first of them was based on certain algebraic properties of the Lebesgue measure λ (namely, on the invariance of this measure under the group of all translations of the real line) and the second one was based on some topological properties of λ (actually, we essentially used the fact that this measure is Radon).

Also, it was mentioned in the previous chapter that the Vitali construction cannot be generalized to the class of all nonzero σ-finite measures given on $\mathbf{R}$ and quasiinvariant with respect to the group of all translations of $\mathbf{R}$ (we recall that a nonzero measure μ given on $\mathbf{R}$ is quasiinvariant under the group of all translations of $\mathbf{R}$ if $dom(\mu)$ and the σ-ideal generated by the family of all μ-measure zero sets are invariant under this group). In other words, the argument used in the Vitali construction does not work for nonzero σ-finite quasiinvariant measures. In this connection, the following two related problems arise:

Let μ be a nonzero σ-finite measure on $\mathbf{R}$ quasiinvariant with respect to the group of all translations of $\mathbf{R}$.

1. Does there exist a subset of $\mathbf{R}$ nonmeasurable with respect to μ?

2. Does there exist a subgroup of $\mathbf{R}$ nonmeasurable with respect to μ?

Our purpose is to establish that the answer to the second question is positive (consequently, the answer to the first question turns out to be positive, too). In the process of establishing this fact we shall essentially use the existence of a Hamel basis of the real line (hence an uncountable form of the Axiom of Choice will be assumed).

In our further consideration, we need several auxiliary statements. These statements are not difficult to prove. We shall see that they directly lead us to the required result.

Lemma 1. *Let G be a commutative group and let G_0 be a subgroup of G. Let μ be a σ-finite G-quasiinvariant measure on G. Suppose also that:*

(1) the factor group G/G_0 is uncountable;

(2) $\mu^*(G_0) > 0$ where μ^* denotes, as usual, the outer measure associated with the initial measure μ.

Then G_0 is nonmeasurable with respect to μ.

Proof. Suppose for a moment that $G_0 \in dom(\mu)$. Then, according to condition (2), we must have the inequality

$$\mu(G_0) > 0.$$

On the other hand, according to condition (1), there is an uncountable family of pairwise disjoint G-translates of G_0 in G. Taking account of the σ-finiteness and G-quasiinvariance of our measure μ, we infer that the equality

$$\mu(G_0) = 0$$

must be true, which contradicts the above-written inequality. The contradiction obtained ends the proof of Lemma 1.

The following exercise contains a slightly more general result than the preceding lemma.

Exercise 2. Let G be an arbitrary group and let X be a subset of G. Suppose that G is equipped with a σ-finite left (respectively, right) G-quasiinvariant measure μ and suppose also that these two relations are fulfilled:

(a) there exists an uncountable family of pairwise disjoint left (respectively, right) G-translates of X in G;

(b) $\mu^*(X) > 0$.

Show that the set X is nonmeasurable with respect to μ.

Lemma 2. Let G be again a commutative group and let $\{G_n \; : \; n \in \omega\}$ be a countable family of subgroups of G. Suppose that the following two conditions are satisfied:

(1) for each natural number n, the factor group G/G_n is uncountable;

(2) $\cup\{G_n \; : \; n \in \omega\} = G$.

Further, let μ be a nonzero σ-finite G-quasiinvariant measure on G. Then there exists at least one $n \in \omega$ such that the corresponding group G_n is nonmeasurable with respect to μ.

Proof. Because μ is not identically equal to zero and the given family of subgroups $\{G_n \; : \; n \in \omega\}$ is a countable covering of G, there exists a natural number n such that

$$\mu^*(G_n) > 0.$$

Now, applying Lemma 1 to the subgroup G_n, we see that this subgroup is necessarily nonmeasurable with respect to μ.

Lemma 3. Let G be a commutative group, let H be a proper subgroup of G and let I be an infinite set of indices. Consider two direct sums of groups $\sum_{i \in I} G_i$ and $\sum_{i \in I} H_i$ where, for each index $i \in I$, the group G_i coincides with G and the group H_i coincides with H. Then we have

$$card((\sum_{i \in I} G_i)/(\sum_{i \in I} H_i)) \geq card(I).$$

In particular, if the set I is uncountable, then the factor group

$$(\sum_{i \in I} G_i)/(\sum_{i \in I} H_i)$$

is uncountable, too.

Proof. Because H is a proper subgroup of G, there exists an element $g \in G \setminus H$. Denote by the symbol D the family of all those elements

$$\{g_i \; : \; i \in I\} \in \sum_{i \in I} G_i$$

that satisfy the relation

$$(\forall i \in I)(g_i \neq 0 \Rightarrow g_i = g).$$

Obviously, we have

$$card(D) = card(I).$$

It is not hard to check that if d' and d'' are any two distinct elements from D, then

$$d' - d'' \notin \sum_{i \in I} H_i.$$

Consequently, we obtain the required inequality

$$card((\sum_{i \in I} G_i)/(\sum_{i \in I} H_i)) \geq card(I),$$

and the proof of Lemma 3 is completed.

Lemma 4. *The additive group of* $\mathbf{R}$ *can be represented in the form*

$$\mathbf{R} = \cup\{G_n : n \in \omega\}$$

where $\{G_n : n \in \omega\}$ *is a countable family of subgroups of* $\mathbf{R}$ *and, for each* $n \in \omega$, *the factor group* $\mathbf{R}/G_n$ *is uncountable.*

Proof. First, let us consider the additive group $\mathbf{Q}$ and show that this group can be represented as the union of a countable increasing (with respect to inclusion) family of its proper subgroups. Indeed, for any natural number n, define

$$Q^{(n)} = \{\frac{k}{n!} : k \in \mathbf{Z}\}.$$

Evidently, $Q^{(n)}$ is a subgroup of $\mathbf{Q}$ and, because there exist prime natural numbers strictly greater than $n!$, this subgroup is proper. Moreover, we may write

$$(\forall n)(\forall m)(n \leq m < \omega \Rightarrow Q^{(n)} \subset Q^{(m)}),$$

i.e., the family of groups $\{Q^{(n)} : n < \omega\}$ is increasing with respect to inclusion. Also, it is clear that

$$\mathbf{Q} = \cup\{Q^{(n)} : n < \omega\}.$$

As demonstrated earlier (with the aid of a Hamel basis), we have a representation of $\mathbf{R}$ in the form of a direct sum

$$\mathbf{R} = \sum_{i \in I} Q_i$$

where I is a set of indices with

$$card(I) = \mathbf{c}$$

and, for each $i \in I$, the group Q_i coincides with $\mathbf{Q}$. Now, for any natural number n, we put

$$G_n = \sum_{i \in I} Q_i^{(n)}.$$

It can easily be shown (by applying Lemma 3) that the family $\{G_n \ : \ n \in \omega\}$ of subgroups of $\mathbf{R}$ is the required one.

The preceding lemmas immediately give us the following generalization of the classical Vitali theorem.

Theorem 1. *For any nonzero σ-finite measure μ on $\mathbf{R}$ quasiinvariant under the group of all translations of $\mathbf{R}$, there exists a subgroup of $\mathbf{R}$ nonmeasurable with respect to μ.*

The next exercise contains a more general result.

Exercise 3. Let E be an uncountable vector space over the field $\mathbf{Q}$. Prove that there exists a countable family $\{E_n \ : \ n \in \omega\}$ of vector subspaces of E, satisfying the following conditions:
(a) $\cup\{E_n : n \in \omega\} = E$;
(b) for each $n \in \omega$, we have $card(E/E_n) > \omega$.
Infer from (a) and (b) that, for any nonzero σ-finite measure μ on E quasiinvariant under the group of all translations of E, there exists at least one $n \in \omega$ for which E_n is nonmeasurable with respect to μ.

Remark 1. By applying some methods of infinitary combinatorics (namely, by using a famous transfinite matrix of Ulam [224]), Theorem 1 and Exercise 3 above can be significantly generalized. More precisely, it can be shown that, for every uncountable commutative group G and for any nonzero σ-finite G-quasiinvariant measure ν on G, there exists a subgroup of G nonmeasurable with respect to ν. Notice that, in order to obtain such a generalized result, we need not only the methods of infinitary combinatorics but also some deep theorems concerning the algebraic structure of infinite commutative groups (for more detalis, see [92] and [107]).

Notice also that similar statements hold true (for uncountable commutative groups again) in terms of the Baire property (see [92]).

On the other hand, there are examples of uncountable noncommutative groups for which analogous results fail to be true.

Let us return to Hamel bases of the real line $\mathbf{R}$.

These bases were found not accidentally but as a tool for solving a concrete question in mathematical analysis. Now, we are going to formulate this question and discuss its close relationships with Hamel bases of $\mathbf{R}$.

Let us consider the class of all functions

$$f \; : \; \mathbf{R} \to \mathbf{R}$$

satisfying the following functional equation:

$$f(x+y) = f(x) + f(y) \qquad (x \in \mathbf{R}, \; y \in \mathbf{R}),$$

which is usually called the Cauchy functional equation. Notice that this equation simply says that f is a homomorphism from the additive group of $\mathbf{R}$ into itself. Also, it is easy to see that any homomorphism f from the additive group of $\mathbf{R}$ into itself satisfies the relation

$$f(qx) = qf(x)$$

for all $x \in \mathbf{R}$ and for all $q \in \mathbf{Q}$. In other words, f can be regarded as a linear mapping when $\mathbf{R}$ is treated as a vector space over the field $\mathbf{Q}$.

The problem is to find all solutions of the Cauchy functional equation.

It is clear that there are very natural solutions of this equation. Namely, every function $f \; : \; \mathbf{R} \to \mathbf{R}$ of the form

$$f(x) = a \cdot x \qquad (x \in \mathbf{R}),$$

where a is an arbitrary real number, is a solution of this equation. Such solutions we shall call trivial ones.

Let us stress that any continuous solution of the Cauchy functional equation is trivial. Moreover, a much stronger result will be presented below stating that any Lebesgue measurable (or having the Baire property) solution of the Cauchy functional equation is necessarily trivial (see Theorem 4 of this chapter).

Hamel bases allow us to construct nontrivial solutions of this equation. Namely, we have the following statement due to Hamel (see [70]).

Theorem 2. *There are nontrivial solutions of the Cauchy functional equation.*

Proof. Let $\{e_i \; : \; i \in I\}$ be an arbitrary Hamel basis of $\mathbf{R}$. As was noticed earlier, each $x \in \mathbf{R}$ can be uniquely represented in the form

$$x = \sum_{i \in I} q_i(x) \cdot e_i.$$

Let us fix an index $i_0 \in I$ and define a function

$$\phi : \mathbf{R} \to \mathbf{R}$$

by the following formula:

$$\phi(x) = q_{i_0}(x) \qquad (x \in \mathbf{R}).$$

It is clear that ϕ satisfies the Cauchy functional equation. Moreover, the range of this function is contained in $\mathbf{Q}$. Also, this function is not constant because

$$\phi(0) = 0, \qquad \phi(e_{i_0}) = 1.$$

So ϕ is not a continuous function (we shall see below that ϕ is not even Lebesgue measurable and does not possess the Baire property). Theorem 2 has thus been proved.

Exercise 4. Show that there exists a nontrivial automorphism of the additive group of $\mathbf{R}$ onto itself. How many such automorphisms are there?

The next statement yields a characterization of nontrivial solutions of the Cauchy functional equation in terms of their graphs.

Theorem 3. *Let f be a solution of the Cauchy functional equation. Then the following two assertions are equivalent:*
(1) the graph of f is dense everywhere in the plane $\mathbf{R}^2$;
(2) f is a nontrivial solution of the Cauchy functional equation.

Proof. Obviously, if f is a trivial solution of the Cauchy functional equation, then the graph of f, being a straight line, is nowhere dense in $\mathbf{R}^2$. Consequently, if a solution of the Cauchy functional equation has the graph everywhere dense in $\mathbf{R}^2$, then this solution must be nontrivial. We thus see that the implication $(1) \Rightarrow (2)$ is valid.

Let us establish the converse implication $(2) \Rightarrow (1)$. In order to do it, suppose that (2) is satisfied but the graph of f is not dense everywhere in $\mathbf{R}^2$. Then there exists a nonempty open rectangle

$$]a, b[\times]c, d[\ \subset \mathbf{R} \times \mathbf{R} = \mathbf{R}^2$$

such that

$$G_f \cap (]a, b[\times]c, d[) = \emptyset,$$

where G_f denotes, as usual, the graph of f. Let us show that at least one of the following two relations holds:

(i) for all $x \in \ ]a, b[$, we have $f(x) \geq c$;
(ii) for all $x \in \ ]a, b[$, we have $f(x) \leq d$.

Indeed, assuming to the contrary that there are two points x_1 and x_2 from the interval $]a, b[$, satisfying the inequalities

$$f(x_1) < c, \qquad f(x_2) > d,$$

we can easily infer that, for some rational numbers $q_1 > 0$ and $q_2 > 0$ with $q_1 + q_2 = 1$, the relation

$$f(q_1 x_1 + q_2 x_2) = q_1 f(x_1) + q_2 f(x_2) \in \ ]c, d[$$

will be fulfilled, i.e.,

$$(q_1 x_1 + q_2 x_2, f(q_1 x_1 + q_2 x_2)) \in \ ]a, b[\times]c, d[,$$

which is impossible. The contradiction obtained yields that at least one of relations (i) and (ii) must be true. We may suppose, without loss of generality, that relation (ii) holds. Now, it is easy to see that there exist a strictly positive real number δ and a real number h, such that

$$]-\delta, \delta[\ + \ h \subset \ ]a, b[.$$

Taking relation (ii) into account and applying the additivity of our function f, we deduce that f is bounded from above on the open interval $]-\delta, \delta[$. Since

$$f(-x) = -f(x) \qquad (x \in \mathbf{R}),$$

we also infer that f is bounded on the same interval and, consequently, is continuous at point 0. Finally, using the additivity of f again, we conclude that f is continuous at all points of $\mathbf{R}$, which contradicts assumption (2). This contradiction establishes the converse implication (2) $\Rightarrow$ (1). Theorem 3 has thus been proved.

Exercise 5. Let E be an arbitrary normed vector space over the field of real numbers (or over the field of complex numbers) and let E have an infinite algebraic dimension. Show that there exists a linear functional defined on E and discontinuous at all points of E.

Compare this result with the fact that every linear functional defined on a finite–dimensional normed vector space is continuous.

Suppose now that a function

$$f : \mathbf{R} \to \mathbf{R}$$

is given satisfying the Cauchy functional equation. Then, as mentioned above, for every $x \in \mathbf{R}$ and for every $q \in \mathbf{Q}$, we have

$$f(q \cdot x) = q \cdot f(x).$$

It immediately follows from this fact that if f is a continuous function at least at one point of $\mathbf{R}$, then

$$f(x) = f(1) \cdot x$$

for each $x \in \mathbf{R}$. This simple result was first established by Cauchy. It is also easy to prove a slightly more general result stating that if f is a solution of the Cauchy functional equation and, in addition, has an upper (or lower) bound on some nonempty open interval, then f is a trivial solution of this equation (cf. the proof of Theorem 3).

Thus, we see that nontrivial solutions of the Cauchy functional equation are very bad from the topological point of view; namely, they are discontinuous at each element of their domain. The next result (due to Frechet) shows us that nontrivial solutions of the Cauchy functional equation are also bad from the points of view of the Lebesgue measure and the Baire property.

Theorem 4. *No nontrivial solution of the Cauchy functional equation is Lebesgue measurable (respectively, possesses the Baire property).*

Proof. There are many proofs of this remarkable theorem. We shall present here a very simple one based on the Steinhaus property for the Lebesgue measure (respectively, for the Baire category). Suppose that a function

$$f \; : \; \mathbf{R} \to \mathbf{R}$$

is given satisfying the Cauchy functional equation and suppose that f is Lebesgue measurable (respectively, has the Baire property). Let us consider the sets of the form

$$f^{-1}([-n, n]) \qquad (n \in \omega).$$

All these sets are Lebesgue measurable (respectively, have the Baire property) and

$$\cup_{n \in \omega} f^{-1}([-n, n]) = \mathbf{R}.$$

Hence there exists a natural number n such that

$$\lambda(f^{-1}([-n, n])) > 0$$

or, respectively,

$$f^{-1}([-n, n]) \in \mathcal{B}a(\mathbf{R}) \setminus \mathcal{K}(\mathbf{R}).$$

Then the Steinhaus property for the Lebesgue measure (respectively, its analogue for the Baire category) implies that the set

$$f^{-1}([-n, n]) - f^{-1}([-n, n])$$

is a neighborhood of the neutral element of $\mathbf{R}$. Let now V be any open interval in $\mathbf{R}$ such that

$$0 \in V \subset f^{-1}([-n, n]) - f^{-1}([-n, n]).$$

Then we get

$$f(V) \subset [-2n, 2n],$$

so the function f is bounded on V. From this fact it immediately follows that f is continuous at point 0. Hence f is a trivial solution of the Cauchy functional equation, and the proof of Theorem 4 is completed.

As mentioned earlier, there are also many other proofs of the preceding theorem which are based on essentially different ideas and methods. For example, one purely analytic proof of this theorem was suggested by Orlicz (see [159]). However, the argument presented above (which used the Steinhaus property for measure and category) seems to be more natural and can be applied in situations where a given Lebesgue measurable function (or function having the Baire property) is not necessarily additive (see, for instance, the proof of Theorem 5 below).

Considering Theorems 2 and 4, we can conclude that the existence of a Hamel basis in $\mathbf{R}$ implies the existence of a Lebesgue nonmeasurable subset of $\mathbf{R}$ and the existence of a subset of $\mathbf{R}$ without the Baire property. More exactly, we see that the proof of these two implications can be done in the theory $\mathbf{ZF}$ & $\mathbf{DC}$.

Exercise 6. Let $\{e_i : i \in I\}$ be a Hamel basis of $\mathbf{R}$. Fix an index $i_0 \in I$ and put

Γ = the vector space (over $\mathbf{Q}$) generated by $\{e_i : i \in I \setminus \{i_0\}\}$.

In other words, Γ is a hyperplane in $\mathbf{R}$ regarded as a vector space over $\mathbf{Q}$. Show in the theory $\mathbf{ZF}$ & $\mathbf{DC}$ that:

(a) the real line can be covered by a countable family of translates of Γ;

(b) Γ does not have the Steinhaus property, i.e., for each $\varepsilon > 0$, there exists a number $q \in \ ]-\varepsilon, \varepsilon[$ such that

$$(q + \Gamma) \cap \Gamma = \emptyset.$$

Deduce from relations (a) and (b) that Γ is not measurable in the Lebesgue sense and does not possess the Baire property. Conclude from this result that the implication

$$(there\ exists\ a\ Hamel\ basis\ of\ \mathbf{R})\ \Rightarrow$$

$$(there\ exists\ a\ subset\ of\ \mathbf{R}\ nonmeasurable\ in\ the\ Lebesgue\ sense$$

$$and\ without\ the\ Baire\ property)$$

is a theorem of the above-mentioned theory.

Exercise 7. By using the theorem on the existence of a Hamel basis of $\mathbf{R}$, describe all solutions of the Cauchy functional equation.

Exercise 8. Describe all continuous functions

$$f\ :\ \mathbf{R} \to \mathbf{R}$$

that satisfy the following functional equation:

$$f(x+y) = f(x) \cdot f(y) \qquad (x \in \mathbf{R},\ y \in \mathbf{R}).$$

Show that there are discontinuous (and Lebesgue nonmeasurable) solutions of this functional equation, too. Describe all such solutions.

One can prove that there are Lebesgue measurable and Lebesgue non-measurable Hamel bases (see exercises below). This fact shows us an essential difference between Hamel bases and Vitali sets from the point of view of Lebesgue measurability. The same is true for the Baire property.

Exercise 9. Prove that there are two sets $A \subset \mathbf{R}$ and $B \subset \mathbf{R}$ both of Lebesgue measure zero and of first category, such that

$$A + B = \mathbf{R}.$$

In particular, we have the equality

$$(A \cup B)\ +\ (A \cup B) = \mathbf{R}$$

where the set $A \cup B$ is also of Lebesgue measure zero and of first category (cf. Exercise 17 from Chapter 1).

Starting with this fact and applying the method of transfinite induction, demonstrate that there exists a Hamel basis in $\mathbf{R}$ contained in the set $A \cup B$. Consequently, that basis is of first category and of Lebesgue measure zero.

In addition, show that if μ is an arbitrary σ–finite measure on $\mathbf{R}$ invariant (or, more generally, quasiinvariant) under the group of all translations of $\mathbf{R}$ and $\{e_i \ : \ i \in I\}$ is an arbitrary Hamel basis in $\mathbf{R}$, then the implication

$$\{e_i \ : \ i \in I\} \in dom(\mu) \Rightarrow \mu(\{e_i \ : \ i \in I\}) = 0$$

holds true.

Exercise 10. By using the method of transfinite induction, prove that there exists a Hamel basis in $\mathbf{R}$, which simultaneously is a Bernstein subset of $\mathbf{R}$. Conclude that such a Hamel basis is λ-nonmeasurable (where λ denotes the standard Lebesgue measure on $\mathbf{R}$) and does not possess the Baire property.

Remark 2. In connection with the results presented in the two preceding exercises, the following question arises naturally: is any Hamel basis of $\mathbf{R}$ totally imperfect in $\mathbf{R}$? It turns out that the answer to this question is negative. Namely, it can be shown that there are nonempty perfect subsets of $\mathbf{R}$ linearly independent over the field $\mathbf{Q}$ (see, for instance, [153] or [227] where a much stronger statement is discussed with its numerous applications). Let P be such a subset. Then, by using the Zorn Lemma, one can easily prove that there exists a Hamel basis of $\mathbf{R}$ containing P. Obviously, this Hamel basis will not be totally imperfect.

The following question also seems to be natural: if μ is a nonzero σ–finite measure on the real line $\mathbf{R}$, invariant under the group of all translations of $\mathbf{R}$, does there exist a Hamel basis in $\mathbf{R}$ nonmeasurable with respect to μ? As pointed out above (see Exercise 10), the answer is positive for the standard Lebesgue measure λ on $\mathbf{R}$. However, in the general case, this question is undecidable in the theory $\mathbf{ZFC}$. The next exercise presents a more precise result concerning the question.

Exercise 11. Prove that the following two statements are equivalent:
(a) the Continuum Hypothesis ($\mathbf{c} = \omega_1$);
(b) for every nonzero σ–finite measure μ given on the real line $\mathbf{R}$ and invariant under the group of all translations of $\mathbf{R}$, there exists a Hamel basis in $\mathbf{R}$ nonmeasurable with respect to μ.

In connection with Exercise 11, see [52] and [190].

In our considerations above, we were primarily concerned with the classical Cauchy functional equation

$$f(x + y) = f(x) + f(y) \qquad (x \in \mathbf{R}, \ y \in \mathbf{R}).$$

Let us now consider the weakest form of the Jensen inequality, which plays a basic role in the theory of real-valued convex functions and is closely connected with the Cauchy functional equation.

We recall that a function

$$f \; : \; \mathbf{R} \to \mathbf{R}$$

satisfies the Jensen inequality (or is midpoint-convex) if

$$f(\frac{x+y}{2}) \leq \frac{f(x)+f(y)}{2}$$

for all $x \in \mathbf{R}$ and $y \in \mathbf{R}$. A more general form of the Jensen inequality which is often used in mathematical analysis is the following one:

$$f(q_1 x + q_2 y) \leq q_1 f(x) + q_2 f(y)$$

for all $x \in \mathbf{R}$ and $y \in \mathbf{R}$ and for all positive real numbers q_1, q_2 whose sum is equal to 1. It is widely known that this more general form of the Jensen inequality describes the class of all convex functions acting from $\mathbf{R}$ into $\mathbf{R}$. This class is very important from the point of view of applications. Also, functions belonging to this class have rather nice properties. For instance, they are continuous and possess a derivative almost everywhere. Moreover, a well-known theorem from analysis states that any convex function possesses a derivative at all points of $\mathbf{R}$ except countably many of them, and possesses also a second derivative at almost all points of $\mathbf{R}$. Indeed, the derivative of a convex function is monotone, so the desired result follows at once from the Lebesgue theorem on the differentiability almost everywhere of any monotone function (see Theorem 2 of Chapter 4).

At the same time, the weakest form of the Jensen inequality written above does not restrict essentially the class of admissible functions. Indeed, if f is a solution of the Cauchy functional equation, then f obviously satisfies the weakest form of the Jensen inequality. But we know that f may be a nontrivial solution of the Cauchy functional equation and, in this case, f is even Lebesgue nonmeasurable (and does not have the Baire property).

In this connection, the following result due to Sierpiński (and generalizing Theorem 4) is of some interest.

Theorem 5. *Suppose that*

$$f \; : \; \mathbf{R} \to \mathbf{R}$$

is a Lebesgue measurable function (respectively, a function with the Baire property) satisfying the inequality

$$f(\frac{x+y}{2}) \leq \frac{f(x)+f(y)}{2} \qquad (x \in \mathbf{R}, \; y \in \mathbf{R}).$$

Then f is continuous and, consequently, f is a convex function in the usual sense.

Proof. The argument is very similar to the one used in the proof of Theorem 4. First of all, we may assume without loss of generality that $f(0) = 0$. Otherwise, it is sufficient to introduce a new function f_1 by the formula

$$f_1(x) = f(x) - f(0) \qquad (x \in \mathbf{R}).$$

Evidently, this new function is also Lebesgue measurable and satisfies the weakest form of the Jensen inequality. So, in our further considerations, we suppose that $f(0) = 0$. In particular, for each $x \in \mathbf{R}$, we have

$$0 = f(0) = f(\frac{x + (-x)}{2}) \leq \frac{f(x) + f(-x)}{2}$$

and, consequently,

$$-f(-x) \leq f(x).$$

Now, for any natural number n, we consider the set

$$X_n = f^{-1}(] - \infty, n]).$$

According to our assumption, all these sets are measurable in the Lebesgue sense (respectively, possess the Baire property). Because

$$\mathbf{R} = \cup\{X_n \; : \; n \in \omega\},$$

at least one of these sets is not of Lebesgue measure zero (respectively, is not of first category). Let X_m be such a set. Then the Steinhaus property implies that the vector sum

$$Z_m = X_m + X_m$$

has a nonempty interior and hence contains some nondegenerate open interval $]a, b[$. Now, the inequality

$$f(\frac{x + y}{2}) \leq \frac{f(x) + f(y)}{2} \qquad (x \in \mathbf{R}, \; y \in \mathbf{R})$$

applied to Z_m immediately yields that

$$f(t) \leq m \qquad (t \in \;]a/2, b/2[),$$

i.e., our function f is bounded from above on the interval $]a/2, b/2[$. In addition, the inequality

$$f(t + r) \leq \frac{f(2t) + f(2r)}{2} \qquad (t \in \ ]a/4, b/4[, \ r \in \mathbf{R})$$

shows us that f is bounded from above on any interval of the form

$$]a/4, b/4[\ + \ r \qquad (r \in \mathbf{R}).$$

Consequently, f is bounded from above on each subinterval of $\mathbf{R}$ with the compact closure and, actually, f turns out to be locally bounded on $\mathbf{R}$.

Let now t be an arbitrary point of $\mathbf{R}$ and let h be a nonzero real number such that $|h| < 1$. Obviously, we may write

$$f(t + h) - f(t) \leq \frac{f(t + 2h) - f(t)}{2}.$$

From this inequality, applying an easy induction on k, we get

$$f(t + h) - f(t) \leq \frac{f(t + 2^k h) - f(t)}{2^k}.$$

But it is clear that, for some natural number k, we have

$$\frac{1}{2^{k+1}} \leq |h| < \frac{1}{2^k}.$$

Let us denote

$$L_0 = L_0(t, f) = sup_{x \in [t-1, t+1]} \ f(x) < +\infty.$$

Then the preceding inequalities imply

$$f(t + h) - f(t) \leq (L_0 - f(t)) \cdot 2|h| = L_1 \cdot |h|,$$

where

$$L_1 = L_1(t, f) \geq 0$$

is some constant depending only on f and t. Using an analogous argument, we easily come to the relation

$$f(t) - f(t + h) \leq L_1 \cdot |h|.$$

Actually, it suffices to apply the inequality

$$f(t) - f(t + h) \leq f(t - h) - f(t)$$

and take into account that

$$f(t - h) - f(t) \leq L_1 \cdot |-h| = L_1 \cdot |h|.$$

Finally, we obtain the relation

$$|f(t + h) - f(t)| \leq L_1 \cdot |h|,$$

which establishes that our function f is continuous at the point t. Hence f is continuous at all points of the real line (because t was taken arbitrarily on this line). In other words, f is a convex function in the usual sense.

Theorem 5 has thus been proved.

An extensive account of the Cauchy functional equation and the weakest form of the Jensen inequality is presented in the monograph by Kuczma [117].

There are also many textbooks and special monographs devoted to the theory of convex sets and convex functions (this theory is usually called convex analysis). Here we wish to present a few additional facts about convex functions interesting from the purely analytical point of view. We shall do it in the next three exercises.

Exercise 12. Let f and g be two functions acting from $\mathbf{R}$ into $\mathbf{R}$. Suppose that, for all points $x \in \mathbf{R}$ and $y \in \mathbf{R}$ and for each number $q \in [0, 1]$, the inequality

$$f(qx + (1 - q)y) \leq qg(x) + (1 - q)g(y)$$

is satisfied. Consider the set

$$\Gamma^*(g) = \{(x, t) \in \mathbf{R}^2 \ : \ g(x) \leq t\}.$$

Let $conv(\Gamma^*(g))$ denote the convex hull of $\Gamma^*(g)$ (recall that, according to the standard definition, the convex hull $conv(A)$ of a set A lying in a vector space E over $\mathbf{R}$ is the smallest convex set in E containing A; actually, $conv(A)$ coincides with the intersection of all convex subsets of E which contain A).

Show that $conv(\Gamma^*(g))$ is identical with the union of the family of all those triangles whose vertices belong to the set $\Gamma^*(g)$. Show also that the boundary of $conv(\Gamma^*(g))$ can be regarded as the graph of some convex function

$$\phi \ : \ \mathbf{R} \to \mathbf{R}$$

for which the relation

$$f(x) \leq \phi(x) \leq g(x) \qquad (x \in \mathbf{R})$$

is fulfilled. In other words, the convex function ϕ separates the given functions f and g.

Exercise 13. Let f be a function acting from $\mathbf{R}$ into $\mathbf{R}$ and let ε be a strictly positive real number. We say that f is an ε-convex function if, for any points $x \in \mathbf{R}$ and $y \in \mathbf{R}$ and for each number $q \in [0, 1]$, we have

$$f(qx + (1 - q)y) \leq qf(x) + (1 - q)f(y) + \varepsilon.$$

By applying the result of the previous exercise, demonstrate that if f is an ε-convex function, then there exists a convex function

$$\phi : \mathbf{R} \to \mathbf{R}$$

such that

$$f(x) \leq \phi(x) \leq f(x) + \varepsilon \qquad (x \in \mathbf{R}).$$

Furthermore, define a function

$$\psi : \mathbf{R} \to \mathbf{R}$$

by the formula

$$\psi(x) = \phi(x) - \varepsilon/2 \qquad (x \in \mathbf{R})$$

and show that

$$|\psi(x) - f(x)| \leq \varepsilon/2 \qquad (x \in \mathbf{R}).$$

In other words, for every ε-convex function f, there exists a convex function ψ such that the norm $||\psi - f||$ is less than or equal to $\varepsilon/2$.

This result is due to Hyers and Ulam (see [75]; some related results and problems are also discussed in [223]).

Exercise 14. Let $\{f_n : n < \omega\}$ be a sequence of real-valued convex functions on $\mathbf{R}$ and suppose that

$$f(x) = lim_{n \to +\infty} f_n(x) \qquad (x \in \mathbf{R}).$$

Show that $f : \mathbf{R} \to \mathbf{R}$ is a convex function, too.

Assuming, in addition, that all functions f_n $(n < \omega)$ and the function f are differentiable, show that

$$f'(x) = lim_{n \to +\infty} f'_n(x) \qquad (x \in \mathbf{R}).$$

Compare this result with Exercise 20 from Chapter 2.

Let now f be a function acting from $\mathbf{R}$ into $\mathbf{R}$ and suppose that, for some $\varepsilon > 0$, we have the relation

$$f(\frac{x+y}{2}) \leq \frac{f(x)+f(y)}{2} + \varepsilon \qquad (x \in \mathbf{R},\ y \in \mathbf{R}).$$

In this case, we cannot assert, in general, that there exists a convex function

$$\psi \ : \ \mathbf{R} \to \mathbf{R}$$

satisfying the inequalities

$$\psi(x) - \varepsilon \leq f(x) \leq \psi(x) + \varepsilon$$

for each $x \in \mathbf{R}$. Indeed, if f is a nontrivial solution of the Cauchy functional equation, then f is a Lebesgue nonmeasurable function and, obviously, it cannot be uniformly approximated by convex functions (which are measurable in the Lebesgue sense).

The next exercise contains some information about the descriptive structure of a Hamel basis of the real line.

Exercise 15. Let H be an arbitrary Hamel basis in $\mathbf{R}$. By starting with the fact that any analytic subset of $\mathbf{R}$ is Lebesgue measurable and applying the Steinhaus property for Lebesgue measurable sets, prove that H is not an analytic subset of $\mathbf{R}$.

This old classical result is due to Sierpiński. In connection with it, let us remark that there are uncountable linearly independent (over $\mathbf{Q}$) subsets of $\mathbf{R}$ having a good descriptive structure. For instance, as mentioned earlier, there are nonempty perfect subsets of $\mathbf{R}$ linearly independent over $\mathbf{Q}$.

Let f be a function acting from $\mathbf{R}$ into $\mathbf{R}$ and let x be a point of $\mathbf{R}$. We recall that f is symmetric with respect to x if

$$f(x - t) = f(x + t)$$

for all $t \in \mathbf{R}$. In other words, f is symmetric with respect to x if its graph is a symmetric subset of $\mathbf{R}^2$ with respect to the straight line $\{x\} \times \mathbf{R}$.

We shall say that a function

$$f \ : \ \mathbf{R} \to \mathbf{R}$$

is almost symmetric with respect to $x \in \mathbf{R}$ if

$$card(\{t \in \mathbf{R} \ : \ f(x - t) \neq f(x + t)\}) < \mathbf{c}.$$

It is easy to see that if f is symmetric with respect to each $x \in \mathbf{R}$, then f is constant. At the same time, by using a Hamel basis of the real line, it can be proved that there exist almost symmetric functions with respect to all $x \in \mathbf{R}$, which are not constant. This result is due to Sierpiński (see [199]). More precisely, Sierpiński established that the following statement is true.

Theorem 6. *There exists a Lebesgue nonmeasurable (and lacking the Baire property) function*

$$f \ : \ \mathbf{R} \to \mathbf{R},$$

which is almost symmetric with respect to all points of $\mathbf{R}$.

In particular, such a function is not equivalent to a constant function, i.e., there exists no set $Z \subset \mathbf{R}$ of Lebesgue measure zero (of first category) such that the restriction of f to $\mathbf{R} \setminus Z$ is constant.

Proof. Let α denote the least ordinal number for which $card(\alpha) = \mathbf{c}$. Let $\{P_\xi \ : \ \xi < \alpha\}$ be the family of all nonempty perfect subsets of $\mathbf{R}$. We may assume without loss of generality that each of the two partial families

$$\{P_\xi \ : \ \xi < \alpha, \ \xi \text{ is an even ordinal}\},$$

$$\{P_\xi \ : \ \xi < \alpha, \ \xi \text{ is an odd ordinal}\}$$

consists of all nonempty perfect subsets of $\mathbf{R}$, too. Now, by applying the method of transfinite recursion, it is not hard to define a family

$$\{e'_\xi \ : \ \xi < \alpha\}$$

of elements of $\mathbf{R}$, such that:

(a) $e'_\xi \in P_\xi$ for each $\xi < \alpha$;

(b) $\{e'_\xi \ : \ \xi < \alpha\}$ is linearly independent over $\mathbf{Q}$.

This construction is very similar to the classical Bernstein construction, so we leave the corresponding details to the reader. Further, we can extend the family $\{e'_\xi \ : \ \xi < \alpha\}$ to some Hamel basis of $\mathbf{R}$. This Hamel basis will be denoted by the symbol $\{e_\xi \ : \ \xi < \alpha\}$. Obviously, we may suppose that each of the sets

$$E_0 = \{e_\xi \ : \ \xi < \alpha, \ \xi \text{ is an even ordinal}\},$$

$$E_1 = \{e_\xi \ : \ \xi < \alpha, \ \xi \text{ is an odd ordinal}\}$$

is a Bernstein subset of $\mathbf{R}$ (hence, these two sets are nonmeasurable in the Lebesgue sense and do not possess the Baire property). Now, for any real number r, we have a unique representation of r in the form

$$r = q_1 e_{\xi_1} + q_2 e_{\xi_2} + \ldots + q_n e_{\xi_n}$$

where $n = n(r)$ is a natural number, all q_1, q_2, ..., q_n are nonzero rational numbers and

$$\xi_1 < \xi_2 < \ldots < \xi_n.$$

We define a mapping

$$\phi : \mathbf{R} \setminus \{0\} \to E_0 \cup E_1$$

by the formula

$$\phi(r) = e_{\xi_n} \qquad (r \in \mathbf{R} \setminus \{0\}).$$

Further, we put

$$X = \{r \in \mathbf{R} : \phi(r) \in E_0\}.$$

Finally, let us denote by f the characteristic function of X. Clearly, f is Lebesgue nonmeasurable and does not possess the Baire property. Fix a point $x \in \mathbf{R}$ and let t be an arbitrary point of $\mathbf{R} \setminus \{0\}$. If $x \neq 0$, then we may write

$$\phi(x) = e_\xi, \qquad \phi(t) = e_\zeta$$

for some ordinal numbers $\xi < \alpha$ and $\zeta < \alpha$. If $\xi < \zeta$, then we have the relation

$$x - t \in X \Leftrightarrow x + t \in X.$$

This relation is also true for $x = 0$. Now, taking account of the equality

$$card(lin(\{e_\zeta : \zeta \leq \xi\})) = card(\xi) + \omega,$$

where the symbol $lin(\{e_\zeta : \zeta \leq \xi\})$ denotes the vector space (over $\mathbf{Q}$) generated by $\{e_\zeta : \zeta \leq \xi\}$, we immediately obtain that

$$card(\{t \in \mathbf{R} : f(x - t) \neq f(x + t)\}) < \mathbf{c},$$

i.e., our function f turns out to be almost symmetric with respect to x. Because x was taken arbitrarily, we conclude that f is almost symmetric with respect to all points of $\mathbf{R}$.

Theorem 6 has thus been proved.

Some additional information around Theorem 6 can be found in [59].

Exercise 16. Let f be a function acting from $\mathbf{R}$ into $\mathbf{R}$ and let $x \in \mathbf{R}$. Let λ denote the standard Lebesgue measure on $\mathbf{R}$. We say that f is λ-almost symmetric with respect to x if f is λ-measurable and

$$\lambda(\{t \in \mathbf{R} \ : \ f(x - t) \neq f(x + t)\}) = 0.$$

Show that if f is λ-almost symmetric with respect to all points belonging to some everywhere dense subset of $\mathbf{R}$, then f is λ-equivalent to a constant function.

Formulate and prove an analogous result in terms of the Baire property.

Hamel bases have many other interesting and important applications and not only in analysis. One of the most beautiful applications may be found in the geometry of Euclidean space, more precisely, in the theory of polyhedra lying in this space. We mean here that part of this theory which is connected with the Hilbert third problem about the non-equivalence (by a finite decomposition) of a three–dimensional unit cube and a regular three–dimensional simplex with the same volume. The Hilbert third problem will be briefly discussed in the next exercise of this chapter. It is reasonable to mention, in this connection, that Hamel was one of Hilbert's many students and worked in the geometry of Euclidean space, as well.

Exercise 17. Let us recall that a (convex) polyhedron in the space $\mathbf{R}^3$ is an arbitrary convex subset of this space which can be represented as the union of a nonempty finite family of closed three–dimensional simplices. For any two polyhedra $X \subset \mathbf{R}^3$ and $Y \subset \mathbf{R}^3$, we say that they are equivalent by a finite decomposition if there exist two finite families

$$\{X_i \ : \ i \in I\}, \qquad \{Y_i \ : \ i \in I\}$$

of polyhedra, such that:

(a) $X = \cup_{i \in I} X_i, \quad Y = \cup_{i \in I} Y_i$;

(b) for all distinct indices $i \in I$ and $i' \in I$, we have

$$int(X_i) \cap int(X_{i'}) = int(Y_i) \cap int(Y_{i'}) = \emptyset$$

where the symbol int denotes, as usual, the operation of taking the interior of a set lying in $\mathbf{R}^3$;

(c) for each index $i \in I$, the polyhedra X_i and Y_i are congruent with respect to the group of all motions (i.e., isometric transformations) of $\mathbf{R}^3$.

Denote by the symbol $\mathcal{P}_3$ the class of all polyhedra in $\mathbf{R}^3$.

Let us consider any solution

$$f \; : \; \mathbf{R} \to \mathbf{R}$$

of the Cauchy functional equation, such that $f(\pi) = 0$.

Starting with this f, we define a functional

$$\Phi_f \; : \; \mathcal{P}_3 \to \mathbf{R}$$

by the formula

$$\Phi_f(X) = \sum_{j \in J} f(\alpha_j) \cdot |b_j| \quad (X \in \mathcal{P}_3),$$

where $(b_j)_{j \in J}$ is the injective family of all edges of the polyhedron X and α_j is the value of the dihedral angle of X corresponding to the edge b_j (i.e., α_j is the value of the dihedral angle which is formed by two faces of X meeting at the edge b_j) and, finally, $|b_j|$ denotes the length of the edge b_j.

Show that the functional Φ_f is invariant under the group of all motions of the space $\mathbf{R}^3$ and has equal values for any two polyhedra equivalent by a finite decomposition.

We say that Φ_f is the Dehn functional on $\mathcal{P}_3$ associated with a solution f of the Cauchy functional equation.

Let α be the value of the dihedral angle corresponding to an edge of a regular three–dimensional simplex.

Show that the two-element set $\{\alpha, \pi\} \subset \mathbf{R}$ is linearly independent over the field $\mathbf{Q}$. Therefore, by applying the Zorn Lemma, this set can be extended to a Hamel basis of $\mathbf{R}$.

Conclude from this fact that there exists a solution $g \; : \; \mathbf{R} \to \mathbf{R}$ of the Cauchy functional equation such that

$$g(\alpha) = 1, \quad g(\pi) = 0.$$

Show that the functional Φ_g assigns a strictly positive value on a regular three–dimensional simplex of volume 1 and, simultaneously, Φ_g assigns the value zero on the closed unit cube in $\mathbf{R}^3$.

Hence these two polyhedra, being of the same volume, are not equivalent by a finite decomposition.

This gives us the solution of the Hilbert third problem, first obtained by his disciple Dehn (see [46]).

Remark 3. The result presented in the above exercise explains why, in the standard course of elementary geometry, we need to use some infinite procedures or limit processes for calculation of the volume of a three–dimensional simplex in the Euclidean space $\mathbf{R}^3$.

Note that for the Euclidean plane $\mathbf{R}^2$ the situation is radically different. Namely, it is well known that any two polygons in $\mathbf{R}^2$ with equal areas are equivalent by a finite decomposition (see, e.g., [19] where a much stronger statement due to Hadwiger and Glur is formulated and proved).

After the Hilbert third problem was solved, another problem arose naturally concerning necessary and sufficient conditions for equivalence by a finite decomposition of two given polyhedra. This essentially more difficult problem was finally solved by Sydler who proved that two given polyhedra X and Y in the space $\mathbf{R}^3$ are equivalent by a finite decomposition if and only if they have the same volume and, for every Dehn functional Φ_f, the equality

$$\Phi_f(X) = \Phi_f(Y)$$

is fulfilled (see the original paper by Sydler [215]; cf. also [80] where a much simpler argument is presented).

This topic and a number of related questions are thoroughly considered in the excellent textbook by Boltjanskii [19].

Exercise 18. Let G be an everywhere dense subgroup of the space $\mathbf{R}^n$ ($n \geq 1$). Starting with the Vitali construction of a nonmeasurable set, show that any two sets $X \subset \mathbf{R}^n$ and $Y \subset \mathbf{R}^n$ with nonempty interiors are countably G-equidecomposable, i.e., there exist countable partitions $\{X_i : i \in I\}$ and $\{Y_i : i \in I\}$ of X and Y, respectively, and a family $\{g_i : i \in I\}$ of elements from G such that

$$(\forall i \in I)(Y_i = g_i + X_i).$$

Infer from this fact that there is no G-invariant measure extending the Lebesgue measure λ_n and defined on the family of all subsets of $\mathbf{R}^n$.

10. Luzin sets, Sierpiński sets, and their applications

In this chapter, our discussion is devoted to the so-called Luzin subsets of the real line $\mathbf{R}$ and the Sierpiński subsets of the same line. These sets are useful in various questions of real analysis and measure theory. Also, they have a number of applications in modern set theory (in particular, in constructing some special models of $\mathbf{ZFC}$).

First of all, we wish to emphasize the fact that the existence of Luzin and Sierpiński subsets of the real line cannot be established within the theory $\mathbf{ZFC}$, so if we want to deal with such subsets, then we need additional set-theoretical axioms (see, for instance, Theorems 1 and 4 below).

We begin our considerations with some properties of Luzin sets. These sets were constructed by Luzin, in 1914, under the assumption of the Continuum Hypothesis. The same sets were constructed by Mahlo one year before Luzin. However, in the mathematical literature the notion of a Luzin set is usually used, probably because Luzin investigated these sets rather deeply and showed a number of their important applications to the theory of real functions and classical measure theory (see, for example, [134]).

We now give the precise definition of Luzin sets.

Let X be a subset of $\mathbf{R}$. We say that X is a Luzin set if:

(1) X is uncountable;

(2) for every first category subset Y of $\mathbf{R}$, the intersection $X \cap Y$ is at most countable.

It is obvious that the family of all Luzin subsets of the real line $\mathbf{R}$ generates the σ–ideal on $\mathbf{R}$ invariant under the group of all those transformations of $\mathbf{R}$, which preserve the σ–ideal of all first category subsets of $\mathbf{R}$ (consequently, the above-mentioned family is invariant with respect to the group of all homeomorphisms of the real line and, in particular, with respect to the group of all translations of this line).

We remark at once that it is impossible to prove in the theory $\mathbf{ZFC}$ the existence of a Luzin set. Namely, assume that the relation

$$(Martin's\ Axiom)\ \&\ (2^\omega > \omega_1)$$

holds and take an arbitrary set $X \subset \mathbf{R}$. If X is at most countable, then it is not a Luzin set. Suppose now that X is uncountable. As known (see, e.g., [119]), Martin's Axiom implies that the union of an arbitrary family $\{Y_i \ : \ i \in I\}$ of first category (respectively, of Lebesgue measure zero) subsets of $\mathbf{R}$, where $card(I) < \mathbf{c}$, is again of first category (respectively, of Lebesgue measure zero). In particular, under Martin's Axiom, each subset of $\mathbf{R}$ with cardinality strictly less than $\mathbf{c}$ is of first category and of Lebesgue measure zero. Let now Y be any subset of X of cardinality ω_1. Then, taking into account the inequality $\omega_1 < \mathbf{c}$, we infer that Y is a first category subset of $\mathbf{R}$ and

$$card(X \cap Y) = card(Y) = \omega_1 > \omega,$$

so we obtain again that X is not a Luzin set.

Fortunately, if we assume the Continuum Hypothesis, then we can prove that Luzin sets exist on the real line (we recall once more that this classical result is due to Luzin and Mahlo).

Theorem 1. *If the Continuum Hypothesis holds, then there are Luzin subsets of* $\mathbf{R}$.

Proof. As we know, the Continuum Hypothesis means the equality

$$\mathbf{c} = \omega_1,$$

which implies, in particular, that the family of all Borel subsets of $\mathbf{R}$ has cardinality ω_1. Let $(X_\xi)_{\xi < \omega_1}$ denote the family of all first category Borel subsets of $\mathbf{R}$. We define, by the method of transfinite recursion, a family $(x_\xi)_{\xi < \omega_1}$ of points from $\mathbf{R}$. Suppose that $\xi < \omega_1$ and that the partial family $(x_\zeta)_{\zeta < \xi}$ has already been constructed. Let us consider the set

$$Z_\xi = (\bigcup_{\zeta < \xi} X_\zeta) \cup \{x_\zeta : \zeta < \xi\}.$$

It is clear that Z_ξ is a first category subset of the real line. Hence, by the classical Baire theorem applied to $\mathbf{R}$, there exists a point

$$x \in \mathbf{R} \setminus Z_\xi.$$

Thus we may put $x_\xi = x$. This ends the construction of the family $(x_\xi)_{\xi < \omega_1}$. Now, we define

$$X = \{x_\xi \ : \ \xi < \omega_1\}.$$

Notice that if $\zeta < \xi < \omega_1$, then $x_\zeta \neq x_\xi$. Therefore, we get

$$card(X) = \omega_1 = \mathbf{c}.$$

Suppose now that Z is an arbitrary first category subset of the real line. Then there exists an ordinal $\xi < \omega_1$ such that $Z \subset X_\xi$ (because the family $(X_\xi)_{\xi < \omega_1}$ forms a base of the σ-ideal of all first category subsets of $\mathbf{R}$). Obviously, we have the relations

$$X \cap Z \subset X \cap X_\xi \subset \{x_\zeta : \zeta \leq \xi\},$$

$$card(X \cap Z) \leq \omega,$$

which enable us to conclude that X is a Luzin set.

This finishes the proof of Theorem 1.

Notice that if X is a Luzin set on $\mathbf{R}$, then the set $X \cup \mathbf{Q}$ is also a Luzin set. Hence, the Continuum Hypothesis implies that there are Luzin sets everywhere dense in $\mathbf{R}$. We can easily get a much stronger result. Namely, a slight modification of the proof presented above gives us a Luzin set X such that, for any set $Y \subset \mathbf{R}$ with the Baire property, we have the relation

$$(card(Y \cap X) \leq \omega) \Leftrightarrow (Y \ is \ of \ first \ category).$$

Exercise 1. By assuming the Continuum Hypothesis and applying the method of transfinite induction, show that there exists a Luzin subset X of $\mathbf{R}$ such that, for each set $Y \subset \mathbf{R}$ possessing the Baire property, the following two relations are equivalent:

(a) $card(Y \cap X) \leq \omega$;

(b) Y is a first category subset of $\mathbf{R}$.

Also, verify whether the assumption that all Y possess the Baire property is essential for the validity of (a) $\Leftrightarrow$ (b).

Remark 1. It is reasonable to mention here that the existence of Luzin subsets of $\mathbf{R}$ is also possible in some cases when the Continuum Hypothesis does not hold. Namely, there are certain Cohen-type models of set theory in which the negation of the Continuum Hypothesis is true and there exist Luzin sets of cardinality $\mathbf{c}$ (for details, see, e.g., [118] and [119]). Notice, in addition, that in those models we also have a subset of $\mathbf{R}$ of cardinality $\omega_1 < \mathbf{c}$, which does not possess the Baire property (cf. the next exercise).

Exercise 2. Let X be a Luzin set on $\mathbf{R}$ with $card(X) = \mathbf{c}$ and let κ be a cardinal number satisfying the inequalities

$$\omega_1 \leq \kappa < \mathbf{c}.$$

Show that no subset Y of X with $card(Y) = \kappa$ possesses the Baire property in $\mathbf{R}$.

Luzin sets have a number of specific features important from the point of view of applications in analysis. The following theorem was also proved by Luzin.

Theorem 2. *Suppose that X is a Luzin set on $\mathbf{R}$. Then X does not possess the Baire property in $\mathbf{R}$ (and, moreover, any uncountable subset of X, being also a Luzin set, does not possess the Baire property). Furthermore, in the space $X \cup \mathbf{Q}$ equipped with the topology induced by the standard topology of $\mathbf{R}$ every first category set is at most countable and, conversely, every at most countable subset of the space $X \cup \mathbf{Q}$ is of first category in $X \cup \mathbf{Q}$.*

Proof. Let X be an arbitrary Luzin set on $\mathbf{R}$. Suppose, for a moment, that X has the Baire property. Because X is uncountable and, for every first category set $Y \subset \mathbf{R}$, we have

$$card(X \cap Y) \leq \omega,$$

we infer that X is not a first category subset of $\mathbf{R}$. But then X contains some uncountable G_δ–subset Z. Let Y be any subset of Z homeomorphic to the classical Cantor discontinuum. Then Y is nowhere dense in $\mathbf{R}$ and $Y \subset X$. We also have

$$card(X \cap Y) = card(Y) = \mathbf{c} \geq \omega_1,$$

which contradicts the definition of a Luzin set. The contradiction obtained yields us that X does not have the Baire property.

The second part of this theorem follows from the fact that $X \cup \mathbf{Q}$ is a Luzin set, too, and, in addition, $X \cup \mathbf{Q}$ is everywhere dense in $\mathbf{R}$.

The proof of Theorem 2 is thus completed.

The result proved above shows us that if the Continuum Hypothesis holds, then there exists a subspace X of $\mathbf{R}$ everywhere dense in $\mathbf{R}$ and such that

$$card(X) = \mathbf{c}, \quad \mathcal{K}(X) = [X]^{\leq \omega},$$

where $\mathcal{K}(X)$ denotes, as usual, the σ–ideal of all first category subsets of the space X and $[X]^{\leq \omega}$ denotes the family of all (at most) countable subsets of X.

The latter equality also implies that

$$\mathcal{B}a(X) = \mathcal{B}(X),$$

where $\mathcal{B}a(X)$ denotes the class of all subsets of X with the Baire property and $\mathcal{B}(X)$ denotes the Borel σ–algebra of X.

By taking this result into account, it is reasonable to introduce the following definition.

A Hausdorff topological space E without isolated points is called a Luzin space if the equality

$$\mathcal{K}(E) = [E]^{\leq \omega}$$

is valid.

Hence we see that, under the Continuum Hypothesis, there exists an everywhere dense Luzin set on the real line $\mathbf{R}$, which can be regarded as an example of a topological Luzin space E with $card(E) = \mathbf{c}$.

In our further considerations, we need one simple auxiliary statement concerning first category supports of σ-finite diffused Borel measures given on separable metric spaces. Recall that a measure μ defined on a σ-algebra of subsets of a set E is diffused (continuous) if, for each $e \in E$, we have $\{e\} \in dom(\mu)$ and $\mu(\{e\}) = 0$.

Lemma 1. *Let E be an arbitrary separable metric space and let μ be a σ-finite diffused Borel measure on E. Then there exists a subset Z of E such that:*

(1) $Z \in \mathcal{K}(E)$ and Z is an F_σ-subset of E;

(2) $\mu(E \setminus Z) = 0$.

In other words, Z is a first category support of μ.

Proof. Without loss of generality, we may assume that μ is a probability measure, i.e., the equality

$$\mu(E) = 1$$

holds true. Let us denote by $\{e_n \ : \ n < \omega\}$ a countable everywhere dense subset of the given space E. Fix a natural number k. Because our μ is a diffused measure, we can find, for each point e_n, an open neighborhood $V_k(e_n)$ such that

$$\mu(V_k(e_n)) < 1/2^{k+n}.$$

Let us put

$$V_k = \cup\{V_k(e_n) \ : \ n < \omega\}.$$

Then V_k is an everywhere dense open subset of E, for which we have

$$\mu(V_k) \leq 1/2^{k-1}.$$

Now, putting

$$Y = \cap\{V_k \ : \ k < \omega\},$$

we obtain a G_δ-subset Y of E of μ-measure zero. According to the definition of Y, the set $Z = E \setminus Y$ is a first category F_σ-subset of E such that

$$\mu(E \setminus Z) = \mu(Y) = 0.$$

Thus, Z is the required support of our measure μ.

Exercise 3. Let E be a topological space satisfying the following conditions:

(a) there exists a countable subset D of E everywhere dense in E;

(b) each one-element set $\{e\}$, where $e \in D$, is a G_δ-subset of E.

Let μ be an arbitrary σ-finite Borel measure on E such that

$$(\forall e \in D)(\mu(\{e\}) = 0).$$

Prove that there exists a first category F_σ-subset Z of E for which

$$\mu(E \setminus Z) = 0.$$

Obviously, this result is a slight generalization of Lemma 1.

The next theorem, also essentially due to Luzin, shows us an interesting connection between Luzin sets on **R** and topological measure theory.

Theorem 3. *Let X be an arbitrary Luzin subset of **R**. Then the following two relations hold:*

*(1) if μ is any σ–finite diffused Borel measure on **R**, then $\mu^*(X) = 0$, where μ^* denotes the outer measure associated with μ;*

(2) if μ is any σ–finite diffused Borel measure on the topological space X, then μ is identically equal to zero.

Proof. It is almost obvious that relations (1) and (2) are equivalent. Hence it is sufficient to prove only the second one. Without loss of generality, we may assume that X is an everywhere dense subset of **R**. Let μ be an arbitrary σ–finite diffused Borel measure given on the topological space X. Because X is a separable metric space, we may apply the preceding lemma on first category supports of diffused Borel measures. According to this lemma, there exists a first category subset of X on which our μ is concentrated. But we know that each first category subset of X is at most countable (see Theorem 2), and our μ is diffused, so we conclude that μ must be identically equal to zero.

Theorems 2 and 3 show us that, on the one hand, from the topological point of view Luzin sets are extremely pathological (because any uncountable subset of a Luzin set does not have the Baire property) but, on the other hand, from the point of view of topological measure theory, Luzin sets are very small because they have measure zero with respect to the completion of any σ–finite diffused Borel measure on $\mathbf{R}$.

Exercise 4. Let E be a topological space such that all one-element sets $\{e\}$, where $e \in E$, are Borel in E. We shall say that E is a universal measure zero space (or universally negligible space) if each σ-finite diffused Borel measure given on E is identically equal to zero.

It immediately follows from Theorem 3 that, under the Continuum Hypothesis, there exist universal measure zero subspaces of the real line, having the cardinality of the continuum (namely, any Luzin subset of $\mathbf{R}$ is such a space). In particular, the Continuum Hypothesis implies that $\mathbf{c}$ is not a real-valued measurable cardinal (for the definition of real-valued measurable cardinals and their properties, see [71], [72], [119], [162], [186], [212], and [224]).

Show that:

(a) each subspace of a universal measure zero space is also a universal measure zero space;

(b) the class of all universal measure zero spaces is closed under finite Cartesian products, but is not closed under countable Cartesian products;

(c) if I is a set of indices, whose cardinality is not real-valued measurable, and $\{E_i : i \in I\}$ is a family of universal measure zero spaces, then the topological sum of $\{E_i : i \in I\}$ is a universal measure zero space, too.

Exercise 5. Let E be a topological space satisfying the assumption of the previous exercise. Prove that the following two assertions are equivalent:

(a) E is a universal measure zero space;

(b) for any topological space E' satisfying the same assumption and containing E as a subspace, and for any σ-finite diffused Borel measure μ on E', we have the equality

$$\mu^*(E) = 0$$

where μ^* denotes, as usual, the outer measure associated with μ.

Deduce from this result that if $\{E_n : n \in \omega\}$ is a countable family of universal measure zero subspaces of a topological space E', then the space $\cup\{E_n : n \in \omega\}$ is universal measure zero, too. In other words, if a space E' is not universal measure zero, then the family of all universal measure zero subspaces of E' forms a σ-ideal in the Boolean algebra of all subsets of E'.

Exercise 6. Let E_1 and E_2 be two topological spaces satisfying the assumption of Exercise 4, and let

$$f : E_1 \to E_2$$

be an injective Borel mapping. Check that if the space E_2 is universal measure zero, then the space E_1 is universal measure zero, too.

Show also that the condition of injectivity of f is essential here. Moreover, show that a continuous image of a universal measure zero space is not, in general, universal measure zero.

Exercise 7. Prove that any universal measure zero subspace X of the real line $\mathbf{R}$ is a Marczewski subset of $\mathbf{R}$ (see Chapter 8 of this book), i.e., for each nonempty perfect set $P \subset \mathbf{R}$, there exists a nonempty perfect set $P' \subset \mathbf{R}$ such that

$$P' \subset P, \quad P' \cap X = \emptyset.$$

As mentioned above, any Luzin subset of the real line is universal measure zero. But the existence of Luzin sets cannot be proved in the theory **ZFC**. On the other hand, it is known that the existence of uncountable universal measure zero subsets of $\mathbf{R}$ can be established within **ZFC** (see, for instance, [120], [136] or [149]). One of the earliest examples of an uncountable universal measure zero subset of $\mathbf{R}$ was constructed by Luzin who applied, in his construction, some specific methods of the theory of analytic sets. In fact, the construction of Luzin yields a universal measure zero set $Y \subset \mathbf{R}$ with $card(Y) = \omega_1$. In particular, we may conclude that it is impossible to establish within the theory **ZFC** that Y is a Luzin set in the sense of the definition presented in this chapter. In addition, let us remark that Luzin's result concerning the cardinality of universal measure zero subsets of $\mathbf{R}$ is rather precise. Namely, as shown by Baumgartner and Laver (cf. [149]), in a certain model of set theory each universal measure zero subspace of $\mathbf{R}$ has cardinality less than or equal to ω_1 and the inequality $\omega_1 < \mathbf{c}$ holds in the same model.

As we see, Luzin sets and universal measure zero sets on $\mathbf{R}$ may be regarded as small subsets of $\mathbf{R}$. There are also many other notions of a small subset of the real line. One of such notions was introduced by Borel in 1919.

Let X be a subset of $\mathbf{R}$. We say that X is small in the Borel sense (or is a strong measure zero set) if, for any sequence $\{\varepsilon_n \ : \ n < \omega\}$ of strictly positive real numbers, there exists a countable covering $\{ \,]a_n, b_n[\ : \ n < \omega\}$ of X by open intervals, such that

$$(\forall n < \omega)(b_n - a_n < \varepsilon_n).$$

It immediately follows from this definition that the family of all subsets of $\mathbf{R}$ small in the Borel sense forms a σ-ideal in the Boolean algebra of all subsets of $\mathbf{R}$.

Exercise 8. Show that any Luzin subset of the real line is small in the Borel sense.

Exercise 9. Let $[a, b]$ be an arbitrary compact subinterval of $\mathbf{R}$ and let μ be a finite diffused Borel measure on $[a, b]$. Show that, for each real $\varepsilon > 0$, there exists a real $\delta > 0$ such that, for any subinterval $[t', t'']$ of $[a, b]$ with $t'' - t' < \delta$, we have $\mu([t', t'']) < \varepsilon$.

Deduce from this fact that any subset of the real line, small in the Borel sense, is universal measure zero (note that the converse assertion is not true; in this connection, see Exercise 25 below).

Assuming the Continuum Hypothesis and taking account of the result of Exercise 8, we have the existence of uncountable subsets of $\mathbf{R}$ small in the Borel sense. The Borel Conjecture is the following set-theoretical statement:

every set on $\mathbf{R}$ *small in the Borel sense is at most countable.*

Thus, under the Continuum Hypothesis, the Borel Conjecture is false. However, Laver demonstrated that there are some models of set theory in which this conjecture holds true (for more detailed information, see [149] and the corresponding references therein).

Let us return to Luzin subsets of the real line. The next two exercises contain some additional information about these sets.

Exercise 10. Let X be a Luzin subset of the real line $\mathbf{R}$ and let μ be an arbitrary σ–finite diffused Borel measure on $\mathbf{R}$. Suppose also that

$$f \ : \ X \to \mathbf{R}$$

is a mapping which has the Baire property. Prove that

$$\mu^*(f(X)) = 0,$$

where μ^* denotes, as usual, the outer measure associated with μ. Hence, $f(X)$ is a universal measure zero subset of $\mathbf{R}$.

Exercise 11. Suppose that the Continuum Hypothesis holds and let X be an uncountable everywhere dense subspace of the real line $\mathbf{R}$, such that

$$\mathcal{B}a(X) = \mathcal{B}(X).$$

Show that X is a Luzin subset of $\mathbf{R}$.

Dual (in a certain sense) objects to Luzin sets are the so-called Sierpiński sets that were constructed by Sierpiński, also under the assumption of the Continuum Hypothesis, in 1924.

Let us introduce the notion of a Sierpiński set and consider some properties of such sets.

Let X be a subset of $\mathbf{R}$. We say that X is a Sierpiński set if:

(1) X is uncountable;

(2) for each Lebesgue measure zero subset Y of $\mathbf{R}$, the intersection $X \cap Y$ is at most countable.

Many facts concerning Sierpiński sets are very similar to the corresponding facts concerning Luzin sets. For example, we have:

(a) the family of all Sierpiński sets generates the σ–ideal of subsets of the real line $\mathbf{R}$, invariant under the group of all those transformations of $\mathbf{R}$ which preserve the σ–ideal of all Lebesgue measure zero subsets of $\mathbf{R}$ (in particular, this family is invariant with respect to the group of all translations of $\mathbf{R}$);

(b) the assumption

$$(Martin's\ Axiom)\ \&\ (2^\omega > \omega_1)$$

implies that there are no Sierpiński sets on the real line.

Analogously, we have the following theorem due to Sierpiński.

Theorem 4. *Assume the Continuum Hypothesis. Then there exist Sierpiński subsets of* $\mathbf{R}$.

The proof of this theorem is very similar to the proof of Theorem 1. The only change is the replacement of the family $(X_\xi)_{\xi<\omega_1}$ of all first category Borel subsets of $\mathbf{R}$ by the family $(Y_\xi)_{\xi<\omega_1}$ of all λ–measure zero Borel subsets of $\mathbf{R}$ (where λ denotes, as usual, the standard Lebesgue measure on $\mathbf{R}$). Also, instead of the classical Baire theorem, we must apply here the trivial fact that λ is not identically equal to zero.

Exercise 12. Give a detailed proof of Theorem 4.

In connection with Theorem 4, we wish to note that the existence of Sierpiński subsets of $\mathbf{R}$ is possible in some situations when the Continuum Hypothesis does not hold. More precisely, there are models of **ZFC** in which the negation of the Continuum Hypothesis is valid and there exist Sierpiński sets of cardinality $\mathbf{c}$ (see, e.g., [118] and [119]). Evidently, in such models we also have Lebesgue nonmeasurable subsets of $\mathbf{R}$ whose cardinality is equal to ω_1 and $\omega_1 < \mathbf{c}$.

Remark 2. There is a general result due to Sierpiński and Erdös, which states that under certain additional set-theoretical hypotheses (in particular, under the Continuum Hypothesis or Martin's Axiom), the σ-ideal of all first category subsets of $\mathbf{R}$ is isomorphic to the σ-ideal of all

Lebesgue measure zero subsets of $\mathbf{R}$. An isomorphism between these two classical σ-ideals is purely set-theoretical and does not have good descriptive properties. However, the existence of such an isomorphism enables us to obtain automatically many theorems for the Lebesgue measure (the Baire category) starting with the corresponding theorems for the Baire category (the Lebesgue measure). In particular, in this way we can easily deduce Theorem 4 from Theorem 1 (and, conversely, Theorem 1 from Theorem 4).

A detailed proof of the Sierpiński-Erdös result mentioned above (the so-called Sierpiński-Erdös Duality Principle) is given in the well-known textbooks [153] and [162] where numerous applications of this principle are presented as well (a general version of the Duality Principle is formulated and proved in [34]).

Exercise 13. Let $\mathcal{T} = \mathcal{T}_d$ denote the density topology on $\mathbf{R}$. Show that a set $Z \subset \mathbf{R}$ is a Sierpiński set in $\mathbf{R}$ if and only if Z is a Luzin set in the space $(\mathbf{R}, \mathcal{T})$ (this means that Z is uncountable and, for every first category set Y in $(\mathbf{R}, \mathcal{T})$, the intersection $Z \cap Y$ is at most countable).

Let us point out another similarity between Luzin and Sierpiński sets.

Theorem 5. *Every Sierpiński set is a first category subset of the real line $\mathbf{R}$. No uncountable subset of a Sierpiński set is Lebesgue measurable.*

Proof. Let X be a Sierpiński set. Let $\mathcal{I}(\lambda)$ denote, as usual, the σ-ideal of all Lebesgue measure zero subsets of $\mathbf{R}$. As we know, the σ–ideals $\mathcal{K}(\mathbf{R})$ and $\mathcal{I}(\lambda)$ are orthogonal, i.e., there exists a partition $\{A, B\}$ of $\mathbf{R}$ such that

$$A \in \mathcal{K}(\mathbf{R}), \quad B \in \mathcal{I}(\lambda).$$

Simultaneously, we have the inequality

$$card(X \cap B) \leq \omega$$

and the inclusion

$$X \subset A \cup (X \cap B),$$

from which we immediately obtain that $X \in \mathcal{K}(\mathbf{R})$.

Suppose now that Y is an uncountable subset of a Sierpiński set X (hence Y is a Sierpiński set, too). Because $X \cap Y$ is uncountable, we observe that $Y \notin \mathcal{I}(\lambda)$. Suppose, to the contrary, that Y is Lebesgue measurable. Then $\lambda(Y) > 0$ and we can find an uncountable set $Z \subset Y$ of Lebesgue measure zero. But then the set $X \cap Z$ is uncountable, so we get a contradiction with the definition of the Sierpiński set X. The contradiction obtained ends the proof of Theorem 5.

Exercise 14. Let X be a Sierpiński subset of $\mathbf{R}$ considered as a topological space with the induced topology. Show that any Borel subset of X is simultaneously an F_σ-set and a G_δ-set in X. In particular, each countable subset of X is a G_δ-set in X.

Exercise 15. Let X be a Sierpiński subset of $\mathbf{R}$. As mentioned above, all countable subsets of X are G_δ-sets in X. Applying this fact, demonstrate that, for any nonempty perfect set $P \subset \mathbf{R}$, the set $X \cap P$ is of first category in P. This result strengthens the corresponding part of Theorem 5.

Exercise 16. Let X be a Sierpiński subset of $\mathbf{R}$. Equip X with the topology induced by the density topology of $\mathbf{R}$. Prove that the topological space X is nonseparable and hereditarily Lindelöf (the latter means that each subspace of X is Lindelöf, i.e., any open covering of a subspace contains a countable subcovering).

Exercise 17. Assume that the Continuum Hypothesis holds. Let X be a Sierpiński set on the real line $\mathbf{R}$. Equip X with the topology induced by the Euclidean topology of $\mathbf{R}$. Prove that

$$\mathcal{A}(X) = \mathcal{B}(X)$$

where $\mathcal{A}(X)$ denotes the class of all analytic subsets of X and $\mathcal{B}(X)$ denotes the class of all Borel subsets of X.

Exercise 18. Let $\mathcal{J}_1$ and $\mathcal{J}_2$ be two orthogonal σ-ideals of subsets of $\mathbf{R}$, each of which is invariant with respect to the group of all translations of $\mathbf{R}$. Let A_1 and A_2 be two subsets of $\mathbf{R}$ satisfying the relations

$$A_1 \notin \mathcal{J}_1, \quad A_2 \notin \mathcal{J}_2.$$

Demonstrate that:
(i) there exists a set $B_1 \in \mathcal{J}_1$ for which we have

$$B_1 - A_2 = \cup\{B_1 - a \ : \ a \in A_2\} = \mathbf{R};$$

(ii) there exists a set $B_2 \in \mathcal{J}_2$ for which we have

$$B_2 - A_1 = \cup\{B_2 - a \ : \ a \in A_1\} = \mathbf{R}.$$

Further, put:
$\mathcal{J}_1 = $ the σ-ideal of all first category subsets of $\mathbf{R}$;
$\mathcal{J}_2 = $ the σ-ideal of all Lebesgue measure zero subsets of $\mathbf{R}$.

Deduce from relations (i) and (ii) that if X is an arbitrary Luzin set on $\mathbf{R}$ and Y is an arbitrary Sierpiński set on $\mathbf{R}$, then the equalities

$$card(X) = card(Y) = \omega_1$$

are fulfilled. We thus conclude that the simultaneous existence in $\mathbf{R}$ of Luzin and Sierpiński sets immediately implies that the cardinality of these sets is as minimal as possible (i.e., is equal to the smallest uncountable cardinal). This result was obtained by Rothberger (see [177]).

It is easy to observe that if we replace the Continuum Hypothesis by Martin's Axiom (which is a much weaker assertion than $\mathbf{CH}$), then we can prove the existence of some analogs of Luzin and Sierpiński sets. Namely, if Martin's Axiom holds, then there exists a set $X \subset \mathbf{R}$ such that:

(1) $card(X) = \mathbf{c}$;

(2) for each set $Y \in \mathcal{K}(\mathbf{R})$, we have

$$card(Y \cap X) < \mathbf{c}.$$

A set X with the above property is usually called a generalized Luzin subset of the real line.

Similarly, if Martin's Axiom holds, then there exists a set $X \subset \mathbf{R}$ such that:

(1') $card(X) = \mathbf{c}$;

(2') for each set $Z \in \mathcal{I}(\lambda)$, we have

$$card(Z \cap X) < \mathbf{c}.$$

A set X with the above property is usually called a generalized Sierpiński subset of the real line.

Let us remark that, for the existence of generalized Luzin sets or generalized Sierpiński sets, we do not need the full power of Martin's Axiom. In fact, the existence of generalized Luzin and Sierpiński sets is implied by some additional set-theoretical assumptions that are much weaker than Martin's Axiom. The next exercise contains the corresponding result for an abstract σ-ideal of subsets of a given infinite set.

Exercise 19. Let E be a set with $card(E) \geq \omega$ and let $\mathcal{J}$ be a σ-ideal of subsets of E, containing in itself all one-element subsets of E. We denote:

$cov(\mathcal{J})$ = the smallest cardinality of a covering of E by sets belonging to $\mathcal{J}$;

$cof(\mathcal{J})$ = the smallest cardinality of a base of $\mathcal{J}$.

Prove that if the equalities

$$cov(\mathcal{J}) = cof(\mathcal{J}) = card(E)$$

are fulfilled, then there exists a subset D of E such that

$$card(D) = card(E)$$

and, for any set $Z \in \mathcal{J}$, we have

$$card(Z \cap D) < card(E).$$

In particular, if our basic set E coincides with the real line $\mathbf{R}$ and $\mathcal{J}$ is the σ-ideal of all first category subsets of $\mathbf{R}$ (respectively, the σ-ideal of all Lebesgue measure zero subsets of $\mathbf{R}$), then we obtain, under Martin's Axiom, the existence of a generalized Luzin subset of $\mathbf{R}$ (respectively, the existence of a generalized Sierpiński subset of $\mathbf{R}$).

Some facts about generalized Luzin sets and generalized Sierpiński sets are presented in the next three exercises.

Exercise 20. Assume that the Continuum Hypothesis holds. Prove that there exists a set $X \subset \mathbf{R}$ satisfying the following conditions:
(a) X is a vector space over the field $\mathbf{Q}$;
(b) X is an everywhere dense Luzin subset of $\mathbf{R}$.
Show also that there exists a set $Y \subset \mathbf{R}$ satisfying the following conditions:
(a') Y is a vector space over the field $\mathbf{Q}$;
(b') Y is an everywhere dense Sierpiński subset of $\mathbf{R}$.
Moreover, by assuming Martin's Axiom, formulate and prove analogous results for generalized Luzin sets and for generalized Sierpiński sets.
In addition, infer from these results, by assuming Martin's Axiom again, that there exist an isomorphism f of the additive group $\mathbf{R}$ onto itself and a generalized Luzin set X in $\mathbf{R}$ such that $f(X)$ is a generalized Sierpiński set in $\mathbf{R}$.

Exercise 21. Suppose that Martin's Axiom holds. Prove that any generalized Luzin subset of $\mathbf{R}$ is universal measure zero. In addition, by using a generalized Luzin set on $\mathbf{R}$, show that there exists a σ-algebra $\mathcal{S}$ of subsets of $\mathbf{R}$, such that:
(a) for each point $x \in \mathbf{R}$, we have $\{x\} \in \mathcal{S}$;
(b) $\mathcal{S}$ is a countably generated σ-algebra, i.e., there exists a countable subfamily of $\mathcal{S}$, which generates $\mathcal{S}$;

(c) there is no nonzero σ-finite diffused measure defined on $\mathcal{S}$.

In particular, we see that Martin's Axiom implies that the cardinal $\mathbf{c}$ is not real-valued measurable.

A result similar to the one presented in Exercise 21 can be proved in the theory **ZFC** if we replace $\mathbf{R}$ by a certain uncountable subspace E of $\mathbf{R}$. Namely, it suffices to take as E a universal measure zero subset of $\mathbf{R}$ with cardinality equal to ω_1. In particular, we immediately obtain from this result that ω_1 is not real-valued measurable (cf. [224]).

Exercise 22. Assume Martin's Axiom. By applying a generalized Luzin set, prove that there exist two σ-algebras $\mathcal{S}_1$ and $\mathcal{S}_2$ of subsets of $\mathbf{R}$, satisfying the following conditions:

(a) $\mathcal{B}(\mathbf{R}) \subset \mathcal{S}_1 \cap \mathcal{S}_2$;

(b) both $\mathcal{S}_1$ and $\mathcal{S}_2$ are countably generated σ-algebras;

(c) there exists a measure μ_1 on $\mathcal{S}_1$ extending the standard Borel measure on $\mathbf{R}$;

(d) there exists a measure μ_2 on $\mathcal{S}_2$ extending the standard Borel measure on $\mathbf{R}$;

(e) there is no nonzero σ-finite diffused measure defined on the σ-algebra of sets, generated by $\mathcal{S}_1 \cup \mathcal{S}_2$.

Note that the result of Exercise 22 will be significantly strengthened in Chapter 11.

We wish to present here one application of a generalized Luzin set to the construction of a function which is extremely bad from the point of view of measure theory. First, we must give the corresponding definition.

Let E be a set (in particular, a topological space) and let f be a function acting from E into $\mathbf{R}$. We shall say that f is absolutely nonmeasurable if, for any nonzero σ-finite diffused measure μ on E, our f is nonmeasurable with respect to μ.

Let us stress that, in this definition, the domain of μ is not a fixed σ-algebra of subsets of E (actually, $dom(\mu)$ may be an arbitrary σ-algebra of subsets of E, containing all singletons).

Theorem 6. *Suppose that Martin's Axiom holds. Then there exists an injective function*

$$f : \mathbf{R} \to \mathbf{R}$$

that simultaneously is absolutely nonmeasurable.

Proof. We know that Martin's Axiom implies the existence of a generalized Luzin subset of $\mathbf{R}$. Let X be such a subset. Because we have

$$card(X) = \mathbf{c} = card(\mathbf{R}),$$

there exists a bijection
$$f \; : \; \mathbf{R} \to X.$$

Obviously, we can consider f as an injection from $\mathbf{R}$ into itself. Let us verify that f is the required function. Suppose, for a moment, that our f is not absolutely nonmeasurable. Then there exists a nonzero σ-finite diffused measure μ on $\mathbf{R}$ such that f is μ-measurable, i.e., for any Borel subset B of $\mathbf{R}$, the relation
$$f^{-1}(B) \in dom(\mu)$$
is satisfied. Equivalently, for any Borel subset B' of X, we have
$$f^{-1}(B') \in dom(\mu).$$

Clearly, without loss of generality, we may assume that μ is a probability measure. Now, for each Borel subset B' of X, we put
$$\nu(B') = \mu(f^{-1}(B')).$$

In this way we obtain a Borel diffused probability measure ν on X, which is impossible since X is a universal measure zero space (see Exercise 21 above).

This contradiction ends the proof of Theorem 6.

Remark 3. The preceding theorem was formulated and proved under the assumption that Martin's Axiom is valid. In this connection, it is reasonable to point out here that we cannot establish the existence of an absolutely nonmeasurable function on $\mathbf{R}$ within the theory **ZFC**. Indeed, if the cardinality of the continuum is real-valued measurable, then such functions do not exist. At the same time, one can easily demonstrate (in **ZFC**) that there exists an absolutely nonmeasurable function

$$f \; : \; \omega_1 \to \mathbf{R}.$$

In order to show this, it suffices to pick a universal measure zero subspace X of $\mathbf{R}$ with
$$card(X) = \omega_1$$
and then to take as f any bijection acting from ω_1 onto X.

Some additive version of Theorem 6 will be discussed in Chapter 11 in connection with invariant extensions of the standard Borel measure on $\mathbf{R}$.

In our further considerations, we shall meet many other applications of Luzin sets and Sierpiński sets (respectively, of generalized Luzin sets and generalized Sierpiński sets). But now we shall use once more Martin's Axiom for giving a construction of a generalized Sierpiński set with the Baire property in the restricted sense.

Theorem 7. *Suppose that Martin's Axiom holds. Then there exists a set $X \subset \mathbf{R}$ such that:*

(1) for every nonempty perfect set $P \subset \mathbf{R}$, the intersection $X \cap P$ is a first category set in P;

(2) for each Lebesgue measurable set $Y \subset \mathbf{R}$ with $\lambda(Y) > 0$, the intersection $X \cap Y$ is nonempty;

(3) X is a generalized Sierpiński subset of $\mathbf{R}$.

Proof. Let α denote the smallest ordinal number whose cardinality is equal to the cardinality of the continuum $\mathbf{c}$. Actually, we may identify α with $\mathbf{c}$ (see the Introduction).

Let $(Z_\xi)_{\xi<\alpha}$ denote the family of all Borel subsets of the real line with a strictly positive Lebesgue measure, i.e.,

$$\{Z_\xi \ : \ \xi < \alpha\} = \mathcal{B}(\mathbf{R}) \setminus \mathcal{I}(\lambda),$$

and let $(T_\xi)_{\xi<\alpha}$ denote the family of all Borel subsets of the real line having Lebesgue measure zero, i.e.,

$$\{T_\xi \ : \ \xi < \alpha\} = \mathcal{B}(\mathbf{R}) \cap \mathcal{I}(\lambda).$$

For each ordinal $\xi < \alpha$, we fix a partition $\{Z_\xi^0, Z_\xi^1\}$ of Z_ξ such that Z_ξ^0 is a Lebesgue measure zero set and Z_ξ^1 is a first category subset of Z_ξ. Notice that the existence of such a partition follows directly from Lemma 1. Now, we define an injective α–sequence $(x_\xi)_{\xi<\alpha}$ of real numbers. Suppose that $\xi < \alpha$ and that the partial sequence $(x_\zeta)_{\zeta<\xi}$ has already been defined. Let us consider the set

$$D_\xi = (\bigcup_{\zeta \leq \xi} Z_\zeta^0) \cup \{x_\zeta \ : \ \zeta < \xi\} \cup (\bigcup_{\zeta \leq \xi} T_\zeta).$$

Martin's Axiom implies that the set D_ξ is also of Lebesgue measure zero. Hence we have

$$Z_\xi \setminus D_\xi \neq \emptyset.$$

So we can choose a point x_ξ from the above-mentioned nonempty difference of sets. In this way, we are able to define the whole α–sequence $(x_\xi)_{\xi<\alpha}$ of points of $\mathbf{R}$. Now, we put

$$X = \{x_\xi \ : \ \xi < \alpha\}$$

and we are going to show that the set X is the required one. Let P be any nonempty perfect subset of $\mathbf{R}$. If its Lebesgue measure is equal to zero, then, for some ordinal $\xi < \alpha$, we may write

$$P = T_\xi.$$

Consequently, from the method of construction of the set X, we immediately obtain

$$card(P \cap X) < \mathbf{c}.$$

Applying Martin's Axiom again, we see that the intersection $P \cap X$ is a first category set in P. Suppose now that $\lambda(P) > 0$. Then, for some ordinal $\xi < \alpha$, we can write

$$P = Z_\xi.$$

Therefore

$$P \cap X \subset Z_\xi^1 \cup \{x_\zeta \ : \ \zeta < \xi\}.$$

Taking account of the fact that the set P does not have isolated points, we obtain from the last inclusion that $P \cap X$ is a first category set in P. Hence condition (1) is satisfied for our set X. Furthermore, since

$$x_\xi \in Z_\xi$$

for each ordinal $\xi < \alpha$, we conclude that condition (2) holds for the set X, too. The validity of condition (3) follows directly from our construction.

Theorem 7 has thus been proved.

Remark 4. This result enables us to conclude that Martin's Axiom implies the existence of a generalized Sierpiński subset of $\mathbf{R}$ that is thick (with respect to the Lebesgue measure λ) and simultaneously has the Baire property in the restricted sense (cf. Exercise 12 from the Introduction).

We want to finish this chapter with some facts and statements concerning universal measure zero sets and strong measure zero sets.

Exercise 23. Check that if A is an arbitrary first category subset of $\mathbf{R}$, then the equality

$$(\mathbf{R} \setminus A) + (\mathbf{R} \setminus A) = \mathbf{R}$$

holds true. By starting with this fact, assuming the Continuum Hypothesis and using the method of transfinite recursion, construct a Luzin set X in $\mathbf{R}$ such that

$$X + X = \mathbf{R}.$$

Formulate and prove an analogous result (under Martin's Axiom) for a generalized Luzin set.

Exercise 24. Let Z be a subset of the Euclidean plane $\mathbf{R}^2$. We say that Z is strong measure zero if, for each sequence $\{\varepsilon_n \ : \ n \in \omega\}$ of strictly positive real numbers, there exists a countable covering $\{V_n \ : \ n \in \omega\}$ of Z by squares, such that

$$(\forall n \in \omega)(diam(V_n) < \varepsilon_n).$$

Now, let l be a straight line in $\mathbf{R}^2$ not parallel to the line $\mathbf{R} \times \{0\}$. Consider the projection

$$pr_{(\mathbf{R},l)} \ : \ \mathbf{R}^2 \to \mathbf{R}$$

of $\mathbf{R}^2$ onto $\mathbf{R}$, canonically associated with the direction l. According to the definition of $pr_{(\mathbf{R},l)}$, for any point $(x,y) \in \mathbf{R}^2$, we have

$$pr_{(\mathbf{R},l)}(x,y) = (x',0),$$

where the vector $(x - x', y)$ is parallel to the line l.

Show that if Z is a strong measure zero subset of $\mathbf{R}^2$, then $pr_{(\mathbf{R},l)}(Z)$ is a strong measure zero subset of $\mathbf{R}$.

In addition, introduce the notion of a Luzin subset (respectively, of a generalized Luzin subset) of the plane $\mathbf{R}^2$ and prove that, under the Continuum Hypothesis (respectively, under Martin's Axiom), there exist Luzin subsets (respectively, generalized Luzin subsets) of $\mathbf{R}^2$. Finally, verify that any Luzin set in $\mathbf{R}^2$ is strong measure zero (and hence universal measure zero).

Exercise 25. Let X denote the Luzin set of Exercise 23 and let

$$\phi \ : \ \mathbf{R}^2 \to \mathbf{R}$$

be a mapping defined by the formula

$$\phi((x,y)) = x + y \qquad ((x,y) \in \mathbf{R}^2).$$

Check that ϕ coincides with the projection $pr_{(\mathbf{R},l)}$ of $\mathbf{R}^2$ onto $\mathbf{R}$, where

$$l = \{(x,y) \in \mathbf{R}^2 \ : \ x + y = 1\}.$$

Show also that

$$\phi(X \times X) = \mathbf{R}.$$

This property of X implies, in particular, that the product $X \times X$ is not a strong measure zero subset of $\mathbf{R}^2$ (cf. the result of Exercise 24).

Conclude from the above-mentioned fact that the Cartesian product of two strong measure zero subsets of $\mathbf{R}$ is not, in general, a strong measure zero subset of $\mathbf{R}^2$.

In this connection, let us recall that the class of all universal measure zero spaces is closed under finite Cartesian products (see Exercise 4 of the present chapter).

Note that the results given in Exercises 23, 24, and 25 are essentially due to Sierpiński (cf. [198]).

The following statement (due to Marczewski) yields a characterization of universal measure zero subsets of $\mathbf{R}$ in terms of Lebesgue measure zero sets.

Theorem 8. *Let X be a subset of $\mathbf{R}$. Then these two assertions are equivalent:*

(1) X is a universal measure zero space;

(2) each homeomorphic image of X lying in $\mathbf{R}$ has Lebesgue measure zero.

Proof. Suppose first that X satisfies relation (1). Let Y be a subset of $\mathbf{R}$ homeomorphic to X. Fix any homeomorphism

$$f \ : \ X \to Y.$$

If $\lambda^*(Y) > 0$, then obviously there exists a nonzero σ-finite diffused Borel measure μ on Y. Putting

$$\nu(Z) = \mu(f(Z)) \quad (Z \in \mathcal{B}(X)),$$

we obtain a nonzero σ-finite diffused Borel measure ν on X. But this is impossible because X is a universal measure zero space. Therefore, the equality $\lambda(Y) = 0$ must be valid. The implication $(1) \Rightarrow (2)$ has thus been proved.

Let now X satisfy relation (2). We are going to demonstrate that relation (1) holds for X, too. Of course, without loss of generality, we may assume that our X is a subset of the unit segment $[0, 1]$. Suppose for a moment that X is not a universal measure zero space. Then there exists a Borel diffused probability measure μ on $[0, 1]$ such that

$$\mu^*(X) > 0.$$

We may also assume that μ does not vanish on any nonempty open subinterval of $[0,1]$ (replacing, if necessary, μ by $(\mu+\lambda)/2$). Now, define a function

$$f \; : \; [0,1] \to [0,1]$$

by the formula

$$f(x) = \mu([0,x]) \qquad (x \in [0,1]).$$

Evidently, f is an increasing homeomorphism from $[0,1]$ onto itself. Let us put

$$\nu(Z) = \mu(f^{-1}(Z)) \qquad (Z \in \mathcal{B}([0,1])).$$

In this way we get a Borel probability measure ν on $[0,1]$ such that

$$\nu^*(f(X)) = \mu^*(X) > 0.$$

Furthermore, it turns out that ν coincides with the standard Borel measure on $[0,1]$. Indeed, for each interval $[a,b] \subset [0,1]$, we may write

$$f^{-1}([a,b]) = \{t \in [0,1] \; : \; \mu([0,t]) \in [a,b]\} = [c,d],$$

where

$$\mu([0,c]) = a, \qquad \mu([0,d]) = b.$$

Then we have

$$\nu([a,b]) = \mu(f^{-1}([a,b])) = \mu([c,d]) =$$
$$= \mu([0,d]) - \mu([0,c]) = b - a.$$

Consequently, the measures ν and λ are identical on the family of all subintervals of $[0,1]$ and hence these two measures coincide on the whole Borel σ-algebra of $[0,1]$. Thus

$$\lambda^*(f(X)) = \nu^*(f(X)) > 0,$$

which contradicts relation (2). The contradiction obtained establishes the implication (2) $\Rightarrow$ (1) and ends the proof of Theorem 8.

Exercise 26. Let X be an uncountable topological space such that all one-element subsets of X are Borel in X. We say that X is a Sierpiński space if X contains no universal measure zero subspace with cardinality equal to $card(X)$. Show that:

(a) any generalized Sierpiński subset of $\mathbf{R}$ is a Sierpiński space;

(b) if X is a Sierpiński space of cardinality ω_1, then $\mathcal{A}(X) = \mathcal{B}(X)$, where $\mathcal{A}(X)$ denotes the class of all analytic subsets of X (i.e., the class of all those sets that can be obtained by applying the (A)-operation to various (A)-systems consisting of Borel subsets of X) and $\mathcal{B}(X)$ denotes, as usual, the class of all Borel subsets of X;

(c) if X_1 and X_2 are two Sierpiński spaces and X is their topological sum, then X is a Sierpiński space, too;

(d) if X is a Sierpiński space, Y is a topological space such that $card(Y) = card(X)$, all one-element subsets of Y are Borel in Y and there exists a Borel surjection from X onto Y, then Y is a Sierpiński space, too. Consequently, if X is a Sierpiński subset of $\mathbf{R}$ and

$$f : X \to \mathbf{R}$$

is a Borel mapping such that $card(f(X)) = card(X)$, then $f(X)$ is a Sierpiński subspace of $\mathbf{R}$.

Exercise 27. By assuming Martin's Axiom and applying the method of transfinite recursion, construct a generalized Sierpiński subset X of $\mathbf{R}$ such that

$$X + X = \mathbf{R}.$$

Infer from this equality that there exists a continuous surjection from the product space $X \times X$ onto $\mathbf{R}$. Further, by starting with this property of X and taking into account assertion (d) of the preceding exercise, show that the product $X \times X$ is not a Sierpiński space. Conclude from this fact that the topological product of two Sierpiński spaces is not, in general, a Sierpiński space.

Exercise 28. Let H be a Hilbert space (over the field $\mathbf{R}$) whose Hilbert dimension is equal to $\mathbf{c}$ (in particular, the cardinality of H equals $\mathbf{c}$, too). Assuming that $\mathbf{c}$ is not a real-valued measurable cardinal, demonstrate that there exists a subset X of H satisfying the following conditions:

(a) $card(X) = \mathbf{c}$;

(b) X is everywhere dense in H (in particular, X is nonseparable);

(c) X is a universal measure zero subspace of H.

Suppose now that $\mathbf{c}$ is not cofinal with ω_1, i.e., $\mathbf{c}$ cannot be represented in the form

$$\mathbf{c} = \sum_{\xi < \omega_1} \kappa_\xi$$

where all cardinal numbers κ_ξ ($\xi < \omega_1$) are strictly less than $\mathbf{c}$.

By starting with the fact that there exists an ω_1-sequence of nowhere dense subsets of H which cover H, show that there is no generalized Luzin subset of H. In other words, show that there is no subset Y of H satisfying the next two relations:

(d) $card(Y) = \mathbf{c}$;

(e) for each first category set $Z \subset H$, the inequality $card(Z \cap Y) < \mathbf{c}$ is fulfilled.

Exercise 29. Consider the Hilbert cube $[0,1]^\omega$. A deep theorem due to Hurewicz states that this cube cannot be covered by countably many zero-dimensional subspaces of it. Also, it is well known that:

(a) any zero-dimensional subset of $[0,1]^\omega$ can be included in some zero-dimensional G_δ-subspace of $[0,1]^\omega$;

(b) any finite-dimensional subset of $[0,1]^\omega$ is contained in a finite union of zero-dimensional subspaces of $[0,1]^\omega$.

For details, see, e.g., [120] (note that (a) is a direct consequence of the Lavrentieff theorem on extensions of homeomorphisms).

Starting with these facts and assuming the Continuum Hypothesis, construct an uncountable Luzin type set $X \subset [0,1]^\omega$ for the σ-ideal generated by the family of all zero-dimensional G_δ-sets in $[0,1]^\omega$.

Show that no uncountable subspace of X is finite-dimensional.

Now, fix a Peano type mapping

$$\phi : C \to [0,1]^\omega,$$

where C denotes the classical Cantor discontinuum on the segment $[0,1]$, and define two functions

$$f_1 : [0,1]^\omega \to C, \quad f_2 : [0,1]^\omega \to C$$

by the following formulas:

$$f_1(x) = min(\phi^{-1}(x)), \quad f_2(x) = max(\phi^{-1}(x)) \quad (x \in [0,1]^\omega).$$

We already know that both these functions are semicontinuous (see Exercise 4 from Chapter 3).

Show that, for any uncountable set $Y \subset X$, the restrictions $f_1|Y$ and $f_2|Y$ are not continuous. Deduce from this circumstance that $f_1|Y$ and $f_2|Y$ are not countably continuous.

Exercise 30. Assume Martin's Axiom. Starting with the existence of a generalized Luzin set in $\mathbf{R}$ and using the topological invariance of the Baire property in the restricted sense (see the Introduction), demonstrate that there exist a function

$$f : \mathbf{R} \to [0,1]$$

and a set $X \subset \mathbf{R}$ having the following properties:

(a) f is upper semicontinuous;
(b) $card(X) = \mathbf{c}$;
(c) for any set $Y \subset X$ with $card(Y) = \mathbf{c}$, the restriction $f|Y$ is not continuous.

Infer from these properties that if f is represented in the form

$$f = \cup\{f_i : i \in I\},$$

where $card(I) < \mathbf{c}$, then at least one partial function f_i is not continuous (in particular, f is not countably continuous).

The result of the last exercise is essentially due to Sierpiński (see [201]).

Rich additional information about Luzin sets and Sierpiński sets (also, about other small subsets of the real line) can be found in [120], [149], [167], and [235].

11. Absolutely nonmeasurable additive functions

In Chapter 10 it was shown that, assuming Martin's Axiom, there exists an injective absolutely nonmeasurable function

$$f : \mathbf{R} \to \mathbf{R}.$$

In other words, it was demonstrated that some functions f acting from $\mathbf{R}$ into $\mathbf{R}$ are extremely bad from the measure-theoretical point of view, i.e., those f are nonmeasurable with respect to any nonzero σ-finite diffused measure defined on a σ-algebra of subsets of $\mathbf{R}$.

In the same chapter it was also pointed out that the existence of absolutely nonmeasurable functions acting from $\mathbf{R}$ into $\mathbf{R}$ cannot be proved within the theory **ZFC**, so necessarily needs some additional set-theoretical axioms.

Here we wish to develop this topic and to establish some interesting connections between absolutely nonmeasurable functions and the measure extension problem raised by Banach.

First, it is natural to ask whether there exist (under **MA**) absolutely nonmeasurable functions $f : \mathbf{R} \to \mathbf{R}$ having important additional (e.g., algebraic or topological) properties.

For instance, one can ask whether there exists an absolutely nonmeasurable homomorphism of the additive group $\mathbf{R}$ into itself.

Exercise 20 from Chapter 10 gives a positive answer to this question (assuming Martin's Axiom). More precisely, if we suppose that there exists a generalized Luzin subset X of $\mathbf{R}$ being simultaneously a vector space over the field $\mathbf{Q}$ of all rational numbers, then the required absolutely nonmeasurable homomorphism from $\mathbf{R}$ into $\mathbf{R}$ can be constructed without any difficulty.

Namely, let us treat $\mathbf{R}$ as a vector space over $\mathbf{Q}$. Because we have

$$card(\mathbf{R}) = card(X) = \mathbf{c},$$

the vector spaces $\mathbf{R}$ and X are isomorphic, so we may take any isomorphism

$$h : \mathbf{R} \to X$$

between these two spaces. Then we may consider h as an injective homomorphism from $\mathbf{R}$ into $\mathbf{R}$. A simple argument given in the same chapter (see Theorem 6 therein) easily yields that h turns out to be an absolutely nonmeasurable function.

Remark 1. We thus conclude that the existence of absolutely nonmeasurable solutions of the Cauchy functional equation can be proved under Martin's Axiom. In this context, it is reasonable to recall that any nontrivial solution of the Cauchy functional equation is necessarily nonmeasurable with respect to the classical Lebesgue measure λ on $\mathbf{R}$ (see Theorem 4 from Chapter 9).

In the present chapter, our main goal is to obtain a certain generalization of the result of Pelc and Prikry [165] concerning the measure extension problem (cf. Exercise 22 from Chapter 10). The existence of absolutely nonmeasurable additive functions acting from $\mathbf{R}$ into $\mathbf{R}$ will be a starting point for our further considerations. Some other questions closely connected with measurability properties of additive functions will be also discussed.

First, we would like to recall several notions and definitions.

Let E be a nonempty set and let μ be a measure defined on a σ-algebra of subsets of E. We recall that μ is diffused (or continuous) if all singletons in E are of μ-measure zero.

Let us emphasize that the measures considered below (in this chapter) are assumed to be diffused.

Let M be a class of measures given on various σ-algebras of subsets of E and let

$$f \; : \; E \to \mathbf{R}$$

be a function. We shall say that f is absolutely nonmeasurable with respect to M if there exists no measure $\mu \in M$ such that f is measurable with respect to μ.

We shall say that f is absolutely nonmeasurable if f is absolutely nonmeasurable with respect to the class of all nonzero σ-finite diffused measures on E.

Remark 2. In the definition above, the real line $\mathbf{R}$ can be replaced by any uncountable Polish topological space (or, more generally, by any uncountable Borel subset of a Polish space). Indeed, since an arbitrary uncountable Borel subset B of a Polish space is Borel isomorphic to $\mathbf{R}$

(see, e.g., [120]), we have a canonical one-to-one correspondence between functions

$$f : E \to \mathbf{R}$$

that are absolutely nonmeasurable with respect to the class M, and functions

$$g : E \to B$$

that are absolutely nonmeasurable with respect to the same M.

Example 1. Let $E = \mathbf{R}$. Take as M the class of all those measures on $\mathbf{R}$, which extend λ and are invariant under the group of all translations of $\mathbf{R}$. Let X be a Vitali set in $\mathbf{R}$ and denote by $f = f_X$ its characteristic function. In view of the Vitali theorem (see also Exercise 4 from Chapter 8), we can assert that f is absolutely nonmeasurable with respect to M.

Exercise 1. Put again $E = \mathbf{R}$ and let M be the class of all those measures on $\mathbf{R}$, which extend λ and are quasiinvariant under the group of all motions of $\mathbf{R}$. Show that there exists a Vitali set Y in $\mathbf{R}$ whose characteristic function f_Y is not absolutely nonmeasurable with respect to M.

Example 2. Let E be an uncountable Polish topological space and let M denote the class of all completions of nonzero σ-finite diffused Borel measures on E. As we know, there exists a Bernstein subset X of E. Let $f = f_X$ be the characteristic function of X. Then we can assert that f is absolutely nonmeasurable with respect to M (see Exercise 9 from Chapter 8).

Let λ_0 stand for the restriction of λ to the Borel σ-algebra $\mathcal{B}(\mathbf{R})$ of $\mathbf{R}$. Pelc and Prikry proved in their work [165] the following statement.

If the Continuum Hypothesis holds, then there exist two σ-algebras $\mathcal{S}_1$ and $\mathcal{S}_2$ of subsets of $\mathbf{R}$, such that:

(1) the Borel σ-algebra of $\mathbf{R}$ is contained in $\mathcal{S}_1 \cap \mathcal{S}_2$;

(2) both σ-algebras $\mathcal{S}_1$ and $\mathcal{S}_2$ are countably generated;

(3) both σ-algebras $\mathcal{S}_1$ and $\mathcal{S}_2$ are invariant under the group Γ of all isometric transformations of $\mathbf{R}$;

(4) there exists a Γ-invariant measure μ_1 on $\mathcal{S}_1$ extending λ_0;

(5) there exists a Γ-invariant measure μ_2 on $\mathcal{S}_2$ extending λ_0;

(6) there is no nonzero diffused σ-finite measure defined on the σ-algebra generated by $\mathcal{S}_1 \cup \mathcal{S}_2$.

The proof of this statement given in [165] essentially utilizes the methods developed in [68] and [82]. Also, in [165] the question is posed whether the statement remains valid assuming Martin's Axiom instead of the Continuum Hypothesis.

By applying the method of Kodaira and Kakutani [111] and by using absolutely nonmeasurable homomorphisms from $\mathbf{R}$ into the one-dimensional unit torus, it will be demonstrated below that an answer to the question of Pelc and Prikry is positive (Theorem 2 of this chapter). Also, the reader will be able to see that our approach essentially differs from the one presented in [165] and leads to a much stronger result in terms of absolutely nonmeasurable additive functions.

Let $\mathbf{T}$ denote the one-dimensional unit torus in the plane $\mathbf{R}^2 = \mathbf{R} \times \mathbf{R}$. Actually, this torus is defined by the equality

$$\mathbf{T} = \{(x, y) \in \mathbf{R}^2 : x^2 + y^2 = 1\}$$

and therefore coincides with the unit circumference in $\mathbf{R}^2$.

We will consider $\mathbf{T}$ as a commutative compact topological group with respect to the natural group operation and topology (the latter is induced by the standard Euclidean topology on $\mathbf{R}^2$).

The group operation in $\mathbf{T}$ will be denoted by $+$ and, accordingly, the neutral element in $\mathbf{T}$ will be denoted by 0.

Being a compact topological group, $\mathbf{T}$ is equipped with the Haar probability measure which will be denoted by ν. In fact, the completion of ν coincides with the standard Lebesgue measure on $\mathbf{T}$.

Further, the group $\mathbf{T}$ is divisible, i.e., for each element $t \in \mathbf{T}$ and for any natural number $n > 0$, there exists an element $z \in \mathbf{T}$ such that

$$t = nz.$$

Some nontrivial subgroups of $\mathbf{T}$ are divisible, too. In particular, consider the subgroup G of $\mathbf{T}$ consisting of all those elements from $\mathbf{T}$ which have finite order; in other words, put

$$G = \{t \in \mathbf{T} : (\exists n < \omega)(n \neq 0 \ \& \ nt = 0)\}.$$

It can easily be verified that G is infinite, countable and divisible.

Exercise 2. Let $(U, +)$ be an arbitrary commutative group and let $(V, +)$ be a divisible commutative group. Suppose that a partial homomorphism

$$\phi : U \to V$$

is given (i.e., ϕ is a homomorphism from some subgroup of U into V). By using the Zorn Lemma, show that ϕ is extendible to a homomorphism from U into V.

Exercise 3. Let $(U, +)$ be an arbitrary commutative group and let W be a divisible subgroup of U. Applying the result of the previous exercise, prove that W is a direct summand in U. In other words, demonstrate that U admits a representation in the form

$$U = W + W' \quad (W \cap W' = \{0\})$$

for some subgroup W' of U (in general, W' is not uniquely determined).

Show also that any direct summand in a divisible commutative group is divisible, too.

By virtue of Exercise 3, the group $\mathbf{T}$ is representable in the form:

$$\mathbf{T} = G + H \quad (G \cap H = \{0\}),$$

where G is the countable group of all those elements in $\mathbf{T}$ that have finite order, and H is a complemented subgroup of $\mathbf{T}$.

It can readily be checked that the following relations are satisfied:

(a) H is a vector space (over $\mathbf{Q}$) isomorphic to $\mathbf{R}$;

(b) H is a ν-thick subset of $\mathbf{T}$, i.e., $\nu^*(H) = 1$;

(c) H is a second category subset of $\mathbf{T}$; moreover, H is thick in the sense of category (i.e., H intersects each second category subset of $\mathbf{T}$ possessing the Baire property).

Exercise 4. Verify the validity of relations (a), (b), and (c).

Taking the above-mentioned facts into account, we come to the following auxiliary proposition.

Lemma 1. *Under Martin's Axiom, there exists a set L in $\mathbf{T}$ such that:*
(1) $L \subset H$;
(2) L is a generalized Luzin subset of $\mathbf{T}$;
(3) L is a vector space over $\mathbf{Q}$.

Proof. Indeed, by using the standard argument (see the proof of Theorem 1 and Exercise 20 from Chapter 10), a generalized Luzin set $L \subset \mathbf{T}$ can be constructed in such a manner that all points of L would be in H and L would be a vector space over $\mathbf{Q}$. We omit the details of this construction (which are not difficult). Let us only remark that, because H is a thick subspace of $\mathbf{T}$ in the sense of category, our L is a generalized Luzin subset of H as well.

The next auxiliary statement plays the key role for our further considerations.

Lemma 2. *Under Martin's Axiom, there exist two functions*

$$\phi : \mathbf{R} \to H, \quad \psi : \mathbf{R} \to H,$$

satisfying the following conditions:
 (1) ϕ and ψ are homomorphisms of vector spaces (over $\mathbf{Q}$);
 (2) the graph of ϕ is $(\lambda \times \nu)$-thick in the product space $\mathbf{R} \times \mathbf{T}$;
 (3) the graph of ψ is $(\lambda \times \nu)$-thick in the product space $\mathbf{R} \times \mathbf{T}$;
 (4) $\phi + \psi$ is an isomorphism between $\mathbf{R}$ and L, where L is a generalized Luzin set of Lemma 1.

Proof. Let α denote the least ordinal of cardinality continuum and let $\{Z_\xi : \xi < \alpha\}$ be the family of all Borel subsets of $\mathbf{R} \times \mathbf{T}$ with strictly positive $(\lambda \times \nu)$-measure. Moreover, fix some partition $\{\Xi_1, \Xi_2, \Xi_3\}$ of $[0, \alpha[$ into three sets of cardinality continuum and assume that every Borel subset of $\mathbf{R} \times \mathbf{T}$ whose $(\lambda \times \nu)$-measure is strictly positive, belongs to each of the partial families:

$$\{Z_\xi : \xi \in \Xi_1\}, \quad \{Z_\xi : \xi \in \Xi_2\}.$$

Let L be as in Lemma 1. Pick a Hamel basis $\{l_\xi : \xi < \alpha\}$ of L. By using the method of transfinite recursion, it is not hard to construct three α-sequences

$$\{x_\xi : \xi < \alpha\}, \quad \{y_\xi : \xi < \alpha\}, \quad \{y'_\xi : \xi < \alpha\},$$

satisfying the relations:
 (a) $\{x_\xi : \xi < \alpha\}$ is a Hamel basis of $\mathbf{R}$;
 (b) $(x_\xi, y_\xi) \in Z_\xi$ for any ordinal $\xi \in \Xi_1$;
 (c) $(x_\xi, y'_\xi) \in Z_\xi$ for any ordinal $\xi \in \Xi_2$;
 (d) $\{y_\xi, y'_\xi\} \subset H$ and $y_\xi + y'_\xi = l_\xi$ for each ordinal $\xi < \alpha$.
Now, we define

$$\phi(x_\xi) = y_\xi, \quad \psi(x_\xi) = y'_\xi \quad (\xi < \alpha).$$

In view of the linear independence of $\{x_\xi : \xi < \alpha\}$, the functions ϕ and ψ can uniquely be extended to homomorphisms

$$\phi : \mathbf{R} \to H, \quad \psi : \mathbf{R} \to H,$$

which also determine the homomorphism

$$\phi + \psi : \mathbf{R} \to H.$$

Since we have

$$(\phi + \psi)(x_\xi) = l_\xi$$

for each $\xi < \alpha$ and $\{l_\xi : \xi < \alpha\}$ is a Hamel basis of L, we claim that $\phi + \psi$ is an isomorphism between $\mathbf{R}$ and L. Finally, by virtue of relations (b) and (c), it is clear that the graphs of ϕ and ψ are $(\lambda \times \nu)$-thick in the product space $\mathbf{R} \times \mathbf{T}$.

This ends the proof of Lemma 2.

Let $f : \mathbf{R} \to \mathbf{T}$ be an arbitrary group homomorphism whose graph is $(\lambda \times \nu)$-thick in $\mathbf{R} \times \mathbf{T}$. Then f can be made measurable with respect to an appropriate invariant extension of λ_0 (see, for instance, the well-known article by Kodaira and Kakutani [111]). Indeed, for each set Z belonging to $dom(\lambda_0 \times \nu)$, denote

$$Z' = \{x \in \mathbf{R} \; : \; (x, f(x)) \in Z\}$$

and consider the family of sets

$$\mathcal{S}' = \{Z' \; : \; Z \in dom(\lambda_0 \times \nu)\}.$$

It is not difficult to verify that $\mathcal{S}'$ is a σ-algebra of subsets of $\mathbf{R}$, containing $dom(\lambda_0)$ and invariant under the group Γ of all isometric transformations of $\mathbf{R}$. Also, we can define a functional μ on $\mathcal{S}'$ by putting

$$\mu(Z') = (\lambda_0 \times \nu)(Z) \qquad (Z' \in \mathcal{S}').$$

It turns out that μ is a measure on $\mathcal{S}'$ extending λ_0 and invariant under Γ (cf. [97], [103], [111]). An easy verification of this fact is left to the reader. Besides, the definition of μ directly implies that the original homomorphism f is measurable with respect to μ.

Exercise 5. Denote by μ' the completion of the above-mentioned measure μ. Show that μ' possesses the uniqueness property, i.e., for any σ-finite Γ-invariant measure θ with $dom(\theta) = dom(\mu')$, there exists a real coefficient $t \geq 0$ such that

$$\theta = t \cdot \mu'.$$

In other words, any σ-finite Γ-invariant measure defined on the domain of μ' is proportional to μ'.

In particular, we see that there are Γ-invariant strong extensions of λ possessing the uniqueness property (similarly to λ). Moreover, it is known that there are even nonseparable Γ-invariant extensions of λ with the same property (in this connection, see [97] and [103]).

Theorem 1. *Assuming Martin's Axiom, there exist group homomorphisms*

$$f_1 \; : \; \mathbf{R} \to \mathbf{T}, \qquad f_2 \; : \; \mathbf{R} \to \mathbf{T}$$

and measures μ_1 and μ_2 on $\mathbf{R}$, such that:
  (1) μ_1 extends λ_0 and is invariant under Γ;
  (2) μ_2 extends λ_0 and is invariant under Γ;
  (3) f_1 is measurable with respect to μ_1;
  (4) f_2 is measurable with respect to μ_2;
  (5) the homomorphism $f_1 + f_2$ is absolutely nonmeasurable.

Proof. It suffices to put

$$f_1 = \phi, \quad f_2 = \psi,$$

where ϕ and ψ are as in Lemma 2. Because the graphs of ϕ and ψ are $(\lambda \times \nu)$-thick subsets of $\mathbf{R} \times \mathbf{T}$, they determine the corresponding Γ-invariant extensions μ_1 and μ_2 of the Borel measure λ_0. At the same time, the group homomorphism $f_1 + f_2$ is injective and its range is a generalized Luzin subset of $\mathbf{T}$. This yields at once that $f_1 + f_2$ is an absolutely nonmeasurable function (cf. Theorem 6 from Chapter 10).

This ends the proof of our statement.

Theorem 2. *Assume again Martin's Axiom. Then there exist two σ-algebras $\mathcal{S}_1$ and $\mathcal{S}_2$ of subsets of $\mathbf{R}$, satisfying the following relations:*
  (1) the Borel σ-algebra of $\mathbf{R}$ is contained in $\mathcal{S}_1 \cap \mathcal{S}_2$;
  (2) both σ-algebras $\mathcal{S}_1$ and $\mathcal{S}_2$ are countably generated;
  (3) both σ-algebras $\mathcal{S}_1$ and $\mathcal{S}_2$ are invariant under the group Γ of all motions of $\mathbf{R}$;
  (4) there exists a Γ-invariant measure μ_1 on $\mathcal{S}_1$ extending λ_0;
  (5) there exists a Γ-invariant measure μ_2 on $\mathcal{S}_2$ extending λ_0;
  (6) there is no nonzero σ-finite diffused measure on the σ-algebra generated by $\mathcal{S}_1 \cup \mathcal{S}_2$.

Proof. Let $\mathcal{B}(\mathbf{R})$ denote, as usual, the Borel σ-algebra of $\mathbf{R}$ and let $\mathcal{B}(\mathbf{R} \times \mathbf{T})$ denote the Borel σ-algebra of $\mathbf{R} \times \mathbf{T}$. Take the homomorphisms f_1 and f_2 of Theorem 1 and define:
  $\mathcal{S}_1 = \{\{x : (x, f_1(x)) \in B\} : B \in \mathcal{B}(\mathbf{R} \times \mathbf{T})\}$;
  $\mathcal{S}_2 = \{\{x : (x, f_2(x)) \in B\} : B \in \mathcal{B}(\mathbf{R} \times \mathbf{T})\}$.
A simple argument shows that these σ-algebras are the required ones.

Indeed, relations (1)–(5) are verified directly in view of Theorem 1. It remains to check the validity of relation (6).

Suppose, to the contrary, that there exists a nonzero σ-finite diffused measure μ on the σ-algebra $\mathcal{S}$ generated by the family $\mathcal{S}_1 \cup \mathcal{S}_2$ (note that $\mathcal{S}$ is also countably generated and invariant under the group of all motions of $\mathbf{R}$). Because the homomorphism f_1 is measurable with respect to $\mathcal{S}_1$ and the homomorphism f_2 is measurable with respect to $\mathcal{S}_2$, we claim that both

f_1 and f_2 are measurable with respect to $\mathcal{S}$ (or, equivalently, with respect to μ).

Consequently, the homomorphism $f_1 + f_2$ must be measurable with respect to μ, too, which contradicts the absolute nonmeasurability of $f_1 + f_2$ (see relation (5) of Theorem 1). The contradiction obtained finishes the proof.

Remark 3. One generalization of Theorem 2 for certain measure type functionals was obtained in [102].

Remark 4. Under Martin's Axiom, Theorems 1 and 2 can be generalized to the case of the n-dimensional Euclidean space $\mathbf{R}^n$ $(n \geq 1)$ equipped with the group Γ_n which is generated by the family of all central symmetries of $\mathbf{R}^n$.

Remark 5. In fact, we do not need the full power of Martin's Axiom for obtaining Theorems 1 and 2. It suffices to utilize the corresponding properties of generalized Luzin sets, which are implied by this axiom.

Exercise 6. Let E be a set and let

$$f : E \to \mathbf{R}$$

be a function. Show that the following two assertions are equivalent:
(a) f is absolutely nonmeasurable;
(b) the set $ran(f)$ is universal measure zero and

$$card(f^{-1}(t)) \leq \omega$$

for all points $t \in \mathbf{R}$.

Deduce from this equivalence that an absolutely nonmeasurable function on E exists if and only if there is a universal measure zero subset X of $\mathbf{R}$ with $card(X) = card(E)$. In particular, if $card(E) > \mathbf{c}$, then no real-valued function on E is absolutely nonmeasurable.

The following example is relevant.

Example 3. Let μ be a nonzero σ-finite measure on $\mathbf{R}$ and let

$$f : \mathbf{R} \to \mathbf{R}$$

be a function such that, for some nonzero σ-finite Borel measure ν on $ran(f)$, the graph of f is $(\mu \times \nu)$-thick in the product set $\mathbf{R} \times ran(f)$. Then f is measurable with respect to an appropriate extension μ' of μ (hence, f is not absolutely nonmeasurable).

Indeed, we may suppose without loss of generality that ν is a Borel probability measure on $ran(f)$ and we may apply once more the method of Kodaira and Kakutani [111]. For each set $Z \in dom(\mu \times \nu)$, let us denote

$$Z' = \{x \in \mathbf{R} \ : \ (x, f(x)) \in Z\}.$$

Furthermore, introduce the family of sets

$$\mathcal{S}' = \{Z' \ : \ Z \in dom(\mu \times \nu)\}.$$

Again, it can easily be seen that $\mathcal{S}'$ is a σ-algebra of subsets of $\mathbf{R}$. In addition, if $X \in dom(\mu)$, then we have

$$X \times ran(f) \in dom(\mu \times \nu),$$

$$X = \{x \in \mathbf{R} \ : \ (x, f(x)) \in X \times ran(f)\},$$

whence it follows that $X \in \mathcal{S}'$. Consequently,

$$dom(\mu) \subset \mathcal{S}'.$$

Now, for any set $Z \in dom(\mu \times \nu)$, let us put

$$\mu'(Z') = (\mu \times \nu)(Z).$$

A straightforward verification shows that the functional μ' is well defined (by virtue of the thickness of the graph of our function f) and that μ' is a measure on $\mathcal{S}'$ extending the initial measure μ. The definition of μ' also implies that f is measurable with respect to μ'.

In particular, if the graph of a function

$$f \ : \ \mathbf{R} \to \mathbf{R}$$

is $(\lambda \times \lambda)$-thick in the plane $\mathbf{R}^2$ (for examples of such functions, see, e.g., [64] or Theorem 5 from Chapter 8), then f can be made measurable with respect to an appropriate extension of λ.

In this context, the following question naturally arises:

Let μ be a measure on $\mathbf{R}$ extending λ and such that there exists a function acting from $\mathbf{R}$ into $\mathbf{R}$, whose graph is $(\mu \times \lambda)$-thick in the plane $\mathbf{R}^2$. Does there exist a group homomorphism from $\mathbf{R}$ into $\mathbf{R}$ whose graph is also $(\mu \times \lambda)$-thick in $\mathbf{R}^2$?

We do not know an answer to this question.

Exercise 7. Let S be an equivalence relation on $\mathbf{R}$ whose all equivalence classes are at most countable. We shall say that a mapping

$$f \; : \; \mathbf{R} \to \mathbf{R}$$

is a Vitali type function for S if $(r, f(r)) \in S$ for each $r \in \mathbf{R}$ and the set $ran(f)$ is a selector of the partition of $\mathbf{R}$ canonically determined by S.

The Vitali theorem (see Chapter 8) states that if V is the classical Vitali equivalence relation on $\mathbf{R}$, i.e.,

$$(x, y) \in V \Leftrightarrow x - y \in \mathbf{Q},$$

then every Vitali type function for V is nonmeasurable with respect to any translation-invariant extension of λ.

Verify that there are additive Vitali type functions for V. Conclude that there exist solutions of the Cauchy functional equation absolutely nonmeasurable with respect to the class of all translation-invariant extensions of λ.

On the other hand, prove that if f is an arbitrary Vitali type function for V, then f is measurable with respect to some measure on $\mathbf{R}$ extending λ.

Exercise 8. Assume Martin's Axiom. Demonstrate that there exist a generalized Luzin set $X \subset \mathbf{R}$ and an equivalence relation $S \subset \mathbf{R} \times \mathbf{R}$ satisfying the following conditions:
  (a) $card(S(r)) = \omega$ for any $r \in \mathbf{R}$;
  (b) X is a selector of the partition $\{S(r) : r \in \mathbf{R}\}$ of $\mathbf{R}$.
Let $h : \mathbf{R} \to \mathbf{R}$ be a Vitali type function for S such that

$$ran(h) = X.$$

Verify that the function h is absolutely nonmeasurable.

Claim from this fact that the validity of the result presented in Exercise 7 is essentially based on specific properties of the Vitali partition of $\mathbf{R}$.

In connection with Exercises 1, 7, and 8, let us note that the following question remains open:

Does there exist a Vitali type function for V nonmeasurable with respect to all measures on $\mathbf{R}$ extending λ and quasiinvariant under the group of all translations of $\mathbf{R}$?

Exercise 9. Prove that there exists a function

$$g : \mathbf{R} \to \mathbf{R}$$

having the following property: for any σ-finite diffused Borel measure μ on $\mathbf{R}$ and for any σ-finite measure ν on $\mathbf{R}$, the graph of g is a $(\mu \times \nu)$-thick subset of the plane $\mathbf{R} \times \mathbf{R}$ (for this purpose, consider a partition $\{X_t : t \in \mathbf{R}\}$ of $\mathbf{R}$ such that all X_t $(t \in \mathbf{R})$ are Bernstein sets in $\mathbf{R}$).

Try to construct a solution of the Cauchy functional equation, possessing the same property.

Exercise 10. Let $E \neq \{0\}$ be a separable Banach space. Demonstrate that E admits a representation in the form:

$$E = X_1 + X_2 \qquad (X_1 \cap X_2 = \{0\}),$$

where
(a) both X_1 and X_2 are vector spaces over $\mathbf{Q}$;
(b) both X_1 and X_2 are Bernstein subsets of E.

Applying such a representation, show that there exists a group homomorphism

$$h : E \to \mathbf{T}$$

possessing the following property: for any σ-finite diffused Borel measure μ on E, the graph of h is a $(\mu \times \nu)$-thick subset of the product group $E \times \mathbf{T}$ (here ν denotes the Haar probability measure on $\mathbf{T}$).

Infer from this property that:
(c) for every σ-finite diffused Borel measure μ on E, the function h becomes measurable with respect to an appropriate extension μ' of μ;
(d) if an initial σ-finite diffused Borel measure μ on E is invariant (quasi-invariant) under a subgroup G of E, then the measure μ' is invariant (quasi-invariant) under the same group G.

The results formulated in Exercises 9 and 10 strengthen some statements given in Chapter 8 (see, for instance, Theorem 5 from that chapter).

Exercise 11. Assuming Martin's Axiom, prove that $\mathbf{R}$ admits a representation in the form:

$$\mathbf{R} = Y_1 + Y_2 \qquad (Y_1 \cap Y_2 = \{0\}),$$

where
(a) both Y_1 and Y_2 are vector spaces over $\mathbf{Q}$;
(b) both Y_1 and Y_2 are generalized Luzin sets (respectively, generalized Sierpiński sets) in $\mathbf{R}$.

Deduce from this fact that there exist two generalized Luzin subsets L_1 and L_2 of $\mathbf{R}$ whose algebraic sum $L_1 + L_2$ is a Bernstein subset of $\mathbf{R}$.

We thus see that (under Martin's Axiom) there are two universal measure zero sets in $\mathbf{R}$ (even two strong measure zero sets in $\mathbf{R}$) whose algebraic sum is absolutely nonmeasurable with respect to the class of the completions of all nonzero σ-finite diffused Borel measures on $\mathbf{R}$.

Let us underline that the last statement can be established only under some additional set-theoretical assumptions (because there exist models of **ZFC** in which $\omega_1 < \mathbf{c}$ and the cardinalities of all universal measure zero subsets of $\mathbf{R}$ do not exceed ω_1).

On the other hand, prove in **ZFC** that there is a Lebesgue measure zero set $X \subset \mathbf{R}$ such that $X + X$ is not measurable in the Lebesgue sense (use the technique of Hamel bases; the existence of X was first obtained by Sierpiński [194]).

Exercise 12. Demonstrate (in **ZFC**) that $\mathbf{R}$ admits a representation in the form:
$$\mathbf{R} = X + Y \qquad (X \cap Y = \{0\}),$$
where
(a) both X and Y are vector spaces over $\mathbf{Q}$;
(b) both X and Y are of Lebesgue measure zero.
This result is due to Erdös, Kunen and Mauldin (see [53]).

Exercise 13. Show that there exists a subset Z of $\mathbf{R}$ satisfying the following relations:
(a) Z is of first category in $\mathbf{R}$;
(b) Z is of Lebesgue measure zero;
(c) for any countable family $\{h_i : i \in I\} \subset \mathbf{R}$, we have

$$\cap\{h_i + Z : i \in I\} \neq \emptyset.$$

Assume the Continuum Hypothesis. Starting with the above-mentioned properties of Z and applying the method of transfinite recursion, construct two subsets X and Y of $\mathbf{R}$ such that:
(d) both X and Y are vector spaces over $\mathbf{Q}$;
(e) $X + Y = \mathbf{R}$ and $X \cap Y = \{0\}$;
(f) both X and Y are of first cateory in $\mathbf{R}$ and have Lebesgue measure zero.

Exercise 14. We know that the real line $\mathbf{R}$ can be represented as $\mathbf{Q} + W$ where W is some vector space over $\mathbf{Q}$ and $\mathbf{Q} \cap W = \{0\}$. Actually, W is a Vitali subset of $\mathbf{R}$ and, simultaneously, a hyperplane in $\mathbf{R}$ regarded as a vector space over $\mathbf{Q}$ (see Chapters 8 and 9). We thus conclude that there exists an infinite countable (hence Borel) subgroup of $\mathbf{R}$ which is a direct summand in $\mathbf{R}$.

Suppose now that

$$\mathbf{R} = X + Y \qquad (X \cap Y = \{0\}),$$

where X and Y are some analytic subgroups of $\mathbf{R}$. Prove that either $X = \{0\}$ or $Y = \{0\}$.

The last exercise of this chapter shows that there are additive Sierpiński-Zygmund functions (cf. Chapter 7).

Exercise 15. Prove that there exists a function

$$f : \mathbf{R} \to \mathbf{R}$$

satisfying the following conditions:

(a) f is a solution of the Cauchy functional equation, i.e., we have

$$f(x + y) = f(x) + f(y)$$

for all reals $x \in \mathbf{R}$ and $y \in \mathbf{R}$;

(b) f is a Sierpiński-Zygmund function (consequently, f turns out to be a nontrivial solution of the Cauchy functional equation);

(c) f is measurable with respect to some $\mathbf{R}$-quasiinvariant extension of the Lebesgue measure λ (in particular, f is not absolutely nonmeasurable).

12. Egorov type theorems

It is well known that one of the earliest important results in real analysis and Lebesgue measure theory was obtained by Egorov [50] who discovered close relationships between the uniform convergence and the convergence almost everywhere of a sequence of real-valued Lebesgue measurable functions. This classical result (known now as the Egorov theorem) has numerous consequences and applications in analysis. For example, it suffices to mention that another classical result in real analysis, the so-called Luzin theorem on the structure of Lebesgue measurable functions, can easily be deduced by starting with the Egorov theorem.

Here we wish to discuss some aspects of the Egorov theorem and, in addition, to show that, for a sequence of nonmeasurable real-valued functions there is no hope of getting a reasonable analogue of this theorem. In other words, we are going to demonstrate in our further considerations that there are some sequences of rather strange real-valued functions, for which even very weak analogues of the Egorov type theorems fail to be true.

First of all, we want to present the Egorov theorem in a form slightly more general than those in which this theorem is usually formulated in the standard course of real analysis and classical measure theory. In order to do this, we need some auxiliary notions and facts.

Let E be a nonempty set and let $\mathcal{S}$ be some class of subsets of E, satisfying the following conditions:

(1) $\emptyset \in \mathcal{S}, \quad E \in \mathcal{S}$;

(2) $\mathcal{S}$ is closed under countable unions and countable intersections.

Suppose also that a functional

$$\nu : \mathcal{S} \to \mathbf{R}$$

is given, such that:

(a) for every increasing (with respect to inclusion) sequence of sets

$$\{X_n : n < \omega\} \subset \mathcal{S},$$

we have

$$\nu(\cup\{X_n : n < \omega\}) \leq sup\{\nu(X_n) : n < \omega\};$$

(b) for every decreasing (with respect to inclusion) sequence of sets

$$\{Y_n \ : \ n < \omega\} \subset \mathcal{S},$$

we have

$$\nu(\cap\{Y_n \ : \ n < \omega\}) \geq inf\{\nu(Y_n) \ : \ n < \omega\}.$$

In this case, we say that $\mathcal{S}$ is an admissible class of subsets of E and ν is an admissible functional on $\mathcal{S}$.

Note that, in analysis, there are many natural examples of admissible functionals. This can be confirmed by the following standard example.

Example 1. Let E be a nonempty set and let $\mathcal{S}$ be some σ-algebra of subsets of E. Then it is obvious that $\mathcal{S}$ is an admissible class. Suppose, in addition, that ν is a finite measure on $\mathcal{S}$. Then it can easily be observed that ν is an admissible functional on E. Actually, in this case, the inequalities of (a) and (b) are reduced to the equalities.

Exercise 1. Give an example of an admissible functional on E, which is not a measure on E.

Similarly to the concept of measurability of real-valued functions with respect to ordinary measures, the concept of measurability of real-valued functions with respect to admissible functionals may be introduced and investigated. Namely, we say that a function

$$f \ : \ E \to \mathbf{R}$$

is measurable with respect to an admissible functional ν on E (or, simply, f is ν-measurable) if, for each open interval $]a, b[\ \subset \mathbf{R}$, we have

$$f^{-1}(]a, b[) \in dom(\nu).$$

Obviously, the same definition can be introduced for partial functions acting from E into $\mathbf{R}$.

The properties of functions (partial functions) measurable with respect to admissible functionals turn out to be rather similar to the properties of functions (partial functions) measurable in the usual sense. The next two exercises vividly illustrate this fact.

Exercise 2. Let ν be an admissible functional on E, let

$$f \ : \ E \to \mathbf{R},$$

$$g \ : \ E \to \mathbf{R}$$

be any two ν-measurable functions and let $t \in \mathbf{R}$. Show that the functions

$$tf, \quad f + g, \quad f \cdot g$$

are ν-measurable, too.

In addition, show that if $g(x) \neq 0$ for all $x \in E$, then the function f/g is also ν-measurable.

Finally, let X be an arbitrary set from $dom(\nu)$. Check that the restriction $f|X$ is a ν-measurable partial function from E into $\mathbf{R}$.

Exercise 3. Let ν be an admissible functional given on a set E and let $\{f_n : n < \omega\}$ be a sequence of ν-measurable functions, pointwise convergent on E. Let us denote

$$f = lim_{n \to +\infty} f_n.$$

Demonstrate that the function f is also measurable with respect to ν.

Now, we are able to formulate and prove a direct analogue of the Egorov theorem for sequences of real-valued functions measurable with respect to an admissible functional.

Theorem 1. *Let E be a basic set and let ν be an admissible functional on E. Suppose, in addition, that a sequence $\{f_n : n < \omega\}$ of ν-measurable functions is given, pointwise convergent on E, and let f denote the corresponding limit function. Then, for each $\varepsilon > 0$, there exists a set $X \in dom(\nu)$ satisfying these two relations:*

(1) $\nu(E) - \nu(X) \leq \varepsilon$;

(2) the sequence of functions $\{f_n|X : n < \omega\}$ converges uniformly to the function $f|X$.

Proof. For any natural number m, let us denote

$$E_{0,m} = \{x \in E : (\forall n \geq m)(|f_n(x) - f(x)| < 1)\}.$$

It is easy to check that the set $E_{0,m}$ belongs to $dom(\nu)$ and

$$E_{0,0} \subset E_{0,1} \subset ... \subset E_{0,m} \subset ...,$$

$$\cup\{E_{0,m} : m < \omega\} = E.$$

Consequently, there exists a natural index m_0 such that

$$\nu(E) - \nu(E_{0,m_0}) < \varepsilon.$$

Let us put
$$X_0 = E_{0,m_0}.$$

Further, for each $m < \omega$, consider the set

$$E_{1,m} = \{x \in X_0 \ : \ (\forall n \geq m)(|f_n(x) - f(x)| < 1/2)\}.$$

Evidently, $E_{1,m}$ belongs to $dom(\nu)$ and

$$E_{1,0} \subset E_{1,1} \subset ... \subset E_{1,m} \subset ...,$$

$$\cup \{E_{1,m} \ : \ m < \omega\} = X_0.$$

Consequently, there exists a natural index m_1 such that

$$\nu(E) - \nu(E_{1,m_1}) < \varepsilon.$$

Let us put
$$X_1 = E_{1,m_1}.$$

Continuing in this manner, we will be able to define by recursion a certain sequence $\{X_k \ : \ k < \omega\}$ of sets belonging to $dom(\nu)$ and satisfying the relations:

(i) $X_0 \supset X_1 \supset ... \supset X_k \supset ...$;

(ii) for any $k < \omega$, we have

$$\nu(E) - \nu(X_k) < \varepsilon; \ \cdot$$

(iii) for any $k < \omega$, there is a natural number m_k such that

$$(\forall n \geq m_k)(\forall x \in X_k)(|f_n(x) - f(x)| < 1/2^k).$$

Finally, we put
$$X = \cap\{X_k \ : \ k < \omega\}.$$

Then, by virtue of the definition of an admissible functional, we get

$$\nu(E) - \nu(X) \leq \varepsilon,$$

and it can easily be verified that the sequence of the restricted functions

$$\{f_n | X \ : \ n < \omega\}$$

converges uniformly to the restricted function $f|X$.

This completes the proof of Theorem 1.

Obviously, the theorem presented above immediately implies the classical Egorov theorem [50]. It suffices to take as ν an arbitrary finite measure on E. In this connection, let us recall that, for σ-finite measures, a direct analogue of the Egorov theorem is not true in general.

Exercise 4. Let λ denote, as usual, the standard Lebesgue measure on the real line $\mathbf{R}$. Give an example of a sequence $\{f_n : n < \omega\}$ of real-valued uniformly bounded λ-measurable (even continuous) functions on $\mathbf{R}$, which is convergent everywhere on $\mathbf{R}$ but there exists no unbounded subset X of $\mathbf{R}$ such that the sequence of the restricted functions $\{f_n|X : n < \omega\}$ is uniformly convergent on X.

Exercise 5. Let E be a normal topological space and let μ be a finite inner regular Borel measure on E, i.e., for each Borel subset Y of E, we have the equality

$$\mu(Y) = sup\{\mu(F) : F \subset Y \ \& \ F \text{ is closed in } E\}.$$

We denote by μ' the usual completion of μ. Let

$$f : E \to \mathbf{R}$$

be an arbitrary μ'-measurable function. By applying the Tietze-Urysohn theorem on the existence of a continuous extension of a continuous real-valued function defined on any closed subset of E, show that, for each $\varepsilon > 0$, there exists a continuous function

$$g : E \to \mathbf{R}$$

satisfying the relation

$$\mu'(\{x \in E : |f(x) - g(x)| \geq \varepsilon\}) < \varepsilon.$$

Deduce from this fact that, for any μ'-measurable function

$$\phi : E \to \mathbf{R},$$

there exists a sequence $\{\phi_n : n < \omega\}$ of continuous real-valued functions on E, convergent to ϕ almost everywhere (with respect to μ').

Exercise 6. Let E be again a normal topological space, let μ be a finite inner regular Borel measure on E and let μ' denote the completion of μ. By starting with the Egorov theorem and applying the result of Exercise 5, prove the following Luzin type theorem: for any μ'-measurable function

$$f : E \to \mathbf{R}$$

and for each real $\varepsilon > 0$, there exists a continuous function

$$g \; : \; E \to \mathbf{R}$$

such that

$$\mu'(\{x \in E \; : \; f(x) \neq g(x)\}) < \varepsilon.$$

The above relation just expresses that the given function f has the so-called (C)-property of Luzin. It is frequently said that all μ'-measurable functions possess this property (and the converse assertion is true, too).

Let us now return to the Egorov theorem. In conformity with it, any convergent sequence of measurable real-valued functions converges uniformly on some large measurable subset of E ("large" means here that the measure of the complement of this subset may be taken arbitrarily small).

In particular, if a given finite Borel measure on a topological space E is nonzero, diffused and inner regular, then we immediately obtain that every convergent sequence of measurable real-valued functions on E converges uniformly on an uncountable closed subset of E. Hence, if E is an uncountable Polish topological space equipped with a nonzero finite diffused Borel measure, then, for any convergent sequence of measurable real-valued functions on E, there exists a nonempty compact perfect subset of E (actually, a subset homeomorphic to the Cantor discontinuum) on which the sequence converges uniformly.

In connection with these observations, it makes sense to consider the following more general situation.

Let E be an arbitrary uncountable complete metric space without isolated points and let $\{f_n \; : \; n < \omega\}$ be a sequence of real-valued Borel functions on E, such that, for some constant $d \geq 0$, we have

$$(\forall n < \omega)(\forall x \in E)(|f_n(x)| \leq d);$$

in other words, our sequence of functions is uniformly bounded. Do there exist a nonempty perfect compact subset P of E and an infinite subset K of ω, for which the partial sequence of functions $\{f_n|P \; : \; n \in K\}$ is uniformly convergent on P?

Evidently, we may restrict our considerations to the case where the given space E is homeomorphic to the Cantor discontinuum (because, according to the well-known theorem from general topology, every complete metric space without isolated points contains a homeomorphic image of the Cantor discontinuum). Actually, we may suppose from the beginning that our E is an uncountable Polish topological space.

Also, in order to get a positive solution to the question formulated above, it suffices to demonstrate that there exist an infinite subset K of ω and a nonempty perfect compact subset P' of E, such that the sequence

$$\{f_n|P' \ : \ n \in K\}$$

converges pointwise on P'. Indeed, suppose that this fact has already been established and equip the set P' with some Borel diffused probability measure μ. Then, evidently, we may apply the Egorov theorem to μ and to the sequence of functions $\{f_n|P' \ : \ n \in K\}$. In accordance with this theorem, there exists a Borel set $X \subset P'$ with $\mu(X) > 1/2$, for which the sequence of functions $\{f_n|X \ : \ n \in K\}$ converges uniformly. Since μ is diffused and $\mu(X) > 0$, we obviously obtain the relation

$$card(X) \geq \omega_1$$

and, consequently,

$$card(X) = \mathbf{c}$$

because X is Borel in P'. It is clear now that X contains a nonempty perfect compact subset P for which the sequence of functions $\{f_n|P \ : \ n \in K\}$ converges uniformly, too.

Mazurkiewicz was the first mathematician to prove that, for any uniformly bounded sequence of real-valued Borel functions given on an uncountable Polish space E, there exists a subset of E homeomorphic to the Cantor discontinuum, on which some subsequence of the sequence converges uniformly (see his remarkable work [144]).

In order to present a detailed proof of this interesting result, we need some auxiliary notions and simple facts.

Let E be an uncountable Polish space and let Φ be a family of real-valued functions defined on E.

We shall say that the family Φ is semicompact if, for each sequence

$$\{\phi_n \ : \ n < \omega\} \subset \Phi$$

and for each nonempty perfect set $P \subset E$, there exist an infinite subset K of ω and a nonempty perfect set P' contained in P, such that the partial sequence of functions $\{\phi_n \ : \ n \in K\}$ converges pointwise on P'.

We shall say that a family $\mathcal{S}$ consisting of some Borel subsets of E is semicompact if the corresponding family of characteristic functions

$$\{f_X \ : \ X \in \mathcal{S}\}$$

is semicompact in the sense of the definition above.

The following auxiliary statement yields a much more vivid description of semicompact families of Borel sets in E.

Lemma 1. *Let S be a family of Borel subsets of an uncountable Polish space E. Then these two assertions are equivalent:*

(1) the family S is semicompact;

(2) for any sequence $\{X_n \ : \ n < \omega\}$ of sets from S and for each nonempty perfect subset P of E, there exists an infinite set $K \subset \omega$ such that either

$$card(\cap\{X_n \ : \ n \in K\} \cap P) > \omega$$

or

$$card(\cap\{E \setminus X_n \ : \ n \in K\} \cap P) > \omega.$$

Proof. Note first that the implication $(2) \Rightarrow (1)$ is almost trivial because if, for example, we have

$$card(\cap\{X_n \ : \ n \in K\} \cap P) > \omega$$

for a nonempty perfect set $P \subset E$ and for some infinite subset K of ω, then the set

$$\cap\{X_n \ : \ n \in K\} \cap P$$

contains a nonempty perfect subset P', and the sequence of characteristic functions

$$\{f_{X_n} \ : \ n \in K\}$$

converges pointwise on the set P' to the characteristic function $f_{P'}$ (in fact, all f_{X_n} $(n \in K)$ are identically equal to 1 on P').

Now, let us establish the implication $(1) \Rightarrow (2)$. Suppose that relation (1) is fulfilled. Let $\{X_n \ : \ n < \omega\}$ be an arbitrary sequence of sets from S and let P be a nonempty perfect subset of E. We may assume, without loss of generality, that

$$P = E.$$

Then, according to (1), there exists an infinite subset K of ω such that the corresponding sequence of characteristic functions $\{f_{X_n} \ : \ n \in K\}$ is convergent on an uncountable Borel subset Y of E. Let us denote

$$f(y) = lim_{n \to +\infty, n \in K} \ f_{X_n}(y) \qquad (y \in Y).$$

Obviously, f is a Borel function on Y, and

$$ran(f) \subset \{0, 1\}.$$

Therefore, at least one of the sets

$$Y_0 = \{y \in Y \ : \ f(y) = 0\},$$

$$Y_1 = \{y \in Y \ : \ f(y) = 1\}$$

is uncountable. Suppose, for example, that $card(Y_1) > \omega$. Then, by taking account of the formula

$$Y_1 = Y_1 \cap limsup\{X_n \ : \ n \in K\} = Y_1 \cap liminf\{X_n \ : \ n \in K\},$$

it can easily be checked that, for some infinite subset K_1 of K, the inequality

$$card(\cap\{X_n \ : \ n \in K_1\}) > \omega$$

is satisfied. If $card(Y_0) > \omega$, then an analogous argument applied to the sequence of characteristic functions

$$\{f_{E \setminus X_n} \ : \ n \in K\}$$

yields the existence of an infinite subset K_0 of K for which the inequality

$$card(\cap\{E \setminus X_n \ : \ n \in K_0\}) > \omega$$

is fulfilled. This establishes the validity of the implication (1) $\Rightarrow$ (2) and finishes the proof of Lemma 1.

The next two auxiliary propositions also are not hard to prove.

Lemma 2. *Let Φ be a semicompact family of real-valued functions defined on an uncountable Polish space E. Then, for any real number $d \geq 0$, the family of functions*

$$\{t\phi \ : \ |t| \leq d, \ \phi \in \Phi\}$$

is semicompact, too.

Lemma 3. *Let Φ_1 and Φ_2 be any two semicompact families of real-valued functions defined on an uncountable Polish space E. Then the family of functions*

$$\{\phi_1 + \phi_2 \ : \ \phi_1 \in \Phi_1, \ \phi_2 \in \Phi_2\}$$

is semicompact, too.

Exercise 7. Give the detailed proofs of Lemma 2 and Lemma 3.

It immediately follows from these lemmas (by using the method of induction) that if $d \geq 0$ and Φ_1, Φ_2,..., Φ_k are some semicompact families of real-valued functions on an uncountable Polish space E, then the family of all those functions ϕ which can be represented in the form

$$\phi = t_1\phi_1 + t_2\phi_2 + ... + t_k\phi_k,$$

where

$$|t_1| \leq d, \quad |t_2| \leq d, \quad ... \quad , \quad |t_k| \leq d,$$

$$\phi_1 \in \Phi_1, \quad \phi_2 \in \Phi_2, \quad ... \quad , \quad \phi_k \in \Phi_k,$$

is also semicompact.

Lemma 4. *Let Φ be a semicompact family of bounded real-valued functions on an uncountable Polish space E, and let Φ^* denote the family of all those functions which are uniform limits of sequences of functions belonging to Φ (in other words, Φ^* is the closure of Φ with respect to the topology of uniform convergence on E). Then the family Φ^* is semicompact, too.*

Proof. Let $\{\phi_n^* \ : \ n < \omega\}$ be an arbitrary sequence of functions from the family Φ^*. In virtue of the definition of Φ^*, for every natural number n, there exists a function $\phi_n \in \Phi$ such that

$$||\phi_n^* - \phi_n|| \leq 1/(n+1).$$

Let us consider the family of functions $\{\phi_n \ : \ n < \omega\}$. According to our assumption, Φ is semicompact. Hence, for each nonempty perfect set $P \subset E$, there exist an infinite subset K of ω and a nonempty perfect subset P' of P, such that the partial sequence of functions

$$\{\phi_n|P' \ : \ n \in K\}$$

converges pointwise on P' to some function ϕ defined on P'. Now, it is easily verified that the corresponding sequence of functions

$$\{\phi_n^*|P' \ : \ n \in K\}$$

also converges pointwise to ϕ. This completes the proof of Lemma 4.

The following auxiliary statement plays the key role in our further considerations.

Lemma 5. *The family of all Borel subsets of an uncountable Polish topological space E is semicompact.*

Proof. Let $\{X_n \ : \ n < \omega\}$ be an arbitrary sequence of Borel subsets of E and let P be a nonempty perfect set in E. Denote by $2^{<\omega}$ the family of all finite sequences whose terms belong to the set $2 = \{0, 1\}$. We are going to construct (by recursion) a dyadic family

$$\{P_\sigma \ : \ \sigma \in 2^{<\omega}\}$$

of nonempty perfect sets in E and a sequence $\{n_k \ : \ k < \omega\}$ of natural numbers, such that:

(a) $P_\emptyset$ is contained in P;

(b) for any $\sigma \in 2^{<\omega}$, we have

$$P_{\sigma 0} \cup P_{\sigma 1} \subset P_\sigma,$$

$$P_{\sigma 0} \cap P_{\sigma 1} = \emptyset;$$

(c) for any nonempty $\sigma \in 2^{<\omega}$, we have

$$diam(P_\sigma) \leq 1/(lh(\sigma))$$

where the symbol $lh(\sigma)$ denotes the length of σ;

(d) the sequence $\{n_k \ : \ k < \omega\}$ is strictly increasing;

(e) for each nonzero $k < \omega$, the inclusion

$$\cup\{P_\sigma \ : \ lh(\sigma) = k\} \subset X_{n_1} \cap X_{n_2} \cap ... \cap X_{n_k}$$

is fulfilled.

Pick a nonempty perfect set $P_\emptyset \subset P$ and a natural number n_0 arbitrarily. Suppose now that the natural numbers

$$n_0 < n_1 < \ ... \ < n_k$$

and the partial family

$$\{P_\sigma \ : \ \sigma \in 2^{<\omega}, \ lh(\sigma) \leq k\}$$

of nonempty perfect sets in E have already been defined.

Only two cases are possible.

1. There exists a natural number $n > n_k$ for which all the sets

$$X_n \cap P_\sigma \qquad (lh(\sigma) = k)$$

are uncountable. In this case, we may put $n_{k+1} = n$ and, for any $\sigma \in 2^{<\omega}$ with $lh(\sigma) = k$, we can construct two nonempty perfect sets $P_{\sigma 0}$ and $P_{\sigma 1}$ satisfying the relations

$$P_{\sigma 0} \cap P_{\sigma 1} = \emptyset,$$

$$P_{\sigma 0} \cup P_{\sigma 1} \subset P_{\sigma} \cap X_{n_{k+1}},$$

$$diam(P_{\sigma 0}) \leq 1/(k+1),$$

$$diam(P_{\sigma 1}) \leq 1/(k+1).$$

So we see that, in this case, the process of our construction can be continued.

2. For each natural number $n > n_k$, there exists a σ from $2^{<\omega}$ with $lh(\sigma) = k$, such that

$$card(P_{\sigma} \cap X_n) \leq \omega.$$

Because the family $\{P_{\sigma} \; : \; lh(\sigma) = k\}$ is finite, we can find an infinite subset M of ω and a $\sigma' \in 2^{<\omega}$ with $lh(\sigma') = k$, such that

$$card(P_{\sigma'} \cap X_n) \leq \omega$$

for all numbers n belonging to M. From the latter relation we obtain

$$card(\cap\{E \setminus X_n \; : \; n \in M\} \cap P_{\sigma'}) > \omega,$$

which immediately gives the desired result (in view of Lemma 1).

Thus, we may restrict our considerations only to case 1. As mentioned above, in this case, the construction just described can be continued and, after ω many steps, it yields a dyadic family

$$\{P_{\sigma} \; : \; \sigma \in 2^{<\omega}\}$$

of nonempty perfect subsets of E. Now, putting

$$P' = \bigcap_{k<\omega} (\cup\{P_{\sigma} \; : \; lh(\sigma) = k\}),$$

we get a nonempty perfect set P' such that

$$P' \subset \cap\{X_{n_k} \; : \; 1 \leq k < \omega\} \cap P.$$

Hence, we may write the inequality

$$card(\cap\{X_{n_k} \; : \; 1 \leq k < \omega\} \cap P) > \omega$$

and, applying once more Lemma 1, we complete the proof of Lemma 5.

Now, taking into account the preceding lemmas, we are able to formulate and prove the following result of Mazurkiewicz.

Theorem 2. *Let Φ be an arbitrary uniformly bounded family of real-valued Borel functions on an uncountable Polish space E. Then Φ is semicompact.*

Proof. Because our Φ is uniformly bounded, there exists a real number $d \geq 0$ such that
$$(\forall \phi \in \Phi)(||\phi|| \leq d).$$
Let us denote by Ψ the family of all those functions ψ that satisfy the following two relations:

(1) $||\psi|| \leq d$;

(2) ψ is a linear combination of characteristic functions of some Borel subsets of E.

Then, according to Lemmas 2, 3 and 5, the family Ψ is semicompact. Also, it is clear that the original family Φ is contained in the closure of Ψ (with respect to the topology of uniform convergence on E). Hence, in view of Lemma 4, the family Φ is semicompact, too.

This ends the proof of Mazurkiewicz's theorem.

Let us observe that if $E = \mathbf{R}$, then Theorem 2 can be extended to an arbitrary family of uniformly bounded real-valued Lebesgue measurable functions on E and to an arbitrary family of uniformly bounded real-valued functions on E having the Baire property. Indeed, in order to obtain the corresponding results, it suffices to apply the following well-known fact: for any Lebesgue measurable (respectively, having the Baire property) real-valued function ϕ on $\mathbf{R}$, there exists a real-valued Borel function ψ on $\mathbf{R}$ such that the set
$$\{x \in \mathbf{R} : \phi(x) \neq \psi(x)\}$$
is of Lebesgue measure zero (respectively, of first category).

Exercise 8. Let E be an uncountable Polish topological space. Prove an analogue of Theorem 2 for real-valued functions on E possessing the Baire property and for real-valued functions on E measurable with respect to the completion of a fixed nonzero σ-finite diffused Borel measure on E.

In addition, give an example of a sequence $\{f_n : n < \omega\}$ of uniformly bounded real-valued Borel functions on $\mathbf{R}$, such that, for each infinite subset M of ω, the corresponding partial sequence $\{f_n : n \in M\}$ is convergent only on a first category subset of $\mathbf{R}$ being simultaneously of Lebesgue measure zero.

As demonstrated above, for any uniformly bounded sequence of real-valued functions possessing good descriptive properties, we have the pointwise convergence (and even the uniform convergence) of an appropriate subsequence on some nonempty perfect set, hence, on some set of cardinality continuum. However, various uniformly bounded sequences of real-valued functions are possible, which are extremely bad from the point of view of convergence pointwise. The following statement (essentially due to Sierpiński) shows that the existence of such sequences can be directly deduced from the existence of a Luzin set Z in an uncountable Polish space E, with

$$card(Z) = \mathbf{c}.$$

In this connection, it is reasonable to recall here that the existence of a Luzin set of cardinality continuum is easily implied by the Continuum Hypothesis (see Theorem 1 from Chapter 10).

Theorem 3. *Let Z be a Luzin subset of the Cantor discontinuum 2^ω, satisfying the equality*

$$card(Z) = \mathbf{c}.$$

Then there exists a sequence $\{X_n \ : \ n < \omega\}$ of subsets of 2^ω such that, for each infinite subset K of ω, the corresponding partial sequence of characteristic functions

$$\{f_{X_n} \ : \ n \in K\}$$

converges pointwise only on a countable subset of 2^ω.

Proof. For every natural number n, let us denote

$$B_n = \{x \in 2^\omega \ : \ x_n = 1\}.$$

The sets B_n $(n < \omega)$ and their complements are clopen in 2^ω and generate a base of the standard product topology on 2^ω. It can easily be seen that, for any infinite subset K of ω, the intersections

$$\cap\{B_n \ : \ n \in K\},$$

$$\cap\{2^\omega \setminus B_n \ : \ n \in K\}$$

are nowhere dense closed subsets of 2^ω. Because Z is a Luzin set in 2^ω, we have

$$card(\cap\{B_n \ : \ n \in K\} \cap Z) \leq \omega,$$

$$card(\cap\{2^\omega \setminus B_n \ : \ n \in K\} \cap Z) \leq \omega.$$

Now, let

$$h \ : \ 2^\omega \to 2^\omega$$

be an injective mapping such that

$$h(2^\omega) = Z.$$

We put

$$X_n = h^{-1}(B_n) \quad (n < \omega).$$

Consider the sequence of characteristic functions

$$\{f_{X_n} \: : \: n < \omega\}.$$

We assert that this sequence is the required one. Indeed, it immediately follows from the definition of the family of sets $\{X_n \: : \: n < \omega\}$ that, for any infinite subset K of ω, the intersections

$$\cap\{X_n \: : \: n \in K\},$$

$$\cap\{2^\omega \setminus X_n \: : \: n \in K\}$$

are at most countable. But from this fact we easily infer that the corresponding partial sequence of characteristic functions $\{f_{X_n} \: : \: n \in K\}$ can be convergent pointwise only on a countable subset of 2^ω.

Theorem 3 has thus been proved.

Exercise 9. Formulate and prove an analogue of Theorem 3 in the situation when there exists a generalized Luzin subset of 2^ω. We recall that the existence of generalized Luzin sets in 2^ω follows, for instance, from Martin's Axiom (see Chapter 10).

In addition, formulate and prove two statements analogous to Theorem 3 under the assumption of the existence of a Sierpiński set in 2^ω (respectively, of a generalized Sierpiński set in 2^ω).

The two results above (namely, Theorem 3 and the corresponding result of Exercise 9) were established under the assumption of the existence of a Luzin set (generalized Luzin set) with cardinality equal to **c**. As we know (see, e.g., Chapter 10), if Martin's Axiom and the negation of the Continuum Hypothesis hold, then there are no Luzin sets on the real line **R** (and, hence, in the Cantor discontinuum 2^ω). So, in such a case, the above argument does not work.

The last theorem of this chapter, presented below, shows us that, under Martin's Axiom, the pointwise convergence of an appropriate subsequence of real-valued functions can be achieved for any subset with small cardinality (i.e., with cardinality strictly less than **c**). Actually, in order to establish the desired result, we do not need the full power of Martin's Axiom. It suffices to apply one purely combinatorial assertion concerning certain families of infinite subsets of ω. This combinatorial assertion is formulated in the following exercise.

Exercise 10. For any two subsets M and K of ω, we shall write

$$M \preceq K$$

if $card(M \setminus K) < \omega$. Obviously, the relation $\preceq$ is a preordering on the family of all subsets of ω.

Suppose that Martin's Axiom holds. Let κ be an infinite cardinal strictly less than $\mathbf{c}$ (as usual, we identify κ with the smallest ordinal number having the same cardinality), and let $\{M_\xi : \xi < \kappa\}$ be a family of infinite sets in ω, such that

$$(\forall \xi)(\forall \zeta)(\xi \leq \zeta < \kappa \Rightarrow M_\zeta \preceq M_\xi).$$

Prove that there exists an infinite subset M of ω satisfying the relation

$$(\forall \xi < \kappa)(M \preceq M_\xi).$$

In addition, show in the theory **ZF** that if $\{M_n : n < \omega\}$ is a sequence of infinite subsets of ω, such that

$$(\forall n)(\forall k)(n \leq k < \omega \Rightarrow M_k \preceq M_n),$$

then there exists an infinite set $M \subset \omega$ satisfying the relation

$$(\forall n < \omega)(M \preceq M_n).$$

Some other combinatorial consequences of Martin's Axiom closely related to the one presented in Exercise 10 are discussed in [71] and [119].

Theorem 4. *Suppose that Martin's Axiom holds. Let E be an arbitrary set of cardinality continuum, let X be a subset of E with*

$$card(X) < card(E),$$

and let $\{f_n : n < \omega\}$ be a uniformly bounded sequence of real-valued functions given on E. Then there exists an infinite subset M of ω such that the partial sequence of functions $\{f_n|X : n \in M\}$ is convergent pointwise on X.

Proof. Let us put $\kappa = card(X)$ and let

$$X = \{x_\xi : \xi < \kappa\}$$

be some enumeration of all elements of X. By applying the result of Exercise 10, it is not hard to define recursively a family $\{M_\xi \ : \ \xi < \kappa\}$ of infinite subsets of ω, satisfying the following conditions:

(a) $M_\zeta \preceq M_\xi$ for $\xi \leq \zeta < \kappa$;

(b) for each $\xi < \kappa$, the partial sequence of real numbers

$$\{f_n(x_\xi) \ : \ n \in M_\xi\}$$

is convergent.

Now, applying the result of Exercise 10 once more, we can define an infinite subset M of ω such that

$$(\forall \xi < \kappa)(M \preceq M_\xi).$$

Then it is easily verified that the partial sequence of functions

$$\{f_n | X \ : \ n \in M\}$$

converges pointwise on the whole set X. This ends the proof of Theorem 4.

In addition, it should be mentioned that the method just described turns out to be rather useful in those questions of analysis, which are concerned with various kinds of convergence of sequences of real-valued functions on a given set E. In fact, this method may be regarded as some generalization of the well-known diagonalization method of Cantor.

Notice, finally, that purely combinatorial arguments (similar to the one presented above) have found numerous applications in real analysis (see, for instance, [30] and [38]).

Exercise 11. Let X be a Polish topological space equipped with a Borel probability measure μ, let

$$F \ : \ X \times \mathbf{R} \to \mathbf{R}$$

be a Borel mapping and suppose that, for each point $x \in X$, there exists a limit

$$lim_{t \to 0} F(x, t) = f(x).$$

Starting with the fact that the projection of a Borel subset of a Polish product space is an analytic set and taking into account the universal measurability of analytic (co-analytic) sets, prove the following parametrized version of the Egorov theorem:

For any real $\varepsilon > 0$, there exists a closed set $Y \subset X$ such that

$$\mu(Y) > 1 - \varepsilon$$

and the equality

$$lim_{t \to 0} F(y, t) = f(y)$$

holds uniformly with respect to a variable $y \in Y$.

This important result is due to Tolstov (see [220]).

13. Sierpiński's partition of the Euclidean plane

In this chapter, we discuss several results and statements that are tightly connected with a Sierpiński partition of the Euclidean plane $\mathbf{R}^2 = \mathbf{R} \times \mathbf{R}$. It turns out that these results and statements can be successfully applied in various fields of mathematics. Especially, they can be applied to certain questions and problems from mathematical analysis, measure theory and general topology.

Let ω denote, as usual, the least infinite ordinal number and let ω_1 denote the least uncountable ordinal number. It is a well-known fact that Sierpiński was the first mathematician who considered, in his classical paper [200], a partition $\{A, B\}$ of the product set $\omega_1 \times \omega_1$, defined as follows:

$$A = \{(\xi, \zeta) : \xi \leq \zeta < \omega_1\},$$

$$B = \{(\xi, \zeta) : \omega_1 > \xi > \zeta\}.$$

He observed that, for any $\xi < \omega_1$ and $\zeta < \omega_1$, the inequalities

$$card(A^\zeta) \leq \omega, \quad card(B_\xi) \leq \omega$$

are true, where

$$A^\zeta = \{\xi : (\xi, \zeta) \in A\},$$
$$B_\xi = \{\zeta : (\xi, \zeta) \in B\}.$$

In other words, each of the sets A and B can be represented as the union of a countable family of "curves" lying in the product set $\omega_1 \times \omega_1$. This property of the partition $\{A, B\}$ implies many interesting and important consequences. For instance, it immediately follows from the existence of $\{A, B\}$ that if the Continuum Hypothesis

$$card(\mathbf{R}) = \mathbf{c} = \omega_1$$

holds, then there exists a partition $\{A', B'\}$ of the Euclidean plane $\mathbf{R}^2$, satisfying the relations:

273

(1) for each straight line L in $\mathbf{R}^2$ parallel to the line $\mathbf{R} \times \{0\}$, the inequality

$$card(A' \cap L) \leq \omega$$

is fulfilled;

(2) for each straight line M in $\mathbf{R}^2$ parallel to the line $\{0\} \times \mathbf{R}$, the inequality

$$card(B' \cap M) \leq \omega$$

is fulfilled.

Moreover, Sierpiński demonstrated that if a covering $\{A', B'\}$ of $\mathbf{R}^2$ with the above-mentioned properties (1) and (2) does exist, then the Continuum Hypothesis is valid.

Indeed, suppose that $\{A', B'\}$ is such a covering of $\mathbf{R}^2$. Choose an arbitrary subset X of $\mathbf{R}$ having cardinality ω_1, and put

$$Z = (X \times \mathbf{R}) \cap B'.$$

Then, according to relation (2), we have

$$card(Z) \leq \omega \cdot \omega_1 = \omega_1.$$

On the other hand, let us show that

$$pr_2(Z) = \mathbf{R}.$$

In order to do this, take an arbitrary point $y \in \mathbf{R}$ and consider the straight line $\mathbf{R} \times \{y\}$. Relation (1) implies that

$$card(A' \cap (\mathbf{R} \times \{y\})) \leq \omega.$$

At the same time, we obviously have

$$card((X \times \mathbf{R}) \cap (\mathbf{R} \times \{y\})) = \omega_1.$$

Hence there exists a point $t \in \mathbf{R}$ such that

$$(t, y) \notin A', \quad (t, y) \in X \times \mathbf{R}.$$

Because $\{A', B'\}$ is a covering of $\mathbf{R}^2$, we infer that $(t, y) \in B'$ and, consequently,

$$(t, y) \in Z, \quad y \in pr_2(Z),$$

which yields the desired equality $pr_2(Z) = \mathbf{R}$. We thus get

$$\mathbf{c} = card(\mathbf{R}) \leq card(Z) \leq \omega_1$$

and, finally, $\mathbf{c} = \omega_1$.

In other words, Sierpiński showed that the Continuum Hypothesis is equivalent to the statement that there exists a partition $\{A', B'\}$ of the Euclidean plane $\mathbf{R}^2$, satisfying relations (1) and (2).

Exercise 1. Let E_1 and E_2 be any two sets, such that

$$card(E_1) = card(E_2) = \mathbf{c}.$$

Check that the following assertions are equivalent:
(i) the Continuum Hypothesis;
(ii) there exists a covering (partition) $\{A, B\}$ of the product set $E_1 \times E_2$, satisfying the relations:

$$(\forall y \in E_2)(card(A \cap (E_1 \times \{y\})) \leq \omega),$$

$$(\forall x \in E_1)(card(B \cap (\{x\} \times E_2)) \leq \omega).$$

Let us mention an important consequence of the existence of a Sierpiński partition $\{A', B'\}$ of $\mathbf{R}^2$. For this purpose, consider the sets

$$A'' = [0,1]^2 \cap A',$$

$$B'' = [0,1]^2 \cap B'.$$

Then we obtain a partition $\{A'', B''\}$ of $[0,1]^2$ with the properties very similar to the ones of $\{A', B'\}$. Let us introduce two functions:
$f = $ the characteristic function of A'';
$g = $ the characteristic function of B''.
It can easily be observed that there exist the iterated integrals

$$\int_0^1 dx \left(\int_0^1 f(x,y)dy \right), \qquad \int_0^1 dy \left(\int_0^1 f(x,y)dx \right),$$

$$\int_0^1 dx \left(\int_0^1 g(x,y)dy \right), \qquad \int_0^1 dy \left(\int_0^1 g(x,y)dx \right),$$

but we have

$$\int_0^1 dx \left(\int_0^1 f(x,y)dy \right) = \int_0^1 dy \left(\int_0^1 g(x,y)dx \right) = 1,$$

$$\int_0^1 dy \left(\int_0^1 f(x,y)dx \right) = \int_0^1 dx \left(\int_0^1 g(x,y)dy \right) = 0$$

and, consequently,

$$\int_0^1 \left(\int_0^1 f(x,y)dy \right)dx \neq \int_0^1 \left(\int_0^1 f(x,y)dx \right)dy,$$

$$\int_0^1 \left(\int_0^1 g(x,y)dy \right)dx \neq \int_0^1 \left(\int_0^1 g(x,y)dx \right)dy.$$

Thus, we infer that the classical Fubini theorem does not hold for each of the functions f and g. But these functions, obviously, are nonnegative and bounded on $[0,1]^2$. Therefore, both f and g are nonmeasurable in the Lebesgue sense.

Remark 1. We see that the Continuum Hypothesis implies the existence of a function f acting from $[0,1]^2$ into $[0,1]$ such that its iterated integrals differ from each other. It is not hard to verify that **CH** is not necessary for this conclusion. For instance, Martin's Axiom also implies the existence of such a function (and, moreover, we do not need here the whole power of **MA**, it suffices to assume that each subset of **R** whose cardinality is strictly less than **c** is measurable in the Lebesgue sense). On the other hand, it was shown in [62] that there are models of set theory in which, for every function

$$g \; : \; [0,1]^2 \rightarrow [0,1],$$

the existence of the iterated integrals

$$\int_0^1 \left(\int_0^1 g(x,y)dx \right)dy, \qquad \int_0^1 \left(\int_0^1 g(x,y)dy \right)dx$$

implies their equality.

For some further statements concerning iterated integrals and tightly connected with a Sierpiński partition of $\mathbf{R}^2$, see [185].

Roughly speaking, we do not have any equivalent of **CH** in terms of iterated integrals. Below, we shall see that there is a beautiful equivalent of **CH** in terms of differentiability.

Exercise 2. Define a function

$$f \; : \; [-1,1]^2 \rightarrow \mathbf{R},$$

by putting:

$f(x,y) = xy/(x^2 + y^2)^2$ if $(x,y) \neq (0,0)$;
$f(x,y) = 0$ if $(x,y) = (0,0)$.
Demonstrate that:

(a) the function f is of first Baire class (hence Lebesgue measurable) but is not Lebesgue integrable;

(b) the iterated integrals for f do exist and

$$\int_{-1}^{1} (\int_{-1}^{1} f(x,y)dx)dy = \int_{-1}^{1} (\int_{-1}^{1} f(x,y)dy)dx = 0.$$

In connection with this exercise, let us note that a much more general result is contained in the old paper by Fichtenholz [58].

We now formulate one statement (also interesting from the view-point of measure theory) which is based on some properties of the Sierpiński partition $\{A, B\}$ of the product set $\omega_1 \times \omega_1$.

If $\mathcal{P}(\omega_1)$ is the σ-algebra of all subsets of ω_1, then the product σ-algebra $\mathcal{P}(\omega_1) \otimes \mathcal{P}(\omega_1)$ coincides with the σ-algebra $\mathcal{P}(\omega_1 \times \omega_1)$ of all subsets of $\omega_1 \times \omega_1$.

In order to establish this result, it is sufficient to consider an arbitrary subset X of the real line $\mathbf{R}$ with $card(X) = \omega_1$ and to apply the well-known fact that the graph of any real-valued function on X is a measurable subset of the product space

$$(X, \mathcal{P}(X)) \times (\mathbf{R}, \mathcal{B}(\mathbf{R})).$$

Exercise 3. Give a detailed proof of the equality

$$\mathcal{P}(\omega_1 \times \omega_1) = \mathcal{P}(\omega_1) \otimes \mathcal{P}(\omega_1).$$

Note that an argument establishing this equality relies essentially on the Axiom of Choice because the existence of an embedding of ω_1 into $\mathbf{R}$ cannot be proved in the theory $\mathbf{ZF}$ & $\mathbf{DC}$. Moreover, as has been shown by Shelah and Raisonnier (see [169], [182]), the existence of such an embedding implies in $\mathbf{ZF}$ & $\mathbf{DC}$ the existence of a subset of $\mathbf{R}$ nonmeasurable in the Lebesgue sense.

From the equality $\mathcal{P}(\omega_1 \times \omega_1) = \mathcal{P}(\omega_1) \otimes \mathcal{P}(\omega_1)$ we can directly obtain the following important statement:

There does not exist a nonzero σ-finite diffused measure μ defined on the whole σ-algebra $\mathcal{P}(\omega_1)$.

Let us recall that this classical statement is due to Ulam [224] who established the nonexistence of such a measure in another way, by applying a transfinite matrix of a special type (for details, see, e.g., [71], [153], [162] or [224]).

In order to prove this statement by using the corresponding properties of the partition $\{A, B\}$, suppose for a moment that such a measure μ does exist and apply the Fubini theorem to the product measure $\mu \times \mu$ and to the sets A and B of the Sierpiński partition. Because all horizontal sections of A are at most countable and all vertical sections of B are also at most countable, we immediately get the equalities

$$(\mu \times \mu)(A) = (\mu \times \mu)(B) = 0$$

and, consequently,

$$(\mu \times \mu)(A \cup B) = (\mu \times \mu)(\omega_1 \times \omega_1) = 0,$$

which yields a contradiction with our assumption that μ is not identically equal to zero. Thus, ω_1 is not a real-valued measurable cardinal.

In fact, the real-valued nonmeasurability of ω_1 is historically the first nontrivial statement, which concerns some important combinatorial properties of uncountable cardinals and which can be established within the theory **ZFC**.

Remark 2. We see, in particular, that if the Continuum Hypothesis holds, then the cardinality of the continuum is not real-valued measurable, either. It is reasonable to recall here that the latter result was first obtained by Banach and Kuratowski in their old joint paper [13]. Actually, the method of [13] gives a more general result. Namely, let us consider the family F of all functions acting from ω into ω. Let f and g be any two functions from F. We put

$$f \preceq g$$

if and only if there exists a natural number $n = n(f, g)$ such that $f(m) \leq g(m)$ for all natural numbers $m \geq n$. Obviously, the relation $\preceq$ is a preordering of F. Now, if the Continuum Hypothesis holds, it is not difficult to define a subset

$$E = \{f_\xi \ : \ \xi < \omega_1\}$$

of F satisfying the following two conditions:

(a) if f is an arbitrary function from F, then there exists an ordinal $\xi < \omega_1$ such that $f \preceq f_\xi$;

(b) for any ordinals ξ and ζ such that $\xi < \zeta < \omega_1$, the relation $f_\zeta \preceq f_\xi$ is not true.

Evidently, each of conditions (a) and (b) implies the equality

$$card(E) = \omega_1.$$

Further, for any two natural numbers m and n, we put

$$E_{m,n} = \{f_\xi \; : \; f_\xi(m) \le n\}.$$

So we get a double family of sets

$$(E_{m,n})_{m<\omega, \; n<\omega},$$

which is usually called a Banach-Kuratowski matrix. It is easy to check that, for each $m < \omega$, we have the inclusions

$$E_{m,0} \subset E_{m,1} \subset \ldots \subset E_{m,n} \subset \ldots$$

and the equality

$$E = \cup\{E_{m,n} \; : \; n < \omega\}.$$

Also, conditions (a) and (b) immediately imply that if f is an arbitrary function from F, then the intersection

$$E_{0,f(0)} \cap E_{1,f(1)} \cap \ldots \cap E_{m,f(m)} \cap \ldots$$

is at most countable.

From these properties of the Banach-Kuratowski matrix it is not hard to deduce that there does not exist a nonzero σ-finite diffused measure on E defined simultaneously for all sets

$$E_{m,n} \quad (m < \omega, \; n < \omega).$$

In addition, it can easily be seen that an analogous result is true for many other functionals (much more general than measures). Namely, let ν be a real-valued positive (i.e., nonnegative) function defined on some class of subsets of E, closed under finite intersections. We say (cf. the corresponding definition presented in Chapter 12) that ν is an admissible functional on E if the following conditions hold:

(1) the family of all countable subsets of E is contained in $dom(\nu)$ and, for any countable set $Z \subset E$, we have the equality $\nu(Z) = 0$;

(2) if $\{X_n \; : \; n < \omega\}$ is an increasing (with respect to inclusion) family of sets, such that $X_n \in dom(\nu)$ for all $n < \omega$, then the set $\cup\{X_n \; : \; n < \omega\}$ also belongs to $dom(\nu)$, and

$$\nu(\cup\{X_n \; : \; n < \omega\}) \le sup\{\nu(X_n) \; : \; n < \omega\};$$

(3) if $\{Y_n \ : \ n < \omega\}$ is a decreasing (with respect to inclusion) family of sets, such that $Y_n \in dom(\nu)$ for all $n < \omega$, then the set $\cap\{Y_n \ : \ n < \omega\}$ also belongs to $dom(\nu)$, and

$$\nu(\cap\{Y_n \ : \ n < \omega\}) \geq inf\{\nu(Y_n) \ : \ n < \omega\}.$$

Evidently, if ν is a finite diffused measure on E, then ν satisfies conditions (1), (2), and (3). But, in general, an admissible functional ν need not have any additive properties similar to the corresponding properties of usual measures. However, the Banach-Kuratowski method works for such functionals, too, and one can conclude that there does not exist a nonzero admissible functional on E defined simultaneously for all sets of a given Banach-Kuratowski matrix.

Exercise 4. Let E be a set with $card(E) > \omega$ for which a Banach-Kuratowski matrix

$$(E_{m,n})_{m<\omega, \ n<\omega}$$

of subsets of E can be constructed. Prove that there does not exist a nonzero admissible functional ν on E satisfying the relation

$$(\forall m < \omega)(\forall n < \omega)(E_{m,n} \in dom(\nu)).$$

Let us return to a Sierpiński partition of the Euclidean plane $\mathbf{R}^2$ and consider some other interesting facts related to it. For instance, we have the following "geometrical" fact:

Assuming the Continuum Hypothesis, there exists a function

$$\phi \ : \ \mathbf{R} \to \mathbf{R}$$

such that

$$\mathbf{R}^2 = \cup\{g_n(\Gamma_\phi) \ : \ n < \omega\}$$

where Γ_ϕ denotes the graph of ϕ and g_n $(n < \omega)$ are some motions of the plane $\mathbf{R}^2$, each of which is either a translation or a rotation (about a point) whose angle is equal to $\pm\pi/2$.

The proof of this result is not difficult and we leave it to the reader.

Exercise 5. By starting with a Sierpiński partition of $\mathbf{R}^2$, give a detailed proof of the above-mentioned result.

Let now X and Y be any two sets. We recall (see, e.g., the Introduction) that a set-valued mapping is an arbitrary function of the type

$$F \ : \ X \to \mathcal{P}(Y)$$

where $\mathcal{P}(Y)$ denotes, as usual, the family of all subsets of Y. According to a well-known definition from general set theory, a subset Z of X is independent with respect to F if, for any two distinct elements $x \in Z$ and $y \in Z$, we have the relations

$$x \notin F(y), \quad y \notin F(x).$$

It can easily be shown that there exists a set-valued mapping

$$F \; : \; \omega_1 \to [\omega_1]^{\leq \omega}$$

such that no two-element subset of ω_1 is independent with respect to F. Actually, the desired set-valued mapping F may be defined as follows:

$$F(\zeta) = A^\zeta \quad (\zeta < \omega_1)$$

where A is the first set of the Sierpiński partition $\{A, B\}$ of $\omega_1 \times \omega_1$.

In connection with this simple fact, let us remark that if a set-valued mapping

$$F \; : \; \omega_1 \to [\omega_1]^{<\omega}$$

is given, then there always exists a subset Ξ of ω_1 satisfying the following two conditions:

(1) $card(\Xi) = \omega_1$;

(2) Ξ is independent with respect to F.

The reader can easily derive this result from the so-called $\triangle$-system lemma (see, e.g., [71] or [119]). It is also reasonable to point out here that the $\triangle$-system lemma is a theorem of the theory **ZF & DC**.

Finally, let us mention that an analogous result (concerning the existence of large independent subsets) holds true for uncountable cardinal numbers, but the proof of this generalized result due to Hajnal and Erdös is more difficult and needs an additional argument.

There are many other interesting statements and facts that are related to the Sierpiński partition of $\omega_1 \times \omega_1$ or can be obtained by using certain properties of this partition (see, e.g., works [35], [38], [60], [92], [113], [151], [152], [190], [191], and [204]).

Here we wish to discuss an application of the Sierpiński partition in real analysis. Namely, we shall present one theorem of Morayne [151] that establishes an interesting connection of this partition with the existence of some strange mappings acting from **R** onto **R** × **R**.

In order to prove the above-mentioned theorem, we need several auxiliary notions and propositions.

Let f be a partial function acting from $\mathbf{R}$ into $\mathbf{R}$, and let X be a subset of $\mathbf{R}$. We say that f satisfies the Banach condition on X if the set

$$\{y \in f(X) \ : \ card(f^{-1}(y) \cap X) > \omega\}$$

is of Lebesgue measure zero.

We also recall that a partial function f acting from $\mathbf{R}$ into $\mathbf{R}$ satisfies the Lipschitz condition if there exists a real $L \geq 0$ such that

$$|f(u) - f(v)| \leq L|u - v|$$

for all u and v belonging to $dom(f)$. In this case, the real L is usually called a Lipschitz constant for f.

It is not hard to prove the following auxiliary statement.

Lemma 1. *Let (X, d) be a metric space, let Y be a subset of X and let f be a function acting from Y into $\mathbf{R}$ and satisfying the Lipschitz condition with a Lipschitz constant*

$$Lip(f) = L \geq 0.$$

Then f can be extended to a function

$$g \ : \ X \to \mathbf{R},$$

which fulfils this condition, too, with the same Lipschitz constant L.

Proof. Assume that $Y \neq \emptyset$ and, for any point $x \in X$, define

$$g(x) = inf\{f(y) + Ld(x, y) \ : \ y \in Y\}.$$

In this way we obtain a mapping g from X into $\mathbf{R}$. Let us check that g is the required extension of f. Fix an arbitrary $x \in Y$. Obviously, for each $y \in Y$, we have

$$g(x) \leq f(y) + Ld(x, y).$$

In particular, putting $y = x$, we get

$$g(x) \leq f(x).$$

On the other hand, the relation

$$|f(x) - f(y)| \leq Ld(x, y) \qquad (y \in Y)$$

implies that

$$f(x) \leq f(y) + Ld(x, y) \qquad (y \in Y)$$

and, hence,

$$f(x) \leq g(x).$$

Consequently, we obtain the equality

$$g(x) = f(x),$$

and claim that g is an extension of f.

Now, let x_1 and x_2 be two arbitrary points from X and let $\varepsilon > 0$. There exist points $y_1 \in Y$ and $y_2 \in Y$ such that

$$f(y_1) + Ld(x_1, y_1) - \varepsilon \leq g(x_1) \leq f(y_2) + Ld(x_1, y_2),$$

$$f(y_2) + Ld(x_2, y_2) - \varepsilon \leq g(x_2) \leq f(y_1) + Ld(x_2, y_1).$$

Then we may write

$$g(x_2) - g(x_1) \leq L(d(x_2, y_1) - d(x_1, y_1)) + \varepsilon \leq Ld(x_1, x_2) + \varepsilon,$$

$$g(x_1) - g(x_2) \leq L(d(x_1, y_2) - d(x_2, y_2)) + \varepsilon \leq Ld(x_1, x_2) + \varepsilon,$$

and, finally,

$$|g(x_1) - g(x_2)| \leq Ld(x_1, x_2) + \varepsilon.$$

Because ε is an arbitrary strictly positive number, we conclude that

$$|g(x_1) - g(x_2)| \leq Ld(x_1, x_2),$$

which finishes the proof of Lemma 1.

Actually, we need only a very special case of this lemma when $X = \mathbf{R}$. In this case, the proof can be done directly.

Lemma 2. *Let f be a partial function acting from $\mathbf{R}$ into $\mathbf{R}$, and suppose that f is differentiable at all points of $dom(f)$, i.e., for each point $t \in dom(f)$, there exists a derivative $f'_{dom(f)}(t)$ relative to the set $dom(f)$. Then the domain of f can be represented in the form*

$$dom(f) = \cup\{P_i \ : \ i \in I\},$$

where

$$card(I) \leq \omega$$

and all sets P_i have the property that $f|P_i$ satisfies the Lipschitz condition.

Proof. First, let us denote

$$D = dom(f)$$

and, for any natural number $n > 0$, define the set $D_n \subset D$ by

$$D_n = \{t \in D \; : \; (\forall t' \in D)(|t' - t| \leq 1/n \Rightarrow |f(t') - f(t)| \leq n|t' - t|)\}.$$

Then, taking into account the assumption that f is differentiable relative to D, it is not hard to check the equality

$$D = \cup\{D_n \; : \; 0 < n < \omega\}.$$

Further, for each natural number $n > 0$, we may write

$$D_n = \cup\{D_{nk} \; : \; k < \omega\}$$

where

$$(\forall k < \omega)(diam(D_{nk}) \leq 1/n).$$

Now, it immediately follows from the definition of D_n that all the restrictions

$$f|D_{nk} \quad (0 < n < \omega, \; k < \omega)$$

satisfy the Lipschitz condition. So we can put

$$\{P_i \; : \; i \in I\} = \{D_{nk} \; : \; 0 < n < \omega, \; k < \omega\},$$

and Lemma 2 is thus proved.

The next auxiliary proposition is well known in real analysis and is due to Banach (see, e.g., [154] or [180]).

Lemma 3. *Let f be a continuous real-valued function of finite variation, defined on some segment $[a, b] \subset \mathbf{R}$. For each $y \in \mathbf{R}$, we put*

$$\phi_f(y) = +\infty$$

if $card(f^{-1}(y)) \geq \omega$, and

$$\phi_f(y) = card(f^{-1}(y))$$

if $card(f^{-1}(y)) < \omega$. Then the function

$$\phi_f \; : \; \mathbf{R} \to \mathbf{R} \cup \{+\infty\}$$

is integrable in the Lebesgue sense, and the relation

$$\int_{\mathbf{R}} \phi_f(y) dy = var_{[a,b]}(f) < +\infty$$

is valid. In particular, for almost all (with respect to the Lebesgue measure λ) points $y \in \mathbf{R}$, we have the inequality

$$card(f^{-1}(y)) < \omega.$$

Exercise 6. Give a proof of Lemma 3.

Lemma 4. *Let f be a function acting from $\mathbf{R}$ into $\mathbf{R}$, and let*

$$D = \{t \in \mathbf{R} \ : \ f'(t) \ exists\}.$$

Then f satisfies the Banach condition on D.

Proof. In view of Lemmas 1 and 2, it suffices to demonstrate that, for each closed bounded interval $[a,b] \subset \mathbf{R}$, any function

$$f^* \ : \ [a,b] \to \mathbf{R}$$

satisfying the Lipschitz condition on $[a,b]$, fulfils the Banach condition on the same interval. But this immediately follows from Lemma 3 because f^* is continuous and of finite variation on $[a,b]$.

Now, we are ready to formulate and prove the result of Morayne [151]. Actually, this result yields a purely analytic equivalent of the Continuum Hypothesis (cf. the material of Chapter 1).

Theorem 1. *The following two assertions are equivalent:*
(1) the Continuum Hypothesis;
(2) there exists a surjection

$$f = (f_1, f_2) \ : \ \mathbf{R} \to \mathbf{R} \times \mathbf{R}$$

such that, for any point $t \in \mathbf{R}$, at least one of the coordinate functions f_1 and f_2 is differentiable at t.

Proof. We first establish the implication (1) $\Rightarrow$ (2). Suppose that the Continuum Hypothesis holds. Then we may consider a Sierpiński type partition $\{A, B\}$ of $\mathbf{R} \times \mathbf{R}$ such that

$$(\forall x \in \mathbf{R})(card(A(x)) \leq \omega),$$

$$(\forall y \in \mathbf{R})(card(B(y)) \leq \omega),$$

where, as usual,

$$A(x) = \{y \in \mathbf{R} \ : \ (x, y) \in A\},$$

$$B(y) = \{x \in \mathbf{R} \ : \ (x, y) \in B\}.$$

Further, we introduce a function

$$\phi \ : \ \mathbf{R} \to \mathbf{R}$$

defined by the formula

$$\phi(t) = t \cdot sin(t) \qquad (t \in \mathbf{R}).$$

It is evident, from the geometrical point of view, that, for any two numbers $u \in \mathbf{R}$ and $v \in \mathbf{R}$, the sets

$$\phi^{-1}(u) \ \cap \ ] - \infty, -1],$$

$$\phi^{-1}(v) \cap [1, +\infty[$$

are countably infinite. So we may write

$$\phi^{-1}(u) \ \cap \ ] - \infty, -1] = \{t_1^u, t_2^u, ..., t_n^u, ...\},$$

$$\phi^{-1}(v) \cap [1, +\infty[\ = \{s_1^v, s_2^v, ..., s_n^v, ...\}.$$

At the same time, we can represent the countable sets

$$A(u) = \{y \in \mathbf{R} \ : \ (u, y) \in A\},$$

$$B(v) = \{x \in \mathbf{R} \ : \ (x, v) \in B\}$$

in the following form:

$$A(u) = \{a_1^u, a_2^u, ..., a_n^u, ...\},$$

$$B(v) = \{b_1^v, b_2^v, ..., b_n^v, ...\}.$$

Let now t be an arbitrary point of $\mathbf{R}$. If $t \in \ ] - \infty, 1[$, then we put

$$f_1(t) = \phi(t).$$

If $t \in [1, +\infty[$, then $t = s_n^v$ for some real v and natural n. In this case, we define

$$f_1(t) = f_1(s_n^v) = b_n^v.$$

Analogously, if $t \in \ ]-1,+\infty[$, then we put

$$f_2(t) = \phi(t).$$

If $t \in \ ]-\infty,-1]$, then $t = t_n^u$ for some real u and natural n. In this case, we define

$$f_2(t) = f_2(t_n^u) = a_n^u.$$

Finally, we introduce a mapping

$$f \ : \ \mathbf{R} \to \mathbf{R} \times \mathbf{R}$$

by the formula

$$f(x) = (f_1(x), f_2(x)) \qquad (x \in \mathbf{R}).$$

Let us verify that f is the required function. For this purpose, take any point $t \in \mathbf{R}$. Because

$$t \in \mathbf{R} = \ ]-\infty,1[\ \cup \ ]-1,+\infty[,$$

there exists an open neighborhood $W(t)$ of t such that

$$W(t) \subset \ ]-\infty,1[\quad \vee \quad W(t) \subset \ ]-1,+\infty[$$

and, hence, at least one of the coordinate functions f_1 and f_2 coincides with the function ϕ on $W(t)$. But ϕ is differentiable everywhere on $\mathbf{R}$. Consequently, at least one of the functions f_1 and f_2 is differentiable on $W(t)$ and, in particular, differentiable at the point $t \in W(t)$.

Also, it can easily be checked that f is a surjection. Indeed, let (u,v) be an arbitrary point of $\mathbf{R}^2$. Because the equality

$$\mathbf{R}^2 = A \cup B$$

holds, the point (u,v) belongs either to A or to B. We may assume, without loss of generality, that $(u,v) \in A$. Then

$$v \in A(u) = \{a_1^u, a_2^u, ..., a_n^u, ...\},$$

$$\phi^{-1}(u) \cap \ ]-\infty,-1] \ = \{t_1^u, t_2^u, ..., t_n^u, ...\}$$

and, for some natural number n, we must have

$$v = a_n^u.$$

Putting $t = t_n^u$ and taking into account the definition of f, we get

$$f(t) = (f_1(t), f_2(t)) = (\phi(t), a_n^u) = (u,v),$$

which shows that our f is a surjection.

In this way we have established the implication $(1) \Rightarrow (2)$.

Suppose now that assertion (2) is valid, i.e., for some surjective mapping

$$f = (f_1, f_2) \ : \ \mathbf{R} \to \mathbf{R} \times \mathbf{R},$$

the equality

$$\mathbf{R} = D_1 \cup D_2$$

is fulfilled, where

$$D_1 = \{t \in \mathbf{R} \ : \ f_1'(t) \ exists\},$$

$$D_2 = \{t \in \mathbf{R} \ : \ f_2'(t) \ exists\}.$$

In conformity with Lemma 4, the coordinate functions f_1 and f_2 satisfy the Banach condition on the sets D_1 and D_2, respectively. Hence the sets

$$K_1 = \{y \in f_1(D_1) \ : \ card(f_1^{-1}(y) \cap D_1) > \omega\},$$

$$K_2 = \{y \in f_2(D_2) \ : \ card(f_2^{-1}(y) \cap D_2) > \omega\}$$

are of Lebesgue measure zero. Let us put

$$M_1 = \mathbf{R} \setminus K_1,$$

$$M_2 = \mathbf{R} \setminus K_2.$$

Then, for M_1 and M_2, we have the equalities

$$card(M_1) = card(M_2) = \mathbf{c}.$$

Now, we define

$$A = f(D_1) \cap (M_1 \times M_2),$$

$$B = f(D_2) \cap (M_1 \times M_2).$$

Taking account of the fact that f is a surjection, we infer

$$A \cup B = M_1 \times M_2.$$

Also, it immediately follows from the definition of the sets M_1 and M_2 that

$$(\forall x \in M_1)(card(A(x)) \leq \omega),$$

$$(\forall y \in M_2)(card(B(y)) \leq \omega).$$

In other words, we obtain a covering $\{A, B\}$ of the product set $M_1 \times M_2$, having properties very similar to those of the Sierpiński partition of $\omega_1 \times$

ω_1. But the product set $M_1 \times M_2$ may be identified (only in the purely set-theoretical sense) with $\mathbf{R} \times \mathbf{R}$. So we conclude that the Continuum Hypothesis must be true (cf. the argument presented in the beginning of this chapter, or Exercise 1). This ends the proof of Theorem 1.

Remark 3. One interesting generalization of the above theorem was obtained in paper [35].

The function f considered in Theorem 1 is singular from the point of view of Lebesgue measurability. The following exercise illustrates this fact.

Exercise 7. Let $f = (f_1, f_2)$ be any surjection from $\mathbf{R}$ onto $\mathbf{R} \times \mathbf{R}$ having the property that, for each $t \in \mathbf{R}$, at least one of the coordinate functions f_1 and f_2 is differentiable at t. Show that f is not measurable in the Lebesgue sense.

In connection with this exercise, let us note that a stronger result is contained in [151].

Exercise 8. Let $n \geq 2$ be a natural number. We consider the n-dimensional Euclidean space $\mathbf{R}^n$ consisting of all n-sequences x of the form

$$x = (x_1, x_2, ..., x_n) \qquad (x_i \in \mathbf{R}, \ i = 1, 2, ..., n).$$

For any $t \in \mathbf{R}$ and for each natural index $i \in [1, n]$, we denote

$$\Gamma_i(t) = \{x \in \mathbf{R}^n \ : \ x_i = t\}.$$

Obviously, $\Gamma_i(t)$ is a coordinate hyperplane in the space $\mathbf{R}^n$.

Demonstrate that the following two assertions are equivalent:
(a) the Continuum Hypothesis;
(b) there exists a partition $\{A_1, A_2, ..., A_n\}$ of $\mathbf{R}^n$ such that, for any $t \in \mathbf{R}$ and for each natural number $i \in [1, n]$, the inequality

$$card(\Gamma_i(t) \cap A_i) \leq \omega$$

is satisfied.

This result generalizes the case $n = 2$ considered above and is also due to Sierpiński.

The following simple exercise shows that, for the infinite-dimensional analogues of an Euclidean space, the situation is essentially different.

Exercise 9. Let us consider the infinite-dimensional topological vector space

$$\mathbf{R}^\omega = R_1 \times R_2 \times ... \times R_n \times ...$$

where
$$(\forall i \in \omega \setminus \{0\})(R_i = \mathbf{R}),$$
and, as above, denote by $\Gamma_i(t)$ the coordinate hyperplane in this space, corresponding to an index $i \in \{1, 2, ..., n, ...\}$ and to a real $t \in \mathbf{R}$.

Show that there does not exist a covering

$$\{A_i \ : \ i \in \omega \setminus \{0\}\}$$

of $\mathbf{R}^\omega$ such that, for any $i \in \omega \setminus \{0\}$ and for each $t \in \mathbf{R}$, the relation

$$card(\Gamma_i(t) \cap A_i) < \mathbf{c}$$

is fulfilled. Consequently, there is no covering

$$\{B_i \ : \ i \in \omega \setminus \{0\}\}$$

of $\mathbf{R}^\omega$ such that, for any $i \in \omega \setminus \{0\}$ and for each $t \in \mathbf{R}$, the relation

$$card(\Gamma_i(t) \cap B_i) \leq \omega$$

is satisfied.

The results presented in the next two exercises are due to Morayne (see his works [151] and [152]).

Exercise 10. Let $n \geq 2$ be a natural number. Prove that the following two assertions are equivalent:
(a) the Continuum Hypothesis;
(b) there exists a surjective mapping

$$f = (f_1, f_2, ..., f_n) \ : \ \mathbf{R} \to \mathbf{R}^n$$

such that, for any point $t \in \mathbf{R}$, at least one of the coordinate functions f_1, f_2, ... , f_n is differentiable at t.

Exercise 11. Show that there is no surjection

$$f = (f_1, f_2, ..., f_n, ...) \ : \ \mathbf{R} \to \mathbf{R}^\omega$$

having the property that, for each point $t \in \mathbf{R}$, at least one of the coordinate functions f_1, f_2, ... , f_n, ... is differentiable at t.

We want to finish this chapter with one statement closely related to the Sierpiński partition of $\omega_1 \times \omega_1$. This statement does not require additional set-theoretical hypotheses and establishes a certain relationship between

Sierpiński type partitions and the nonmeasurability in the Lebesgue sense (cf., e.g., [84]).

Theorem 2. *In the theory* **ZF** & **DC**, *the assertion*

there exists a bijection from **R** *onto a well ordered set*

implies the assertion

there exists a subset of **R** *nonmeasurable in the Lebesgue sense.*

Proof. Obviously, the existence of a bijection between **R** and some well ordered set is equivalent to the existence of a well-ordering of **R** and, actually, means that **R** can be represented as an injective family of points

$$\mathbf{R} = \{x_\xi \ : \ \xi < \alpha\},$$

where α denotes some ordinal number of cardinality continuum. Also, it is clear that if we want to prove the existence of Lebesgue nonmeasurable subsets of **R**, it suffices for us to establish the existence of subsets of the plane $\mathbf{R}^2$, nonmeasurable with respect to the standard two-dimensional Lebesgue measure on $\mathbf{R}^2$. Let

$$\lambda_2 = \lambda \times \lambda$$

denote the latter measure (where, as usual, λ is the standard Lebesgue measure on **R**). Let $\beta \leq \alpha$ be the least ordinal for which

$$\lambda^*(\{x_\xi \ : \ \xi < \beta\}) > 0,$$

where λ^* is the outer measure associated with λ. If the set $\{x_\xi \ : \ \xi < \beta\}$ is nonmeasurable with respect to λ, then we are done. Otherwise, we may write

$$\{x_\xi \ : \ \xi < \beta\} \in dom(\lambda),$$

$$\lambda(\{x_\xi \ : \ \xi < \beta\}) > 0$$

and, according to the definition of β, for each ordinal $\gamma < \beta$, we have

$$\lambda(\{x_\xi \ : \ \xi < \gamma\}) = 0.$$

Consider now a subset Z of $\mathbf{R}^2$ defined as follows:

$$Z = \{(x_\xi, x_\zeta) \ : \ \xi \leq \zeta < \beta\}.$$

We assert that Z is nonmeasurable with respect to λ_2. Indeed, suppose for a moment that $Z \in dom(\lambda_2)$. Then, considering the vertical and horizontal sections of Z and applying the classical Fubini theorem to Z, we get, on the one hand, the relation

$$\lambda_2(Z) > 0$$

and, on the other hand, the equality

$$\lambda_2(Z) = 0.$$

Because this is impossible, we conclude that Z is not λ_2-measurable, which also implies the existence of a Lebesgue nonmeasurable subset of the real line.

Exercise 12. By applying a certain topological version of the classical Fubini theorem, i.e., the so-called Kuratowski-Ulam theorem (see, e.g., [120], [153], or [162]), prove an analogue of Theorem 2 for the Baire property. More precisely, demonstrate in the theory **ZF** & **DC** that the assertion

there exists a bijection from **R** *onto a well ordered set*

implies the assertion

there exists a subset of **R** *without the Baire property.*

In other words, Theorem 2 and the result of the preceding exercise show us that the existence of a well-ordering of the real line **R** immediately yields the existence of subsets of **R** having very bad descriptive structure (from the points of view of Lebesgue measurability and the Baire property).

In this connection, let us recall that the existence of a totally imperfect subset of **R** of cardinality continuum also implies (in the same theory **ZF** & **DC**) the existence of a Lebesgue nonmeasurable subset of **R** and the existence of a subset of **R** without the Baire property (see Exercise 5 from Chapter 8).

Exercise 13. Let Z be a subset of the Euclidean plane $\mathbf{R}^2$ such that

$$card(Z \cap L) < \omega$$

for each straight line L of the plane, which is parallel to the line $\mathbf{R} \times \{0\}$. Let μ be an arbitrary σ-finite measure on $\mathbf{R}^2$ quasiinvariant under the group of all translations of $\mathbf{R}^2$.

Prove that if $Z \in dom(\mu)$, then $\mu(Z) = 0$.

Exercise 14. Demonstrate that there exists no covering $\{X, Y\}$ of the plane $\mathbf{R}^2$ satisfying the following conditions:

(a) $card(X \cap L) < \omega$ for all straight lines L in $\mathbf{R}^2$ that are parallel to the line $\mathbf{R} \times \{0\}$;

(b) $card(Y \cap M) < \mathbf{c}$ for all straight lines M in $\mathbf{R}^2$ that are parallel to the line $\{0\} \times \mathbf{R}$.

Give a simple direct proof of this fact by applying the well-known theorem of set theory, stating that the cardinality of the continuum is not cofinal with ω.

Also, give another proof of this fact by starting with the result of the previous exercise and using a certain measure on $\mathbf{R}^2$ which extends λ_2 and is invariant under the group of all translations of $\mathbf{R}^2$.

Exercise 15. Show that the following two assertions are equivalent:

(i) the Continuum Hypothesis;

(ii) there exists a partition $\{X, Y, Z\}$ of the Euclidean space $\mathbf{R}^3$ satisfying the relations:

(a) $card(X \cap L) < \omega$ for all straight lines L in $\mathbf{R}^3$ that are parallel to the line $\mathbf{R} \times \{0\} \times \{0\}$;

(b) $card(Y \cap M) < \omega$ for all straight lines M in $\mathbf{R}^3$ that are parallel to the line $\{0\} \times \mathbf{R} \times \{0\}$;

(c) $card(Z \cap N) < \omega$ for all straight lines N in $\mathbf{R}^3$ that are parallel to the line $\{0\} \times \{0\} \times \mathbf{R}$.

This equivalence was first established by Sierpiński (see, e.g., [190]).

Exercise 16. Let $(G, +)$ be an arbitrary uncountable commutative group and let 0 denote the neutral element of G.

Prove that there exist three subgroups G_1, G_2, G_3 of G satisfying the relations:

(a) $card(G_1) = card(G_2) = card(G_3) = \omega_1$;

(b) $G_1 \cap (G_2 + G_3) = \{0\}$;

(c) $G_2 \cap (G_1 + G_3) = \{0\}$;

(d) $G_3 \cap (G_1 + G_2) = \{0\}$.

In other words, this result shows that any uncountable commutative group G contains a direct sum of three uncountable subgroups of G.

Moreover, prove that any uncountable commutative group G contains a direct sum of ω_1-sequence of uncountable subgroups of G.

Exercise 17. Let $(G, +)$ be a commutative group and let X be a subset of G. We say that X is G-negligible in G if the following relations hold:

(a) there exists a nonzero σ-finite G-invariant measure μ on G such that $X \in dom(\mu)$;

(b) for every σ-finite G-invariant measure ν on G, we have

$$X \in dom(\nu) \Rightarrow \nu(X) = 0.$$

Starting with the results of Exercises 15 and 16, demonstrate that if $(G, +)$ is uncountable, then it admits a representation in the form

$$G = X \cup Y \cup Z,$$

where X, Y, and Z are some G-negligible subsets of G.

14. Bad functions defined on second category sets

If E is an arbitrary topological space that is not of first category on itself, then we say, in short, that E is of second category (see the Introduction).

Analogously, if X is a subset of E and X is not of first category in E, then we simply say that X is of second category (in E).

Obviously, if $X \subset E$ is of first category in E and

$$f : X \to \mathbf{R}$$

is an arbitrary function, then f can be extended to a function

$$f^* : E \to \mathbf{R}$$

possessing the Baire property. Indeed, it suffices to put $f^*(x) = 0$ for all points $x \in E \setminus X$.

In this chapter we are interested in the following question:

Let E be a topological space without isolated points and let $X \subset E$ be of second category in E. Does there exist a function $f : X \to \mathbf{R}$ that cannot be extended to a function $f^* : E \to \mathbf{R}$ having the Baire property?

Naturally, it is reasonable first to examine this question for the classical case $E = \mathbf{R}$ and then to try to consider a more general situation.

It turns out that, for $E = \mathbf{R}$, the answer is positive and this result is essentially due to Novikov (see [158]).

In his above-mentioned work Novikov dealt only with the question of the existence of a function $f : X \to \mathbf{R}$ that cannot be extended to a Borel function $f^* : \mathbf{R} \to \mathbf{R}$, where X is an arbitrary second category subset of $\mathbf{R}$. However, slightly modifying his argument, it is not difficult to obtain the required result for the Baire property.

Now, we are going to discuss thoroughly this remarkable result of descriptive set theory (let us emphasize that it does not appeal to extra set-theoretical axioms, i.e., belongs to **ZFC** theory).

We need several auxiliary statements. First, let us recall the precise formulation of the Kuratowski-Ulam theorem that can be regarded as a topological version of the Fubini theorem (see, e.g., [120] or [162]).

Theorem 1. *Let E_1 and E_2 be two topological spaces, let E_2 possess a countable base and let $Z \subset E_1 \times E_2$. Then:*

(1) if Z is of first category in the product space $E_1 \times E_2$, then for almost all points $x \in E_1$ the section

$$Z(x) = \{y \in E_2 : (x, y) \in Z\}$$

is of first category in the space E_2;

(2) if Z possesses the Baire property in the product space $E_1 \times E_2$, then for almost all points $x \in E_1$ the section $Z(x)$ possesses the Baire property in the space E_2;

(3) if Z possesses the Baire property in $E_1 \times E_2$ and, for almost all $x \in E_1$, the set $Z(x)$ is of first category in E_2, then Z is of first category in $E_1 \times E_2$.

As pointed out in preceding chapters, the Kuratowski-Ulam theorem has many interesting consequences and applications in set-theoretical topology and mathematical analysis. One nontrivial application of this theorem will be given below, in Chapter 18, where some questions about the descriptive structure of generalized derivatives are discussed.

Exercise 1. Let E be a topological space. A family $\{U_i : i \in I\}$ of nonempty open subsets of E is called a π-base in E if, for each nonempty open set $U \subset E$, there exists an index $i \in I$ such that $U_i \subset U$.

Evidently, any base of E is also a π-base in E but the converse assertion is not true, in general.

Let E_1 be an arbitrary topological space and let E_2 be a topological space with a countable π-base. For a set $Z \subset E_1 \times E_2$, prove the validity of relations (1), (2), and (3) of Theorem 1.

Exercise 2. Let E be a topological space of second category, satisfying the Suslin condition, and let $\{U_i : i \in I\}$ be some π-base in E. Show that there exists a base $\mathcal{B}$ of the σ-ideal $\mathcal{K}(E)$ of all first category subsets of E, such that

$$card(\mathcal{B}) \leq (card(I))^\omega.$$

In this chapter we need only the following corollary of the Kuratowski-Ulam theorem for the Euclidean plane $\mathbf{R}^2 = \mathbf{R} \times \mathbf{R}$.

Lemma 1. *Let Z be a subset of the plane $\mathbf{R} \times \mathbf{R}$ such that:*
(1) $pr_1(Z)$ is of second category in $\mathbf{R}$;
(2) for any $x \in pr_1(Z)$, the section $Z(x)$ is of second category in $\mathbf{R}$.
Then the set Z is of second category in $\mathbf{R} \times \mathbf{R}$.

Obviously, this lemma is a direct consequence of Theorem 1.

Lemma 2. *Let A be an arbitrary subset of the plane $\mathbf{R} \times \mathbf{R}$. Denote*

$$A^* = \{(x,y) \in \mathbf{R} \times \mathbf{R} : y \in cl(A(x))\},$$

where $cl(A(x))$ stands for the closure of $A(x)$ in $\mathbf{R}$.
If A is analytic in $\mathbf{R} \times \mathbf{R}$, then A^* is also analytic in $\mathbf{R} \times \mathbf{R}$.
In particular, if A is Borel in $\mathbf{R} \times \mathbf{R}$, then A^* is analytic in $\mathbf{R} \times \mathbf{R}$.

Proof. According to the definition of A^*, we can write

$$(x,y) \in A^* \Leftrightarrow (\forall n < \omega)(\exists z \in \mathbf{R})(|y - z| < \frac{1}{n+1} \ \& \ (x,z) \in A).$$

For any $n < \omega$, define the set D_n in the space $\mathbf{R}^3$ by the formula

$$D_n = \{(x,y,z) : |y - z| < \frac{1}{n+1} \ \& \ (x,z) \in A\}.$$

It can easily be seen that D_n is an analytic subset of $\mathbf{R}^3$. This implies at once that the set

$$A^* = \bigcap_{n < \omega} pr_{\mathbf{R} \times \mathbf{R}} D_n$$

is analytic in $\mathbf{R} \times \mathbf{R}$, and the proof is completed.

Lemma 3. *Let B be a Borel subset of $\mathbf{R} \times \mathbf{R}$ such that*

$$(\forall x \in \mathbf{R})(card(B(x)) \leq \omega).$$

Let us denote

$$B' = \{x \in \mathbf{R} \ : \ B \cap (\{x\} \times \mathbf{R}) \text{ is nowhere dense in } \{x\} \times \mathbf{R}\}.$$

Then the set B' is also Borel in $\mathbf{R}$.

Proof. Fix a countable base $\{U_n : n < \omega\}$ of open subsets of $\mathbf{R}$. For any $n < \omega$, define the set

$$I(n) = \{m < \omega : U_m \subset U_n\}.$$

Clearly, we can write

$$x \in B' \Leftrightarrow (\forall n < \omega)(\exists m \in I(n))((\{x\} \times U_m) \cap B = \emptyset).$$

Or, equivalently,

$$x \in B' \Leftrightarrow (\forall n < \omega)(\exists m \in I(n))\neg(\exists y)(y \in U_m \,\&\, (x,y) \in B).$$

It suffices to establish that all sets

$$\{x \in \mathbf{R} : \neg(\exists y)(y \in U_m \,\&\, (x,y) \in B)\} \qquad (m < \omega)$$

are Borel in $\mathbf{R}$. We observe that, for each $m < \omega$, the set

$$T_m = \{x \in \mathbf{R} : (\exists y)((x,y) \in (\mathbf{R} \times U_m) \cap B)\}$$

coincides with the projection on $\mathbf{R}$ of some Borel subset of $\mathbf{R}^2$ whose all vertical sections are at most countable. Consequently (see the Introduction), the set T_m is also Borel in $\mathbf{R}$. Clearly, the same is true for the set $\mathbf{R} \setminus T_m$, which yields the required result.

Lemma 4. *Let $X \subset \mathbf{R}$ be a set of second category in $\mathbf{R}$. There exists a set $Z \subset \mathbf{R} \times \mathbf{R}$ such that:*
(1) $pr_1(Z)$ is a subset of X and is also of second category in $\mathbf{R}$;
(2) for any $x \in pr_1(Z)$, the set $Z(x)$ is nowhere dense in $\mathbf{R}$;
(3) Z is of second category in $\mathbf{R} \times \mathbf{R}$.

Proof. The argument presented below is very similar to that one given in the proof of Theorem 2 from Chapter 13. Let us represent our set X in the form of an injective transfinite α-sequence of points:

$$X = \{x_\xi : \xi < \alpha\},$$

where $\alpha \geq \omega_1$ is some ordinal number. Let $\beta \leq \alpha$ denote the least ordinal number such that the set $\{x_\xi : \xi < \beta\}$ is of second category. Then we have

$$(\forall \xi < \beta)(\text{the set } \{x_\zeta : \zeta < \xi\} \text{ is of first category in } \mathbf{R}).$$

Let us put

$$X_0 = \{x_\xi : \xi < \beta\},$$

$$S = \{(x_\xi, x_\zeta) : \zeta \leq \xi < \beta\}.$$

Evidently, we have

$$pr_1(S) = X_0, \quad pr_2(S) = X_0,$$

and, for any $x \in pr_1(S)$, the set $S(x)$ is of first category. Applying the classical Sierpiński argument to S (cf. Chapter 13) and taking into account

Lemma 1, we claim that S does not possess the Baire property, hence S is of second category. At the same time, it is evident that S admits a representation:

$$S = \cup\{Z_n : n < \omega\},$$

where each set Z_n $(n < \omega)$ has the property:

$$(\forall x \in pr_1(Z_n))(Z_n(x) \text{ is nowhere dense in } \mathbf{R}).$$

Obviously, there exists $n_0 < \omega$ such that Z_{n_0} is of second category in $\mathbf{R} \times \mathbf{R}$. We define $Z = Z_{n_0}$. It is not difficult to verify that relations (1), (2), and (3) of the lemma are satisfied for Z.

We now are ready to prove the following statement.

Theorem 2. *Let $X \subset \mathbf{R}$ be a set of second category in $\mathbf{R}$. Then there exists a function*

$$f : X \to \mathbf{R}$$

that does not admit an extension $f^* : \mathbf{R} \to \mathbf{R}$ possessing the Baire property.

Proof. In view of the preceding lemma, a set $Z \subset \mathbf{R} \times \mathbf{R}$ can be found such that:

(1) $pr_1(Z) \subset X$ and $pr_1(Z)$ is of second category in $\mathbf{R}$;
(2) Z is of second category in $\mathbf{R} \times \mathbf{R}$;
(3) for any $x \in pr_1(Z)$, the set $Z(x)$ is nowhere dense in $\mathbf{R}$.
It suffices to show that there exists a function

$$f : pr_1(Z) \to \mathbf{R},$$

which does not admit an extension $f^* : \mathbf{R} \to \mathbf{R}$ having the Baire property.

For each point $x \in pr_1(Z)$, denote by $D(x)$ a subset of $Z(x)$ that is at most countable and everywhere dense in $Z(x)$. Further, consider the set

$$T = \bigcup_{x \in pr_1(Z)} (\{x\} \times D(x)).$$

Note that $pr_1(T) = pr_1(Z)$, the set

$$T^* = \{(x, y) \in \mathbf{R} \times \mathbf{R} : y \in cl(T(x))\}$$

contains Z and hence is of second category in $\mathbf{R} \times \mathbf{R}$. Obviously, we can represent T as a countable union of the graphs of functions acting from $pr_1(Z)$ into $\mathbf{R}$. In other words, we can write

$$T = \cup\{f_n : n < \omega\},$$

where
$$f_n \; : \; pr_1(Z) \to \mathbf{R}$$
for each $n < \omega$.

Now, we assert that at least one function f_n cannot be extended to a function $f_n^* : \mathbf{R} \to \mathbf{R}$ possessing the Baire property.

Suppose to the contrary that every f_n admits such an extension f_n^*. Then, according to the well-known theorem from general topology (see, e.g., [120], [162] or Exercise 13 from the Introduction), for any $n < \omega$, we can find an F_σ-set $P_n \subset \mathbf{R}$ of first category, such that the restriction $f_n^*|(\mathbf{R} \setminus P_n)$ is continuous. Let us denote

$$P = \cup\{P_n : n < \omega\}.$$

The set P is also of type F_σ and of first category in $\mathbf{R}$. In addition, all restrictions $f_n^*|(\mathbf{R} \setminus P)$, where $n < \omega$, are continuous. Let us also point out that the set

$$K = pr_1(Z) \setminus P$$

is of second category in $\mathbf{R}$, the set

$$B = \bigcup_{x \in \mathbf{R} \setminus P} (\{x\} \times \{f_0^*(x), f_1^*(x), ..., f_n^*(x), ...\})$$

is Borel in $\mathbf{R} \times \mathbf{R}$ and, in view of the definition of all functions f_n^*, we have

$$\bigcup_{x \in K} (\{x\} \times D(x)) \subset B.$$

Putting

$$H = \bigcup_{x \in K} (\{x\} \times D(x))$$

and taking into account the equality

$$T = H \cup (\bigcup_{x \in pr_1(Z) \cap P} (\{x\} \times D(x))),$$

we see that the set

$$H^* = \{(x, y) \in \mathbf{R} \times \mathbf{R} : y \in cl(H(x))\}$$

is of second category in $\mathbf{R} \times \mathbf{R}$. By virtue of Lemma 3, the set B' is Borel in $\mathbf{R}$ and the inclusion $K \subset B'$ holds. We thus claim that

$$H \subset B \cap (B' \times \mathbf{R}).$$

Observe now that the set

$$L = B \cap (B' \times \mathbf{R})$$

is Borel in $\mathbf{R}^2$ and, by Lemma 2, the set L^* is analytic in $\mathbf{R}^2$. Also, all vertical sections of L^* are nowhere dense. Remembering that L^* has the Baire property (see the Introduction), we conclude by the Kuratowski-Ulam theorem that L^* is of first category in $\mathbf{R} \times \mathbf{R}$. But this is impossible since $H^* \subset L^*$ and, as mentioned above, H^* is of second category.

The contradiction obtained finishes the proof.

Let us underline once more that Theorem 2 is a result of **ZFC** theory and essentially relies on deep facts from classical descriptive set theory. Obviously, we may replace the real line $\mathbf{R}$ by any nonempty Polish space E without isolated points, and prove an analogous result for E.

Exercise 3. Let $A \subset \mathbf{R} \times \mathbf{R}$ be an analytic set. Demonstrate that, for every natural number n, the set

$$A_n = \{x \in \mathbf{R} : card(A(x)) \geq n\}$$

is analytic in $\mathbf{R}$.

Deduce from this fact that the set

$$A_\omega = \{x \in \mathbf{R} : card(A(x)) \geq \omega\}$$

is also analytic in $\mathbf{R}$.

Exercise 4. Let $f : \mathbf{R} \to \mathbf{R}$ be a Sierpiński-Zygmund function. Assuming that each set in $\mathbf{R}$ of cardinality strictly less than $\mathbf{c}$ is of first category, show that for any second category set $X \subset \mathbf{R}$, the function $f|X$ cannot be extended to a function $f^* : \mathbf{R} \to \mathbf{R}$ possessing the Baire property.

Now, let us turn our attention to the general situation where a topological space E without isolated points is given with its second category subset X. Does it always exist a function $f : X \to \mathbf{R}$ that is not extendible to a function possessing the Baire property? It turns out that we cannot positively answer this question within **ZFC** theory. Indeed, assuming the existence of some large cardinal, Kunen proved the consistency with **ZFC** of the following statement:

There exists a topological space K satisfying the relations:

(1) $card(K) = \omega_1$;

(2) K is Hausdorff and contains no isolated points;

(3) K is a Baire space;

(4) no nonempty open set $U \subset K$ admits a representation in the form of the union of two disjoint subsets each of which is everywhere dense in U.

We recall (see Exercise 5 from Chapter 2) that a topological space E is resolvable if it admits a representation

$$E = A \cup B,$$

where A and B are disjoint and everywhere dense in E. Clearly, if E is resolvable, then it does not contain isolated points.

Relation (4) above says that no nonempty open subspace of the Kunen space K is resolvable.

Exercise 5. Suppose that in a topological space E no nonempty open set U admits a representation in the form of the union of two disjoint everywhere dense subsets of U.

Show that any subset X of E with $int(X) = \emptyset$ is nowhere dense in E.

Deduce from this fact that in such a space E every subset can be represented as the union of an open set and a nowhere dense set. Consequently, each subset of E has the Baire property in E.

We thus see that any subset of the Kunen space K possesses the Baire property. Hence, any real-valued function defined on K has the Baire property. Therefore, the answer to the question posed above is trivially negative for such a space.

So we must introduce some natural restrictions on a general topological space E if we want our question for E would be solved positively. Those restrictions can be formulated in purely topological terms, but it is desirable to formulate some conditions only in terms of the structure of the σ-ideal $\mathcal{K}(E)$, not touching the inner properties of the topology of E. Note that, in many cases, those properties are not preserved under taking subspaces.

Exercise 6. Demonstrate that there exist a Baire space E without isolated points and satisfying the Suslin condition (even separable) and a subspace X of E of second category in E, such that the Suslin condition does not hold for X.

The following definition is useful for our further considerations.

Let E be a nonempty set equipped with a σ-algebra $\mathcal{S}$ of its subsets and let $\mathcal{I}$ be a σ-ideal of subsets of E such that $\mathcal{I} \subset \mathcal{S}$. The members of $\mathcal{I}$ are usually called small sets (or negligibles) with respect to the measurable structure $(E, \mathcal{S})$. The triple $(E, \mathcal{S}, \mathcal{I})$ is called a measurable space with negligibles (cf. [61]). Of course, we have the following two widely known examples of such $(E, \mathcal{S}, \mathcal{I})$.

Example 1. Let E be a topological space of second category, $\mathcal{S} = \mathcal{B}a(E)$ be the σ-algebra of all subsets of E having the Baire property and let $\mathcal{I} = \mathcal{K}(E)$ be the σ-ideal of all first category subsets of E. Then the triple $(E, \mathcal{S}, \mathcal{I})$ can be regarded as a measurable space with negligibles.

Example 2. Let E be a set, $\mathcal{S}$ be a σ-algebra of subsets of E and let μ be a nonzero σ-finite complete measure with $dom(\mu) = \mathcal{S}$. Denote by $\mathcal{I} = \mathcal{I}(\mu)$ the σ-ideal of all μ-measure zero sets in E. Then the triple $(E, \mathcal{S}, \mathcal{I})$ is a measurable space with negligibles.

It is easy to see that the question posed in the beginning of this chapter can be reformulated in the following more general form:

Let $(E, \mathcal{S}, \mathcal{I})$ be a measurable space with negligibles and let X be a subset of E not belonging to $\mathcal{I}$. Does there exist a function $f : X \to \mathbf{R}$ which cannot be extended to an $\mathcal{S}$-measurable function $f^* : E \to \mathbf{R}$?

A natural approach to this question is contained in the next statement.

Theorem 3. *Let $(E, \mathcal{S}, \mathcal{I})$ be a measurable space with negligibles, X be a subset of E and let*

$$\mathcal{S}_X = \{Y \cap X : Y \in \mathcal{S}\}.$$

Suppose that $\mathcal{S}_X \neq \mathcal{P}(X)$. Then there exists a function $g : X \to \mathbf{R}$ which cannot be extended to an $\mathcal{S}$-measurable function $g_1 : E \to \mathbf{R}$.

In particular, if $\mathcal{S}_X \neq \mathcal{P}(X)$ for all nonsmall sets $X \subset E$, then there are real-valued functions on any nonsmall set in E, which do not admit an $\mathcal{S}$-measurable extension defined on E.

Proof. The argument is very easy. Because $\mathcal{S}_X \neq \mathcal{P}(X)$, there exists a set $Z \subset X$ not belonging to $\mathcal{S}_X$. Let

$$g_Z \ : \ X \to \{0, 1\}$$

denote the characteristic function of Z. Clearly, g_Z is not $\mathcal{S}_X$-measurable. We assert that $g = g_Z$ is the required one. Indeed, suppose to the contrary that there exists a function

$$g_1 : E \to \mathbf{R}$$

extending g_Z and measurable with respect to $\mathcal{S}$. Then, for any open set $U \subset \mathbf{R}$, we can write

$$g_Z^{-1}(U) = g_1^{-1}(U) \cap X.$$

Because $g_1^{-1}(U) \in \mathcal{S}$, we get $g_Z^{-1}(U) \in \mathcal{S}_X$. Therefore, g_Z must be $\mathcal{S}_X$-measurable, which contradicts the definition of g_Z. The contradiction obtained ends the proof.

Actually, the same argument establishes the following much stronger statement: g_Z cannot be extended to a partial $\mathcal{S}$-measurable real-valued function.

Returning to the problem of the existence of a real-valued partial function on a topological space E, which does not admit an extension defined on the whole E and possessing the Baire property, we will see below that this problem can be solved positively in the case when we are able to prove that any second category subset X of E includes a set without the Baire property (with respect to X).

The following exercises yield some results in this direction. They are based on Theorem 3 proved above.

Exercise 7. Let E be an arbitrary topological space, A be a subset of E and let e be a point in E. We say that A is of second category at e if, for every neighborhood $U(e)$ of e, the set $U(e) \cap A$ is of second category in E.

Let X be an arbitrary second category subset of E. Show that there exists a set $Y \subset X$ satisfying the following relations:

(a) Y is of second category at each point $y \in Y$;

(b) $X \setminus Y$ is of first category in E.

Also, demonstrate that:

(c) $\mathcal{K}(Y) = \{Z \cap Y : Z \in \mathcal{K}(E)\}$;

(d) $\mathcal{B}a(Y) = \{Z \cap Y : Z \in \mathcal{B}a(E)\}$.

Exercise 8. Let E be an infinite set and let $\{X_j : j \in J\}$ be a family of subsets of E such that:

(a) $card(J) \leq card(E)$;

(b) $(\forall j \in J)(card(X_j) = card(E))$.

Prove that there exists a family $\{Y_i : i \in I\}$ of subsets of E satisfying the relations:

(c) $card(I) > card(E)$;

(d) $card(Y_i \cap Y_{i'}) < card(E)$ for any $i \in I$, $i' \in I$, $i \neq i'$ (in other words, the family $\{Y_i : i \in I\}$ is almost disjoint);

(e) $(\forall i \in I)(\forall j \in J)(card(X_j \cap Y_i) = card(E))$.

Use the method of transfinite induction (cf. Exercise 7 from Chapter 8).

This result generalizes the classical Sierpiński theorem on almost disjoint families of sets (see [190]).

It follows from the said above that there exists an almost disjoint family $\{Y_i : i \in I\}$ of infinite subsets of ω with $card(I) > \omega$.

Strengthen this fact and prove (within **ZF**) that there exists an almost disjoint family of infinite subsets of ω, whose cardinality is equal to **c** (the last result is also due to Sierpiński).

Exercise 9. Let E be a topological space satisfying the conditions:
(1) $card(E) = \omega_1$;
(2) E is of second category and has no isolated points;
(3) there exists a base of $\mathcal{K}(E)$ whose cardinality does not exceed ω_1.
Show that there exists a subset of E which does not possess the Baire property. Moreover, prove that, for each second category subset X of E, there exists a set $Y \subset X$ without the Baire property (in E).

For this purpose, take into account the results of the two preceding exercises and apply the Banach theorem on open sets of first category (see Exercise 30 from the Introduction).

Further, show that, for any second category subset X of E, there exists a function $g : X \to \mathbf{R}$ not extendible to a function $g^* : E \to \mathbf{R}$ possessing the Baire property.

In particular, one can claim that an analogue of Theorem 2 holds true (under **CH**) for any topology on $\mathbf{R}$ extending the standard topology of $\mathbf{R}$ and such that the σ-ideal of all first category sets possesses a base whose cardinality does not exceed **c**.

It is useful to compare the above exercise with the Kunen result formulated in this chapter.

Exercise 10. Let E be a set with $card(E) = \omega_1$ and let μ be a nonzero σ-finite complete diffused measure defined on some σ-algebra of subsets of E. Denote by $\mathcal{S}$ the domain of μ and by $\mathcal{I} = \mathcal{I}(\mu)$ the σ-ideal of all μ-measure zero sets in E. Applying to the measurable space with negligibles $(E, \mathcal{S}, \mathcal{I})$ the classical Ulam theorem on the non-real-valued measurability of ω_1, show that for any set $X \notin \mathcal{I}$, there exists a function $g : X \to \mathbf{R}$ not extendible to an $\mathcal{S}$-measurable function $g^* : E \to \mathbf{R}$.

Exercise 11. Let E be a topological space of second category and suppose that the following conditions are satisfied:
(a) for any family $\mathcal{F} \subset \mathcal{K}(E)$ with $card(\mathcal{F}) < card(E)$, we have

$$\cup \mathcal{F} \in \mathcal{K}(E),$$

i.e., the σ-ideal $\mathcal{K}(E)$ is $card(E)$-additive;
(b) there exists a base of $\mathcal{K}(E)$ whose cardinality does not exceed $card(E)$.
Let $\{X_i : i \in I\} \subset \mathcal{K}(E)$ be a point-finite family of sets, i.e.,

$$(\forall x \in E)(card(\{i \in I : x \in X_i\}) < \omega),$$

and suppose, in addition, that $\cup\{X_i : i \in I\} = E$.

Prove that there exist two disjoint sets $I_1 \subset I$ and $I_2 \subset I$ such that both sets $\cup\{X_i : i \in I_1\}$ and $\cup\{X_i : i \in I_2\}$ do not possess the Baire property in E.

Exercise 12. Let E be a topological space satisfying the conditions of Exercise 11, let E_1 be a metric space and let

$$\Phi : E \to \mathcal{P}(E_1)$$

be a set-valued mapping such that $\Phi(x)$ is nonempty and compact for any $x \in E$. Suppose also that, for each open set $U \subset E_1$, the set

$$\{x \in E : \Phi(x) \cap U \neq \emptyset\}$$

possesses the Baire property in E.

Fix a real $\varepsilon > 0$. Demonstrate that there exists a ball $B \subset E_1$ with $diam(B) \leq \varepsilon$, such that the set

$$\{x \in E : \Phi(x) \cap B \neq \emptyset\}$$

is of second category in E.

Assume, in addition, that E satisfies the Suslin condition. Prove that there exists a selector f of Φ possessing the Baire property, i.e., there exists a function

$$f : E \to E_1$$

possessing the Baire property and such that

$$f(x) \in \Phi(x) \qquad (x \in E).$$

The following exercise yields a nontrivial application of almost disjoint families of sets to a problem arising in the theory of Banach spaces.

Exercise 13. Let E be a Banach space and let E_1 be a Banach subspace of E. We recall that E_1 admits a complemented space if there exists a Banach subspace E_2 of E such that E is a direct topological sum of E_1 and E_2. In this case, E_1 is also called a direct topological summand in E.

For example, any Hilbert subspace E_1 of a given Hilbert space E is a direct topological summand in E.

Let l_∞ denote the Banach space of all bounded real-valued sequences and let c_0 denote the Banach subspace of l_∞ consisting of all sequences tending to zero. Note that l_∞ is nonseparable and c_0 is separable.

Demonstrate that c_0 does not admit a complemented space in l_∞ (the old result of Sobczyk).

Do this in the following manner.

Start with the observation that the conjugate space $(l_\infty)^*$ contains a countable subfamily separating the points in l_∞. Consequently, each vector subspace F of l_∞ possesses the same property, i.e., F^* contains a countable subfamily which separates the points in F.

Assuming to the contrary that F is a complemented space for c_0, infer first that F must be isomorphic to the Banach factor-space l_∞/c_0.

On the other hand, show that $(l_\infty/c_0)^*$ does not contain a countable subfamily separating the points in l_∞/c_0.

For this purpose, take an arbitrary uncountable almost disjoint family $\{A_i : i \in I\}$ of infinite subsets of ω (see Exercise 8) and consider the characteristic functions

$$f_{A_i} : \omega \to \{0, 1\} \quad (i \in I)$$

of these subsets. Observe that all functions f_{A_i} belong to l_∞. Let

$$\phi : l_\infty \to l_\infty/c_0$$

denote the canonical surjective homomorphism of Banach spaces and let

$$x_i = \phi(f_{A_i}) \quad (i \in I).$$

Further, for any functional $g \in (l_\infty/c_0)^*$ and for any index $i \in I$, we put $a_i = sgn(g(x_i))$. Check that if a set $I_0 \subset I$ is finite, then

$$\left\| \sum_{i \in I_0} a_i x_i \right\| \leq 1$$

in view of the almost disjointness of $\{A_i : i \in I\}$. Deduce from this fact that

$$\|g\| \geq |g(\sum_{i \in I_0} a_i x_i)| = \sum_{i \in I_0} |g(x_i)|,$$

which implies that the family $\{|g(x_i)| : i \in I\}$ of reals is summable.

Consequently, there are only countably many members of this family which differ from zero.

Finally, conclude that if $\{g_k : k < \omega\}$ is an arbitrary countable family of functionals from $(l_\infty/c_0)^*$, then there exists an index $i_0 \in I$ satisfying the relations

$$g_k(x_{i_0}) = 0 \quad (k < \omega).$$

In other words, $\{g_k : k < \omega\}$ does not separate the points $x_{i_0} \neq 0$ and 0 in the space l_∞/c_0.

The obtained contradiction shows that c_0 is not a direct topological summand in l_∞.

Exercise 14. Observe that if a Banach space E_1 is a direct topological summand in a Banach space E and a Banach space E_2 is a direct topological summand in E_1, then E_2 is a direct topological summand in E. Deduce from this fact and the result of the previous exercise that the Banach space c consisting of all convergent real-valued sequences does not admit a complemented space in l_∞.

Exercise 15. Applying Exercise 8 of this chapter, prove that if E is a set and μ is a nonatomic probability measure defined on the family $\mathcal{P}(E)$ of all subsets of E, then μ is not perfect (cf. Exercise 1 from Chapter 7).

On the other hand, show that if ν is a two-valued probability measure defined on the same family $\mathcal{P}(E)$, then ν is perfect.

15. Sup-measurable and weakly sup-measurable functions

It is well known that the notion of measurability of sets and functions plays an important role in various fields of classical and modern analysis (also, in probability theory and general topology). For functions of several variables, a related notion of sup-measurability was introduced and investigated (see, e.g., [3], [8], [38], [41], [67], [96], [101], [104], [106], [115], [116], [176], [187], [188] and the references given therein). It turned out that this notion can successfully be applied to some topics from analysis and, in particular, to the theory of ordinary differential equations (for more information concerning applications of sup-measurable mappings in the above-mentioned theory, see [116] and Chapter 17 of this book).

Here we shall introduce and examine the following three classes of functions acting from $\mathbf{R}^2 = \mathbf{R} \times \mathbf{R}$ into $\mathbf{R}$:

(1) the class of sup-continuous mappings;

(2) the class of sup-measurable mappings;

(3) the class of weakly sup-measurable mappings.

We begin with the definitions of sup-continuous and sup-measurable functions.

We shall say that a mapping

$$\Phi \ : \ \mathbf{R} \times \mathbf{R} \to \mathbf{R}$$

is sup-continuous (sup-measurable) with respect to the second variable y if, for every continuous (Lebesgue measurable) function

$$\phi \ : \ \mathbf{R} \to \mathbf{R},$$

the superposition

$$\Phi_\phi \ : \ \mathbf{R} \to \mathbf{R}$$

given by the formula

$$\Phi_\phi(x) = \Phi(x, \phi(x)) \qquad (x \in \mathbf{R})$$

309

is also continuous (Lebesgue measurable).

Let us mention that, actually, the first notion yields nothing new: it turns out that the class of all sup-continuous mappings coincides with the class of all continuous mappings acting from $\mathbf{R} \times \mathbf{R}$ into $\mathbf{R}$. For the sake of completeness, we present here the proof of this simple (and probably well-known) fact.

Theorem 1. *Let Φ be a mapping acting from $\mathbf{R} \times \mathbf{R}$ into $\mathbf{R}$. Then the following two assertions are equivalent:*

(1) Φ is continuous;

(2) Φ is sup-continuous.

Proof. The implication (1) $\Rightarrow$ (2) is trivial. So it remains to establish only the converse implication (2) $\Rightarrow$ (1). Let Φ be sup-continuous, and suppose that Φ is not continuous. Then there exist a point (x_0, y_0) of $\mathbf{R} \times \mathbf{R}$, a real number $\varepsilon > 0$ and a sequence of points

$$\{(x_n, y_n) \ : \ n \in \mathbf{N}, \ n > 0\} \subset \mathbf{R} \times \mathbf{R}$$

such that:

(a) $lim_{n \to +\infty} (x_n, y_n) = (x_0, y_0)$;

(b) $|\Phi(x_n, y_n) - \Phi(x_0, y_0)| > \varepsilon$ for all $n \in \mathbf{N} \setminus \{0\}$.

We may assume, without loss of generality, that the sequence of points

$$\{x_n \ : \ n \in \mathbf{N}, \ n > 0\} \subset \mathbf{R}$$

is injective and $x_n \neq x_0$ for each $n \in \mathbf{N} \setminus \{0\}$. Indeed, if

$$f_n \ : \ \mathbf{R} \to \mathbf{R} \qquad (n = 1, 2, ...)$$

denotes the function identically equal to y_n, then the function

$$\Phi_{f_n} \ : \ \mathbf{R} \to \mathbf{R}$$

is continuous and $\Phi_{f_n}(x_n) = \Phi(x_n, y_n)$. Therefore, for some strictly positive real number $\delta = \delta(x_n)$ and for all points x belonging to the open interval $]x_n - \delta, \ x_n + \delta[$, we have the inequality

$$|\Phi_{f_n}(x) - \Phi(x_0, y_0)| > \varepsilon$$

or, equivalently,

$$|\Phi(x, y_n) - \Phi(x_0, y_0)| > \varepsilon.$$

From this fact it immediately follows that the above-mentioned sequence of points $\{x_n \; : \; n \in \mathbf{N}, \; n > 0\}$ can be chosen injective and satisfying the relation

$$(\forall n \in \mathbf{N} \setminus \{0\})(x_n \neq x_0).$$

Now, it is not difficult to define a continuous function

$$f \; : \; \mathbf{R} \to \mathbf{R}$$

such that

$$(\forall n \in \mathbf{N})(f(x_n) = y_n).$$

For this function f, we get the continuous superposition

$$\Phi_f \; : \; \mathbf{R} \to \mathbf{R}.$$

Because

$$lim_{n \to +\infty} \, x_n = x_0,$$

we must have the equality

$$lim_{n \to +\infty} \, \Phi_f(x_n) = \Phi_f(x_0)$$

and, consequently,

$$lim_{n \to +\infty} \, \Phi(x_n, y_n) = \Phi(x_0, y_0),$$

which is impossible. This contradiction finishes the proof of Theorem 1.

A completely different situation can be observed for sup-measurable mappings.

On the one hand, simple examples show that if

$$\Phi \; : \; \mathbf{R} \times \mathbf{R} \to \mathbf{R}$$

is a Lebesgue measurable mapping, then it need not be sup-measurable.

Exercise 1. Give an example of a function

$$\Phi \; : \; \mathbf{R} \times \mathbf{R} \to \mathbf{R}$$

that is Lebesgue measurable but is not sup-measurable. More precisely, demonstrate that the existence of such examples follows directly from the widely known fact that the composition of Lebesgue measurable functions (acting from $\mathbf{R}$ into $\mathbf{R}$) need not be Lebesgue measurable.

On the other hand, it turns out that there exist (under some additional set-theoretical axioms) various sup-measurable mappings which are not measurable in the Lebesgue sense. In order to present this result, let us first formulate and prove one simple auxiliary statement.

Lemma 1. *Suppose that* Ψ *is a mapping acting from* $\mathbf{R} \times \mathbf{R}$ *into* $\mathbf{R}$. *Then the following two assertions are equivalent:*

(1) Ψ *is sup-measurable;*

(2) for every continuous function ψ : $\mathbf{R} \to \mathbf{R}$, *the function* Ψ_ψ *is Lebesgue measurable.*

Proof. The implication (1) $\Rightarrow$ (2) is trivial. Let us show that the converse implication (2) $\Rightarrow$ (1) is true, too. Let Ψ satisfy (2) and let ψ be an arbitrary Lebesgue measurable function acting from $\mathbf{R}$ into $\mathbf{R}$. Applying the classical theorem of Luzin, we can find a countable disjoint covering $\{X_k : k < \omega\}$ of $\mathbf{R}$ and a countable family $\{\psi_k : k < \omega\}$ of functions acting from $\mathbf{R}$ into $\mathbf{R}$, such that:

(a) all sets X_k $(1 \leq k < \omega)$ are closed in $\mathbf{R}$ and X_0 is of Lebesgue measure zero;

(b) all functions ψ_k $(1 \leq k < \omega)$ are continuous;

(c) for each index $k < \omega$, the restriction of ψ to X_k coincides with the restriction of ψ_k to X_k.

Let us denote by f_k the characteristic function of X_k. Then it is not difficult to check the equality

$$\Psi_\psi = \sum_{k<\omega} f_k \cdot \Psi_{\psi_k}.$$

According to our assumption, all superpositions Ψ_{ψ_k} $(1 \leq k < \omega)$ are Lebesgue measurable. In addition, the function $f_0 \cdot \Psi_{\psi_0}$ is equivalent to zero. Thus, we easily conclude that the superposition Ψ_ψ is Lebesgue measurable, too.

Exercise 2. Let Φ be a mapping acting from $\mathbf{R} \times \mathbf{R}$ into $\mathbf{R}$. Suppose that this mapping satisfies the so-called Carathéodory conditions, i.e.,

(1) for each $x \in \mathbf{R}$, the partial function

$$y \to \Phi(x, y) \qquad (y \in \mathbf{R})$$

is continuous;

(2) for each $y \in \mathbf{R}$, the partial function

$$x \to \Phi(x, y) \qquad (x \in \mathbf{R})$$

is Lebesgue measurable.

Show that:

(a) Φ is measurable with respect to the usual two-dimensional Lebesgue measure λ_2 on the plane $\mathbf{R}^2$ (more precisely, Φ is measurable with respect to the product of the σ-algebras $dom(\lambda)$ and $\mathcal{B}(\mathbf{R})$);

(b) Φ is sup-measurable.

Exercise 3. Let Φ be a mapping acting from $\mathbf{R} \times \mathbf{R}$ into $\mathbf{R}$. Suppose that the following two conditions hold:

(1) for each $x \in \mathbf{R}$, the partial function

$$y \rightarrow \Phi(x, y) \qquad (y \in \mathbf{R})$$

is a continuous mapping from $\mathbf{R}$ into $\mathbf{R}$;

(2) for each $y \in \mathbf{R}$, the partial function

$$x \rightarrow \Phi(x, y) \qquad (x \in \mathbf{R})$$

is a Borel mapping from $\mathbf{R}$ into $\mathbf{R}$.

Show that Φ is a Borel mapping from $\mathbf{R} \times \mathbf{R}$ into $\mathbf{R}$.

By using the method of transfinite induction, give an example of a function Ψ acting from $\mathbf{R} \times \mathbf{R}$ into $\mathbf{R}$ such that:

(a) for any points $x_0 \in \mathbf{R}$ and $y_0 \in \mathbf{R}$, the partial functions

$$y \rightarrow \Psi(x_0, y) \qquad (y \in \mathbf{R}),$$

$$x \rightarrow \Psi(x, y_0) \qquad (x \in \mathbf{R})$$

are Borel (more precisely, are upper semicontinuous);

(b) Ψ is not measurable with respect to the two-dimensional Lebesgue measure λ_2 on $\mathbf{R} \times \mathbf{R}$.

Exercise 4. Let $n > 1$ be a natural number and let Φ be a mapping acting from the n-dimensional Euclidean space $\mathbf{R}^n$ into $\mathbf{R}$. Suppose also that Φ satisfies the following condition: for each natural index $i \in [1, n]$ and for any points

$$x_1 \in \mathbf{R}, \quad ..., \quad x_{i-1} \in \mathbf{R}, \quad x_{i+1} \in \mathbf{R}, \quad ..., \quad x_n \in \mathbf{R},$$

the partial function

$$x \rightarrow \Phi(x_1, ..., x_{i-1}, x, x_{i+1}, ..., x_n) \qquad (x \in \mathbf{R})$$

is continuous. Prove, using induction on n, that Φ is a Borel mapping from $\mathbf{R}^n$ into $\mathbf{R}$. More precisely, prove that the Baire order of Φ is less than or equal to $n - 1$ (cf. Theorem 4 of Chapter 2).

The next exercise shows that, for some standard infinite-dimensional spaces, the situation is essentially different from the one described in the previous exercise.

Exercise 5. Let $\mathbf{T}$ denote the one-dimensional unit torus, i.e., the set

$$\mathbf{T} = \{(x,y) \in \mathbf{R} \times \mathbf{R} \;:\; x^2 + y^2 = 1\}$$

is regarded as a commutative compact topological group with respect to the usual group operation and the Euclidean topology. We denote by e the neutral element of this group. Now, consider the product group $\mathbf{T}^\omega$. It is a commutative compact topological group, too. Equip this product group with the Haar probability measure μ. In fact, μ is the product measure of a countable family of measures, each of which coincides with the Haar probability measure on $\mathbf{T}$. Further, denote by G the subset of $\mathbf{T}^\omega$ consisting of all those elements $\{x_n \;:\; n \in \omega\} \in \mathbf{T}^\omega$ for which we have

$$card(\{n \in \omega \;:\; x_n \neq e\}) < \omega.$$

Obviously, G is an everywhere dense Borel subgroup of $\mathbf{T}^\omega$. Finally, let $\{D_i \;:\; i \in I\}$ be the injective family of all G-orbits in $\mathbf{T}^\omega$. Check that

$$card(I) = \mathbf{c},$$

where $\mathbf{c}$ denotes, as usual, the cardinality of the continuum. Prove that there exists a subset J of I such that the set

$$D = \cup\{D_j \;:\; j \in J\}$$

is not measurable with respect to the completion of the Haar measure μ. Deduce from this fact that there exists a mapping

$$\Psi \;:\; \mathbf{T}^\omega \to \mathbf{R}$$

satisfying the relations:
(a) $ran(\Psi) = \{0,1\}$;
(b) Ψ is constant with respect to each variable x_n $(n \in \omega)$; in particular, Ψ is continuous with respect to each x_n;
(c) Ψ is nonmeasurable with respect to the completion of μ.

Exercise 6. Show that the result presented in the previous exercise has a direct analogue in terms of the Baire property.

Now, we wish to formulate and prove the following statement.

Theorem 2. *Let* λ *denote the standard Lebesgue measure on* $\mathbf{R}$ *and let* $[\mathbf{R}]^{<\mathbf{c}}$ *be the family of all subsets of* $\mathbf{R}$, *whose cardinalities are strictly less than* $\mathbf{c}$. *There exists a subset* Z *of* $\mathbf{R} \times \mathbf{R}$ *such that:*

(1) no three distinct points of Z *belong to a straight line (in other words,* Z *is a set of points in general position);*

(2) Z *is a Lebesgue nonmeasurable subset of* $\mathbf{R} \times \mathbf{R}$;

(3) if the inclusion $[\mathbf{R}]^{<\mathbf{c}} \subset dom(\lambda)$ *holds, then the characteristic function of* Z *is sup-measurable.*

Proof. The argument is very similar to the one applied in the construction of a Sierpiński-Zygmund function (see Chapter 7; cf. also Lemma 2 from Chapter 8).

Obviously, we can identify $\mathbf{c}$ with the first ordinal number α such that $card(\alpha) = \mathbf{c}$. Let λ_2 denote the standard two-dimensional Lebesgue measure on the plane $\mathbf{R} \times \mathbf{R}$ and let $\{Z_\xi : \xi < \alpha\}$ be the family of all Borel subsets of $\mathbf{R} \times \mathbf{R}$ having a strictly positive λ_2-measure. In addition, let $\{\phi_\xi : \xi < \alpha\}$ be the family of all continuous functions acting from $\mathbf{R}$ into $\mathbf{R}$. As usual, we identify any function from $\mathbf{R}$ into $\mathbf{R}$ with its graph lying in the plane $\mathbf{R} \times \mathbf{R}$. Now, using the method of transfinite recursion, we are going to define an α-sequence of points

$$\{(x_\xi, y_\xi) : \xi < \alpha\} \subset \mathbf{R} \times \mathbf{R},$$

satisfying the following conditions:

(a) if $\xi < \alpha$, $\zeta < \alpha$ and $\xi \neq \zeta$, then $x_\xi \neq x_\zeta$;

(b) for each $\xi < \alpha$, the point (x_ξ, y_ξ) belongs to the set Z_ξ;

(c) for each $\xi < \alpha$, the point (x_ξ, y_ξ) does not belong to the union of the family $\{\phi_\zeta : \zeta \leq \xi\}$;

(d) for each $\xi < \alpha$, no three distinct points of the set $\{(x_\zeta, y_\zeta) : \zeta \leq \xi\}$ belong to a straight line.

Suppose that, for an ordinal $\xi < \alpha$, the partial ξ-sequence of points

$$\{(x_\zeta, y_\zeta) : \zeta < \xi\} \subset \mathbf{R} \times \mathbf{R}$$

has already been defined. Let us consider the set Z_ξ. We have

$$\lambda_2(Z_\xi) > 0.$$

According to the classical Fubini theorem, we can write

$$\lambda(\{x \in \mathbf{R} : Z_\xi(x) \in dom(\lambda) \ \& \ \lambda(Z_\xi(x)) > 0\}) > 0$$

where $Z_\xi(x)$ denotes the section of Z_ξ corresponding to a point $x \in \mathbf{R}$. Taking account of the latter formula, we see that there exists an element

$$x_\xi \in \mathbf{R} \setminus \{x_\zeta \ : \ \zeta < \xi\}$$

for which $\lambda(Z_\xi(x_\xi)) > 0$. In particular, we get the equality

$$card(Z_\xi(x_\xi)) = \mathbf{c}.$$

Consequently, there exists an element

$$y_\xi \in Z_\xi(x_\xi) \setminus \cup\{\phi_\zeta(x_\xi) \ : \ \zeta \leq \xi\}.$$

Moreover, y_ξ can be chosen in such a way that the corresponding point (x_ξ, y_ξ) does not belong to the union of all straight lines having at least two common points with the set $\{(x_\zeta, y_\zeta) \ : \ \zeta < \xi\}$.

We have thus defined the point $(x_\xi, y_\xi) \in \mathbf{R} \times \mathbf{R}$. Proceeding in this manner, we are able to construct the α-sequence $\{(x_\xi, y_\xi) \ : \ \xi < \alpha\}$ satisfying conditions (a), (b), (c), and (d). Finally, let us put

$$Z = \{(x_\xi, y_\xi) \ : \ \xi < \alpha\}$$

and let Φ denote the characteristic function of Z (obviously, Z is considered as a subset of the plane $\mathbf{R} \times \mathbf{R}$). Notice that Z can also be regarded as the graph of a partial function acting from $\mathbf{R}$ into $\mathbf{R}$. Hence the inner λ_2-measure of Z is equal to zero. On the other hand, the construction of Z immediately yields that Z is a λ_2-thick subset of the plane. Consequently, Z is nonmeasurable in the Lebesgue sense and the same is true for its characteristic function Φ. It remains to check that Φ is a sup-measurable mapping under the assumption

$$[\mathbf{R}]^{<\mathbf{c}} \subset dom(\lambda).$$

Let us take an arbitrary continuous function $\phi \ : \ \mathbf{R} \to \mathbf{R}$. Then $\phi = \phi_\xi$ for some ordinal $\xi < \alpha$. Now, we can write

$$\{x \in \mathbf{R} \ : \ \Phi(x, \phi_\xi(x)) \neq 0\} = \{x \in \mathbf{R} \ : \ (x, \phi_\xi(x)) \in Z\}$$

and it easily follows from condition (c) that

$$card(\{x \in \mathbf{R} \ : \ (x, \phi_\xi(x)) \in Z\}) \leq card(\xi) + \omega < \mathbf{c}.$$

Because the inclusion $[\mathbf{R}]^{<\mathbf{c}} \subset dom(\lambda)$ holds, we obtain that the function $\Phi_{\phi_\xi} = \Phi_\phi$ almost vanishes (with respect to λ) and, in particular, Φ_ϕ is λ-measurable. Applying Lemma 1, we conclude that Φ is sup-measurable.

Remark 1. It is reasonable to stress here that the set Z (and, consequently, its characteristic function Φ) is defined within the theory **ZFC**. We used an additional set-theoretical hypothesis only to prove that Φ is a sup-measurable mapping. Let us also recall that the first construction of a Lebesgue nonmeasurable subset of the Euclidean plane, no three points of which belong to a straight line, is due to Sierpiński (see, for instance, [162]).

Remark 2. The preceding theorem was proved in the theory **ZFC**. In connection with this theorem, the question naturally arised whether it is possible to establish within **ZFC** the existence of a sup-measurable mapping that is not measurable in the Lebesgue sense (see, e.g., [96]). This question was solved negatively by Shelah and Roslanowski. Namely, they have shown in [176] that the statement "all sup-measurable mappings are Lebesgue measurable" is consistent with **ZFC**.

Exercise 7. By assuming Martin's Axiom and using an argument similar to the proof of Theorem 2, demonstrate that there exists a mapping

$$\Phi : \mathbf{R} \times \mathbf{R} \to \mathbf{R}$$

satisfying the following conditions:

(a) for every Lebesgue measurable function $\phi : \mathbf{R} \to \mathbf{R}$, the superpositions

$$x \to \Phi(\phi(x), x) \quad (x \in \mathbf{R}),$$

$$x \to \Phi(x, \phi(x)) \quad (x \in \mathbf{R})$$

are also Lebesgue measurable;

(b) Φ is not measurable in the Lebesgue sense.

Remark 3. In fact, for the existence of a function Φ of the previous exercise, we do not need the whole power of Martin's Axiom. It suffices to apply a certain set-theoretical hypothesis weaker than Martin's Axiom (cf. Theorem 3 below).

On the other hand, it is not difficult to prove that if a mapping

$$\Psi : \mathbf{R} \times \mathbf{R} \to \mathbf{R}$$

has the property that, for any two Borel functions f and g acting from $\mathbf{R}$ into $\mathbf{R}$, the superposition

$$x \to \Psi(f(x), g(x)) \quad (x \in \mathbf{R})$$

is Lebesgue measurable, then Ψ is Lebesgue measurable, too (see, e.g., [96]). The next exercise contains a slightly more general result.

Exercise 8. Let Ψ be a function acting from $\mathbf{R} \times \mathbf{R}$ into $\mathbf{R}$, such that the superposition

$$x \to \Psi(f(x), g(x)) \qquad (x \in \mathbf{R})$$

is Lebesgue measurable for all continuous functions f and g acting from $\mathbf{R}$ into $\mathbf{R}$. Show that Ψ is Lebesgue measurable (cf. Lemma 1 of this chapter).

We now introduce the notion of a weakly sup-measurable function.

Let Φ be a mapping acting from $\mathbf{R} \times \mathbf{R}$ into $\mathbf{R}$. We shall say that Φ is weakly sup-measurable if, for any continuous function $\phi : \mathbf{R} \to \mathbf{R}$ differentiable almost everywhere (with respect to λ), the superposition Φ_ϕ is Lebesgue measurable.

Evidently, from the view-point of the theory of ordinary differential equations, the notion of a weakly sup-measurable mapping is more important than the notion of a sup-measurable mapping, because any solution of an ordinary differential equation must be continuous everywhere and differentiable almost everywhere.

Clearly, Theorem 2 can be formulated in terms of weakly sup-measurable mappings. In this connection, the following question arises naturally: does there exist a weakly sup-measurable mapping that is not sup-measurable? In order to give a partial answer to this question, we need one auxiliary statement due to Jarník (see [77]).

Lemma 2. *There exists a continuous function $f : \mathbf{R} \to \mathbf{R}$, which is nowhere approximately differentiable.*

We recall that Lemma 2 was proved in Chapter 6 of our book, devoted to one special construction of a nowhere approximately differentiable function.

We also want to recall that, in fact, Jarník proved in [77] that the set of all those functions from the Banach space $C([0,1])$, which are nowhere approximately differentiable, is residual in $C([0,1])$, i.e., is the complement of a first category set. Nevertheless, in our further considerations, we need only one such function.

Theorem 3. *Suppose that:*
(1) $[\mathbf{R}]^{<\mathbf{c}} \subset dom(\lambda)$;
(2) for any cardinal number $\kappa < \mathbf{c}$ and for any family $\{X_\xi : \xi < \kappa\}$ of λ-measure zero subsets of $\mathbf{R}$, we have

$$\cup\{X_\xi : \xi < \kappa\} \neq \mathbf{R}.$$

Then there exists a weakly sup-measurable mapping $\Phi : \mathbf{R} \times \mathbf{R} \to \mathbf{R}$ that is not sup-measurable.

Proof. We can identify $\mathbf{c}$ with the first ordinal number α such that $card(\alpha) = \mathbf{c}$. Let f be a function from Lemma 2. Let $\{B_\xi : \xi < \alpha\}$ be some Borel base of the σ-ideal of all Lebesgue measure zero subsets of $\mathbf{R}$ and let $\{\phi_\xi : \xi < \alpha\}$ be the family of all continuous functions acting from $\mathbf{R}$ into $\mathbf{R}$ and differentiable almost everywhere in $\mathbf{R}$. We are going to construct (by transfinite recursion) an injective α-sequence

$$\{(x_\xi, y_\xi) : \xi < \alpha\} \subset \mathbf{R} \times \mathbf{R}$$

of points belonging to the graph of f. Suppose that, for an ordinal $\xi < \alpha$, the partial ξ-sequence $\{(x_\zeta, y_\zeta) : \zeta < \xi\}$ has already been defined. Notice that, for each $\zeta \leq \xi$, the closed set

$$P_\zeta = \{x \in \mathbf{R} : \phi_\zeta(x) = f(x)\}$$

is of Lebesgue measure zero. Indeed, assuming otherwise, i.e., $\lambda(P_\zeta) > 0$, we can find a density point x of P_ζ belonging to P_ζ such that there exists an approximate derivative $f'_{ap}(x) = \phi'_\zeta(x)$. But this is impossible in view of the property of f. Consequently, $\lambda(P_\zeta) = 0$ for all $\zeta \leq \xi$, and the set

$$\mathbf{R} \setminus ((\cup\{B_\zeta : \zeta \leq \xi\}) \cup (\cup\{P_\zeta : \zeta \leq \xi\}) \cup \{x_\zeta : \zeta < \xi\})$$

is not empty. Let x_ξ be an arbitrary point from this set and let $y_\xi = f(x_\xi)$.

Proceeding in such a manner, we are able to define the required family of points $\{(x_\xi, y_\xi) : \xi < \alpha\}$. Now, we put

$$Z = \{(x_\xi, y_\xi) : \xi < \alpha\}, \qquad X = \{x_\xi : \xi < \alpha\}$$

and denote by Φ the characteristic function of Z. Then it can easily be seen that Φ is a weakly sup-measurable mapping (cf. the proof of Theorem 2). On the other hand, let us consider the superposition Φ_f. Obviously, we have

$$\Phi_f(x) = 1 \Leftrightarrow (x, f(x)) \in Z \Leftrightarrow x \in X.$$

It follows from our construction that X is a Sierpiński type subset of the real line $\mathbf{R}$ (for the definition and various properties of Sierpiński sets, see, e.g., [120], [149], [153], [162] or Chapter 10 of our book which is specially devoted to Luzin and Sierpiński sets). In particular, as shown in that chapter, X is not measurable in the Lebesgue sense and, therefore, Φ_f is not Lebesgue measurable, either. We thus conclude that Φ is not a sup-measurable mapping. This completes the proof of the theorem.

Remark 4. It is well known that assumptions (1) and (2) of Theorem 3 are logically independent (see, for instance, [118]). Slightly changing the argument presented above, one can show (under the assumptions of Theorem 3) that there exists a weakly sup-measurable mapping, which is not sup-measurable and, in addition, is not Lebesgue measurable.

We do not know whether the assertion of Theorem 3 (i.e., the existence of a weakly sup-measurable mapping which is not sup-measurable) can be proved in the theory **ZFC**.

Remark 5. Evidently, the notion of sup-measurability can be formulated in terms of the Baire property instead of measurability in the Lebesgue sense. More precisely, we say that a function

$$\Phi \; : \; \mathbf{R} \times \mathbf{R} \to \mathbf{R}$$

is sup-measurable in the sense of the Baire property if, for any function

$$\phi \; : \; \mathbf{R} \to \mathbf{R}$$

possessing the Baire property, the superposition

$$x \to \Phi(x, \phi(x)) \qquad (x \in \mathbf{R})$$

possesses the Baire property, too. In a similar way, the notion of weak sup-measurability (in the sense of the Baire property) can be introduced.

It is not difficult to verify that, for functions with the Baire property, a direct analogue of Theorem 2 holds true. The corresponding analogue of Theorem 3 also holds (in this case, we do not need Lemma 2; it suffices to apply the existence of a continuous nowhere differentiable function acting from **R** into **R**). The corresponding details are left to the reader as a useful exercise.

Exercise 9. Prove an analogue of Theorem 2 for the Baire property.

Exercise 10. Prove an analogue of Theorem 3 for the Baire property.

Exercise 11. Let k be a strictly positive integer. Suppose that:

(a) any subset X of **R** with $card(X) < \mathbf{c}$ has the Baire property, i.e., is of first category;

(b) for any family $\{X_i \; : \; i \in I\}$ with $card(I) < \mathbf{c}$, consisting of first category subsets of **R**, we have

$$\mathbf{R} \neq \cup\{X_i \; : \; i \in I\}.$$

Show that there exists a mapping

$$\Phi \; : \; \mathbf{R} \times \mathbf{R} \to \mathbf{R}$$

satisfying the following relations:
    (1) for every k-times continuously differentiable function

$$g \; : \; \mathbf{R} \to \mathbf{R},$$

the superposition

$$x \to \Phi(x, g(x)) \qquad (x \in \mathbf{R})$$

has the Baire property;
    (2) there is a $(k - 1)$-times continuously differentiable function

$$h \; : \; \mathbf{R} \to \mathbf{R}$$

such that the superposition

$$x \to \Phi(x, h(x)) \qquad (x \in \mathbf{R})$$

does not have the Baire property.
    Formulate and prove an analogous result in terms of the Lebesgue measure.

Exercise 12. Let Φ be a function acting from $\mathbf{R} \times \mathbf{R}$ into $\mathbf{R}$. Show that the following two assertions are equivalent:
    (a) Φ is sup-measurable in the sense of the Baire property;
    (b) for every Borel function ϕ acting from $\mathbf{R}$ into $\mathbf{R}$, the superposition

$$x \to \Phi(x, \phi(x)) \qquad (x \in \mathbf{R})$$

has the Baire property.
    Also, determine the precise Baire order of functions ϕ that is sufficient for the equivalence of these two assertions.

    As mentioned above (see Exercise 2 of this chapter), the functions of two variables, satisfying the Carathéodory conditions, are good from the point of view of sup-measurability. In addition, such functions play an important role in the theory of ordinary differential equations. The following definition introduces a slightly more general class of functions.
    Let $(X, \mathcal{S}, \mu)$ be a space with a complete probability measure, let Y be a topological space and let

$$f \; : \; X \times Y \to \mathbf{R}$$

be a function. We say that f almost satisfies the Carathéodory conditions if:

(1) for almost all (with respect to μ) points $x \in X$, the partial function $f(x, \cdot)$ is continuous on Y;

(2) for all points $y \in Y$, the partial function $f(\cdot, y)$ is μ-measurable.

Obviously, if f satisfies the Carathéodory conditions, then it almost satisfies these conditions. The converse assertion is not true, in general. However, it can easily be verified that if f almost satisfies the Carathéodory conditions, then there exists a function

$$g \ : \ X \times Y \to \mathbf{R}$$

satisfying these conditions, such that

$$f(x, \cdot) = g(x, \cdot)$$

for almost all points $x \in X$.

Some useful additional information about functions which almost satisfy the Carathéodory conditions is given in the exercises below.

Exercise 13. Let $(X, \mathcal{S}, \mu)$ be a space with a complete probability measure, let Y be a topological space with a countable base and let

$$f \ : \ X \times Y \to [0, 1]$$

be a function satisfying the Carathéodory conditions. Pick a countable base $\{U_n \ : \ n \in \mathbf{N}\}$ of Y and consider the family F of all those functions ϕ which can be represented in the form

$$\phi = q_n \cdot \psi_{U_n},$$

where

$$n \in \mathbf{N}, \quad q_n \in \mathbf{Q} \cap [0, 1]$$

and ψ_{U_n} is the characteristic function of the set U_n. Because the family F is countable, we may write

$$F = \{\phi_k \ : \ k \in \mathbf{N}\}.$$

Further, for any $k \in \mathbf{N}$, let us define

$$X_k = \{x \in X \ : \ (\forall y \in Y)(\phi_k(y) \leq f(x, y))\}.$$

Fix a countable subset Y_0 of Y everywhere dense in Y and, for each $k \in \mathbf{N}$, put

$$X_k' = \{x \in X \ : \ (\forall y \in Y_0)(\phi_k(y) \leq f(x, y))\}.$$

Check that the set X'_k is μ-measurable and the equality $X_k = X'_k$ holds.

Therefore, X_k is a μ-measurable subset of X.

Show that
$$f(x, y) = sup_{k \in \mathbf{N}} \ \psi_{X_k}(x)\phi_k(y)$$
for all $x \in X$ and $y \in Y$ (here ψ_{X_k} denotes the characteristic function of X_k). Deduce from this fact that:

(a) the function f is measurable with respect to the product of the σ-algebras S and $B(Y)$, where $B(Y)$ denotes, as usual, the Borel σ-algebra of Y;

(b) the function f is sup-measurable, i.e., for any μ-measurable mapping
$$h \ : \ X \to Y,$$
the superposition
$$x \to f(x, h(x)) \qquad (x \in X)$$
is μ-measurable, too.

Suppose, in addition, that X is a Hausdorff topological space and μ is the completion of a Radon probability measure on X. By applying a Luzin type theorem on the structure of μ-measurable real-valued functions, prove that, for each $\varepsilon > 0$, there exists a compact set $P \subset X$ with $\mu(P) > 1 - \varepsilon$, such that the restriction of f to the product set $P \times Y$ is lower semicontinuous.

Extend the results presented above to the functions which almost satisfy the Carathéodory conditions.

Exercise 14. Let X be a Hausdorff topological space, let μ be the completion of a Radon probability measure on X and let Y be a topological space with a countable base. By using the results of Exercise 13, show that, for a function
$$f \ : \ X \times Y \to [0, 1],$$
the following two assertions are equivalent:

(a) f almost satisfies the Carathéodory conditions;

(b) for any $\varepsilon > 0$, there exists a compact set $P \subset X$ with $\mu(P) > 1 - \varepsilon$, such that the restriction of f to the product set $P \times Y$ is continuous.

Note that the equivalence of these two assertions is usually called the Scorza Dragoni theorem.

Exercise 15. Let (X, S, μ) be a space with a complete σ-finite measure, let Y be a locally compact topological space with a countable base and let Z be a subset of the product space $X \times Y$, measurable with respect to the product of the σ-algebras S and $B(Y)$. Demonstrate that the set
$$pr_1(Z) = \{x \in X \ : \ (\exists y \in Y)((x, y) \in Z)\}$$

is measurable with respect to μ. For this purpose, apply Choquet's theorem on capacities (see, e.g., [47], [156], or [34]).

Exercise 16. Let $(X, \mathcal{S}, \mu)$ be a space with a complete probability measure, let Y be a nonempty compact metric space and let

$$f \: : \: X \times Y \to [0, 1]$$

be a function satisfying the Carathéodory conditions. For each point $x \in X$, denote

$$F(x) = \{y \in Y \: : \: f(x, y) = inf_{t \in Y} f(x, t)\}.$$

Check that $F(x)$ is a nonempty closed subset of Y. Hence we have a set-valued mapping

$$F \: : \: X \to \mathcal{P}(Y).$$

Prove that, for any open set $U \subset Y$, the set

$$\{x \in X \: : \: F(x) \cap U \neq \emptyset\}$$

is μ-measurable. Derive from this fact, by using the theorem of Kuratowski and Ryll-Nardzewski on the existence of measurable selectors (see [121] or [125]), that there exists a μ-measurable mapping

$$h \: : \: X \to Y$$

such that

$$f(x, h(x)) = inf_{y \in Y} f(x, y)$$

for all points $x \in X$.

Formulate and prove an analogous statement for functions almost satisfying the Carathéodory conditions.

Let us present a result, which is much deeper than that given in the exercise above. For this purpose, we need the notion of a C-set. This notion was introduced (by Luzin and Kolmogorov) many years ago and was thoroughly investigated by several authors (see, e.g., [132], [157], and [181]).

Let E be a metric space. We define the family of all C-sets in E as the smallest class containing all open subsets of E and closed under the operation of taking the complement and under the (A)-operation. Because the (A)-operation includes in itself countable unions and countable intersections, we see that the class of all C-sets in E forms a certain σ-algebra containing all analytic subsets of E (in particular, the Borel σ-algebra of E is contained in the class of all C-subsets of E). It immediately follows from the definition that all C-sets are universally measurable and possess the Baire property in the restricted sense (see the Introduction).

Now, let E' be another metric space and let

$$f : E \to E'$$

be a mapping. We say that f is C-measurable if, for each open subset U of E', the preimage $f^{-1}(U)$ is a C-set in E.

From the definition of the (A)-operation it follows that the preimage of the result of this operation over a given family of sets coincides with the result of the same operation over the family of preimages of sets. Taking this fact into account, we may easily infer that, for any C-measurable mapping f from E into E' and for any C-set Z in E', the preimage $f^{-1}(Z)$ is a C-set in E.

In particular, we claim that the composition of two C-measurable functions is a C-measurable function, too (this is an important feature of C-measurable functions, which has no analogue, e.g., in the class of Lebesgue measurable functions).

Let now X and Y be any two Polish spaces and let

$$f : X \times Y \to [0,1]$$

be a Borel mapping such that

$$(\forall x \in X)(\exists y \in Y)(f(x,y) = inf_{t \in Y} f(x,t)).$$

Let us define
$$f^*(x) = inf_{t \in Y} f(x,t) \qquad (x \in X).$$

Note that, for each $a \in \mathbf{R}$, we have

$$\{x \in X : f^*(x) < a\} = \{x \in X : (\exists t \in Y)(f(x,t) < a)\}.$$

Because the original function f is Borel, we conclude that the first set in the equality above is analytic, i.e., the function f^* is measurable with respect to the σ-algebra generated by all analytic subsets of X (hence f^* is C-measurable as well). In the product space $X \times \mathbf{R} \times Y$ consider the set

$$B = \{(x,z,y) : f(x,y) = z\}.$$

Obviously, this set is Borel (as a homeomorphic image of the graph of a Borel function). According to the classical uniformization theorem of Luzin,

Jankov, and von Neumann (which is an easy consequence of the theorem of Kuratowski and Ryll-Nardzewski), there exists a function

$$h \ : \ pr_{X \times \mathbf{R}}(B) \to Y$$

such that:

(1) h is measurable with respect to the σ-algebra generated by all analytic subsets of $dom(h)$;

(2) the graph of h is contained in B.

Further, our assumption on f and the definition of f^* imply the relations:

(a) $pr_X(dom(h)) = X$;

(b) for any $x \in X$, we have $(x, f^*(x)) \in dom(h)$.

Consequently, we may write

$$(\forall (x,z) \in dom(h))((x, z, h(x,z)) \in B),$$

$$(\forall (x,z) \in dom(h))(f(x, h(x,z)) = z),$$

$$(\forall x \in X)(f(x, h(x, f^*(x))) = f^*(x)).$$

Now, define a function

$$g \ : \ X \to \mathbf{R}$$

by the formula

$$g(x) = h(x, f^*(x)) \qquad (x \in X).$$

Then g is C-measurable (as a composition of two C-measurable functions) and we get

$$f(x, g(x)) = f^*(x) = inf_{t \in Y} f(x, t)$$

for all points $x \in X$.

Exercise 17. Let us consider the diagonal

$$\{(x, y) \ : \ x \in \mathbf{R}, \ y \in \mathbf{R}, \ x = y\}$$

of the Euclidean plane $\mathbf{R}^2$ and let Z be a subset of this diagonal, nonmeasurable with respect to the standard one-dimensional Lebesgue measure on it. Define a function

$$f \ : \ \mathbf{R}^2 \to \mathbf{R},$$

by putting $f(x, y) = 0$ if $(x, y) \notin Z$, and $f(x, y) = -1$ if $(x, y) \in Z$.

Show that:

(a) for each point $x \in \mathbf{R}$, the partial function $f(x, \cdot) \ : \ \mathbf{R} \to \mathbf{R}$ is lower semicontinuous;

(b) for each point $y \in \mathbf{R}$, the partial function $f(\cdot, y) \, : \, \mathbf{R} \to \mathbf{R}$ is lower semicontinuous;

(c) the function

$$x \to f(x, x) \qquad (x \in \mathbf{R})$$

is not Lebesgue measurable (consequently, f is not sup-measurable);

(d) f is measurable with respect to the two-dimensional Lebesgue measure λ_2 on $\mathbf{R}^2$.

Exercise 18. Suppose that there exists a Luzin subset of the Euclidean plane $\mathbf{R}^2$. Demonstrate that, in this case, there exists a mapping

$$\Phi \, : \, \mathbf{R}^2 \to \mathbf{R}$$

satisfying the following relations:

(a) Φ does not have the Baire property;

(b) for each function $\phi \, : \, \mathbf{R} \to \mathbf{R}$ whose graph is a first category subset of the plane (in particular, for each ϕ possessing the Baire property), the superposition

$$x \to \Phi(x, \phi(x)) \qquad (x \in \mathbf{R})$$

possesses the Baire property.

Deduce an analogous result (in terms of the Lebesgue measurability) from the existence of a Sierpiński subset of the plane.

Exercise 19. Let us denote by the symbol $M_0 = M_0[0, 1]$ the family of all Lebesgue measurable functions acting from $[0, 1]$ into $\mathbf{R}$. Obviously, we have a canonical equivalence relation $\equiv$ in M_0 defined by the formula

$$f \equiv g \Leftrightarrow f \text{ and } g \text{ coincide almost everywhere on } [0, 1].$$

We denote by M the factor set with respect to this equivalence relation (i.e., M is the family of all equivalence classes with respect to $\equiv$). If f is an arbitrary function from the original family M_0, then the symbol $[f]$ will denote the class of all those functions that are equivalent to f. The natural algebraic operations in M_0 are compatible with the relation $\equiv$ and, consequently, induce the corresponding algebraic operations in M. Therefore, M becomes a linear algebra over the field $\mathbf{R}$.

Further, for any two elements $[f] \in M$ and $[g] \in M$, we put

$$d([f], [g]) = \int_{[0, 1]} \frac{|f(t) - g(t)|}{1 + |f(t) - g(t)|} dt.$$

Check that this definition is correct (i.e., it does not depend on the choice of f and g) and that the function

$$d \; : \; M \times M \to \mathbf{R}$$

obtained in this way turns out to be a metric on M.

Show that:

(a) the pair (M, d) is a Polish topological vector space;

(b) there exists no nonzero linear continuous functional defined on the entire space M (in other words, the conjugate space M^* is trivial).

Now, let us extend the equivalence relation $\equiv$ introduced above onto the family of all Lebesgue measurable partial functions acting from $[0, 1]$ into $\mathbf{R}$. Namely, for any two such functions f and g, we put

$$(f \equiv g) \Leftrightarrow (\lambda(dom(f) \triangle dom(g)) = 0 \ \&$$

f coincides with g almost everywhere on $dom(f) \cap dom(g))$.

Further, we denote by the symbol M' the family of all equivalence classes with respect to $\equiv$. Evidently, we have the canonical embedding

$$j \; : \; M \to M',$$

so we may identify M with the subset $j(M)$ of M'.

Suppose now that some mapping (operator) $H \; : \; M' \to M'$ is given and satisfies the following conditions:

(1) for each element $[f] \in M'$ and for any Lebesgue measurable partial function g such that $[g] = H([f])$, we have the equality

$$\lambda(dom(f) \triangle dom(g)) = 0;$$

(2) if $[f] \in M'$, all partial functions f_n $(n < \omega)$ are the restrictions of f to pairwise disjoint Lebesgue measurable subsets of $\mathbf{R}$ whose union coincides with $dom(f)$ and

$$H([f_n]) = [g_n] \qquad (n < \omega),$$

$$dom(g_n) \cap dom(g_m) = \emptyset \qquad (n < \omega, \ m < \omega, \ n \neq m),$$

then we have the equality

$$H([f]) = [g],$$

where g denotes the common extension of all partial functions g_n $(n < \omega)$ with

$$dom(g) = \cup \{dom(g_n) \; : \; n < \omega\}.$$

In this case, we say that H is an admissible operator acting from the family M' into itself.

Suppose that such an operator H is given.

Under the assumption that the σ-ideal $\mathcal{I}(\lambda)$ of all Lebesgue measure zero subsets of $\mathbf{R}$ is $\mathbf{c}$-additive (i.e., $\cup\{X_\xi : \xi < \beta\} \in \mathcal{I}(\lambda)$ whenever $card(\beta) < \mathbf{c}$ and $X_\xi \in \mathcal{I}(\lambda)$ for each $\xi < \beta$), demonstrate the existence of a mapping

$$\Phi : [0,1] \times \mathbf{R} \to \mathbf{R}$$

satisfying the following condition: for every Lebesgue measurable partial function

$$f : [0,1] \to \mathbf{R},$$

the superposition

$$\Phi_f : [0,1] \to \mathbf{R}$$

is a Lebesgue measurable partial function, too, and the equality

$$H([f]) = [\Phi_f]$$

holds true. In other words, any admissible operator $H : M' \to M'$ is representable in the form of some superposition operator Φ (this result is essentially due to Krasnoselskii and Pokrovskii). Note, in addition, that the descriptive properties of this Φ can be very bad. In particular, Φ can be nonmeasurable with respect to the standard two-dimensional Lebesgue measure on $\mathbf{R}^2$ (cf. Theorem 2 of this chapter).

Finally, formulate and prove an analogue of the preceding result for partial functions acting from $[0,1]$ into $\mathbf{R}$ and having the Baire property (under the assumption that the σ-ideal $\mathcal{K}(\mathbf{R})$ of all first category subsets of $\mathbf{R}$ is $\mathbf{c}$-additive).

Exercise 20. In this exercise a question about the existence of extensions of partial functions of two variables, satisfying the Carathéodory conditions, is considered.

Let X be a set equipped with a complete σ-finite measure μ, let Y be a compact metric space and let

$$\Phi_0 : X \times Y \to \mathbf{R}$$

be a partial mapping measurable with respect to the product of the σ-algebras $dom(\mu)$ and $\mathcal{B}(Y)$.

Applying Exercise 15 of this chapter, the Tietze-Urysohn theorem on extensions of continuous real-valued functions and the theorem of Kuratowski and Ryll-Nardzewski on measurable selectors, prove that the following two assertions are equivalent:

(a) for each element $x \in X$, the partial function $\Phi_0(x, \cdot)$ is uniformly continuous on its domain;

(b) there exists a mapping $\Phi : X \times Y \to \mathbf{R}$ extending Φ_0 and satisfying the Carathéodory conditions.

Exercise 21. Let X be a topological space equipped with a σ-finite inner regular Borel measure μ and let Y be a compact metric space. Denote:

$\mu' =$ the completion of μ;

$\mathcal{S}' = dom(\mu')$;

$\mathcal{F}(Y) =$ the family of all nonempty closed subsets of Y.

Let $\Psi : X \to \mathcal{F}(Y)$ be a set-valued mapping. Prove that the following assertions are equivalent:

(a) Ψ is μ'-measurable as a function acting from the measure space $(X, \mathcal{S}', \mu')$ into the space $\mathcal{F}(Y)$ endowed with the Hausdorff metric;

(b) Ψ is lower $\mathcal{S}'$-measurable, i.e.,

$$\{x \in X : \Psi(x) \cap U \neq \emptyset\} \in \mathcal{S}'$$

for all open sets $U \subset Y$;

(c) Ψ is upper $\mathcal{S}'$-measurable, i.e.,

$$\{x \in X : \Psi(x) \cap F \neq \emptyset\} \in \mathcal{S}'$$

for all closed sets $F \subset Y$.

16. Generalized step-functions and superposition operators

The previous chapter was devoted to some properties of sup-measurable and weakly sup-measurable functions of two variables. In this chapter, for a given σ-ideal of sets, we introduce the notion of a real-valued generalized step-function and investigate generalized step-functions in connection with the problem of sup-measurability of certain functions of two variables, regarded as superposition operators.

Let $\mathbf{R}$ denote, as usual, the real line and let

$$\Phi : \mathbf{R} \times \mathbf{R} \to \mathbf{R}$$

be a function of two variables. Then (see Chapter 15) this Φ can be treated as a superposition operator defined as follows: for any function $f : \mathbf{R} \to \mathbf{R}$, we put

$$(\Phi(f))(x) = \Phi(x, f(x)) \qquad (x \in \mathbf{R}).$$

Sometimes, Φ is also called the Nemytskii superposition operator (cf. [114], [188]).

Let λ denote the standard Lebesgue measure on $\mathbf{R}$. In many cases, it is important to know whether a given superposition operator Φ preserves the class $L(\mathbf{R}, \mathbf{R})$ of all real-valued Lebesgue measurable functions on $\mathbf{R}$ (i.e., $\Phi(f)$ is λ-measurable whenever f is λ-measurable). There are various sufficient conditions under which Φ maps $L(\mathbf{R}, \mathbf{R})$ into itself.

In particular, we know (see Chapter 15) that if Φ is λ-measurable with respect to the first variable and continuous with respect to the second variable (the so-called Carathéodory classical conditions), then Φ preserves $L(\mathbf{R}, \mathbf{R})$ or, in short, Φ is sup-measurable. In such a case, Φ is also λ_2-measurable where

$$\lambda_2 = \lambda \times \lambda$$

stands for the two-dimensional Lebesgue measure on the plane $\mathbf{R}^2 = \mathbf{R} \times \mathbf{R}$.

Other conditions for the sup-measurability of Φ can be found, e.g., in [187].

331

We have already shown in Chapter 15 that, under some additional set-theoretical axioms, there exist sup-measurable operators Φ, which are not λ_2-measurable.

In this chapter our attention is focused on the following problem:

Give a characterization of all those functions $f \in L(\mathbf{R}, \mathbf{R})$ for which there exists a superposition operator Φ having rather good descriptive properties and such that $\Phi(f)$ does not belong to $L(\mathbf{R}, \mathbf{R})$.

In order to present a solution of this problem, we need several auxiliary notions and propositions. First of all, let us formulate the following classical statement from descriptive set theory.

Lemma 1. *Let E be a Polish topological space, E' be a metric space and let $\phi : E \to E'$ be a continuous mapping whose range is uncountable. Then there exists a set $C \subset E$ homeomorphic to the Cantor discontinuum, such that the restriction $\phi|C$ is injective (consequently, $\phi|C$ is a homeomorphism between C and $\phi(C)$).*

For the proof of Lemma 1, see, e.g., [120]. In fact, this lemma directly implies that the cardinality of any uncountable analytic subset of a Polish space is equal to the cardinality of the continuum.

Recall that $f \in L(\mathbf{R}, \mathbf{R})$ is a step-function if

$$card(ran(f)) \leq \omega,$$

i.e., the range of f is at most countable. In this case, there exists a countable partition $\{X_j : j \in J\}$ of $\mathbf{R}$ consisting of Lebesgue measurable sets such that, for any $j \in J$, the restriction $f|X_j$ is a constant function.

We shall say that $f \in L(\mathbf{R}, \mathbf{R})$ is a generalized step-function if there exists at least one step-function $g \in L(\mathbf{R}, \mathbf{R})$ such that f and g are equivalent with respect to the measure λ, i.e.,

$$\{x \in \mathbf{R} : f(x) \neq g(x)\} \in \mathcal{I}(\lambda).$$

The next lemma yields a characterization of generalized step-functions. This characterization will be useful below.

Lemma 2. *If $f \in L(\mathbf{R}, \mathbf{R})$, then the following two assertions are equivalent:*

(1) f is not a generalized step-function;

(2) there exists a set $Y \subset \mathbf{R}$ with $\lambda^*(Y) > 0$ such that the restriction $f|Y$ is injective.

Proof. The implication $(2) \Rightarrow (1)$ is trivial. Let us prove the implication $(1) \Rightarrow (2)$. Suppose that $f \in L(\mathbf{R}, \mathbf{R})$ satisfies (1). Let us denote

$$T_0 = \{t \in ran(f) : \lambda(f^{-1}(t)) > 0\}.$$

In view of the σ-finiteness of λ, we have the inequality

$$card(T_0) \leq \omega.$$

Because f is not a generalized step-function, we also have

$$\lambda(\mathbf{R} \setminus f^{-1}(T_0)) > 0.$$

Moreover, applying to f the classical Luzin theorem, we claim that there exists a closed set $P \subset \mathbf{R} \setminus f^{-1}(T_0)$ with $\lambda(P) > 0$ for which the restriction $f|P$ is continuous and

$$card(ran(f|P)) > \omega.$$

Let us put

$$h = f|P, \quad T = ran(h).$$

Then $\lambda(h^{-1}(t)) = 0$ for each point $t \in T$. Denote by α the least ordinal number of cardinality continuum and let $(P_\xi)_{\xi < \alpha}$ be an injective family of all closed subsets of P having strictly positive λ-measure. Construct, by using the method of transfinite recursion, a family of points

$$\{y_\xi : \xi < \alpha\} \subset P.$$

Namely, take an ordinal $\xi < \alpha$ and suppose that the partial family $\{y_\zeta : \zeta < \xi\}$ has already been defined. Keeping in mind Lemma 1, it is not difficult to check that

$$P_\xi \setminus \cup\{h^{-1}(h(y_\zeta)) : \zeta < \xi\} \neq \emptyset.$$

Hence, there exists a point y belonging to $P_\xi \setminus \cup\{h^{-1}(h(y_\zeta)) : \zeta < \xi\}$. We put $y_\xi = y$. By proceeding in this manner, the required family of points $\{y_\xi : \xi < \alpha\}$ will be constructed. Denote now

$$Y = \{y_\xi : \xi < \alpha\}.$$

It immediately follows from our construction that Y is a partial selector of the disjoint family of sets $\{h^{-1}(t) : t \in T\}$. This implies that the restriction $h|Y$ (consequently, the restriction $f|Y$) is injective. Moreover, since

$$P_\xi \cap Y \neq \emptyset$$

for each $\xi < \alpha$, we easily infer that

$$\lambda^*(Y) = \lambda(P) > 0.$$

This completes the proof of Lemma 2.

Lemma 3. *If $f \in L(\mathbf{R}, \mathbf{R})$ is not a generalized step-function, then there exists a λ-nonmeasurable set $X \subset \mathbf{R}$ for which the restriction $f|X$ is injective.*

Proof. According to Lemma 2, there exists a set $Y \subset \mathbf{R}$ with $\lambda^*(Y) > 0$ such that $f|Y$ is an injection. If Y is not measurable in the Lebesgue sense, then we are done. Suppose now that $Y \in dom(\lambda)$ and hence $\lambda(Y) > 0$. It is well known (see, e.g., Chapter 8 of this book) that Y contains a subset nonmeasurable with respect to λ. Take any such subset and denote it by X. Clearly, $f|X$ is an injection and the proof is completed.

Theorem 1. *Let $f \in L(\mathbf{R}, \mathbf{R})$ and suppose that f is not a generalized step-function. Then there exists a superposition operator*

$$\Phi : \mathbf{R} \times \mathbf{R} \to \mathbf{R}$$

satisfying the following relations:
  (1) $ran(\Phi) = \{0, 1\}$;
  (2) for any $x \in \mathbf{R}$, the partial function $\Phi(x, \cdot)$ is lower semicontinuous;
  (3) for any $y \in \mathbf{R}$, the partial function $\Phi(\cdot, y)$ is lower semicontinuous;
  (4) Φ is a λ_2-measurable operator;
  (5) the function $\Phi(f)$ is not λ-measurable.

Proof. According to Lemma 3, there exists a λ-nonmeasurable set $X \subset \mathbf{R}$ for which the restriction $f|X$ is injective. Define the required superposition operator Φ as follows:

$$\Phi(x, y) = 0 \qquad (x \in X, \ y = f(x)),$$

$$\Phi(x, y) = 1 \qquad (x \in \mathbf{R} \setminus X, \ y = f(x)),$$

$$\Phi(x, y) = 1 \qquad (x \in \mathbf{R}, \ y \in \mathbf{R}, \ y \neq f(x)).$$

For this Φ, relations (1), (2) and (3) are verified directly. Further, because the graph of f is a λ_2-measure zero subset of $\mathbf{R}^2$, we claim that Φ is equivalent to a constant function and, consequently, Φ is λ_2-measurable. Finally, we have

$$\Phi(x, y) = 0 \Leftrightarrow (x \in X \ \& \ y = f(x))$$

whence it follows that
$$(\Phi(f))^{-1}(0) = X$$
and, therefore, $\Phi(f)$ is not λ-measurable. This ends the proof.

Theorem 2. *Let $f \in L(\mathbf{R}, \mathbf{R})$ and suppose that f is not a generalized step-function. Then there exists a superposition operator*

$$\Psi : \mathbf{R} \times \mathbf{R} \to \mathbf{R}$$

such that:
  (1) $ran(\Psi) = \{1, 2\}$,
  (2) for any $x \in \mathbf{R}$, the partial function $\Psi(x, \cdot)$ is lower semicontinuous;
  (3) for any $y \in \mathbf{R}$, the partial function $\Psi(\cdot, y)$ is lower semicontinuous;
  (4) Ψ is a λ_2-nonmeasurable operator;
  (5) the function $\Psi(f)$ is λ-nonmeasurable.

Proof. By using the method of transfinite recursion and applying the standard argument (cf. Chapter 8), an injective function

$$g : \mathbf{R} \to \mathbf{R}$$

can be defined whose graph is λ_2-thick in $\mathbf{R}^2$ and does not intersect the graph of f. Let χ_g denote the characteristic function of the graph of g (of course, this graph is considered as a subset of $\mathbf{R}^2$). We put

$$\Psi = \Phi + 1 - \chi_g,$$

where Φ is the superposition operator of Theorem 1. It is easy to verify that Ψ is the required superposition operator, i.e., Ψ satisfies all relations (1)–(5) of Theorem 2.

Remark 1. If a superposition operator

$$\Phi : \mathbf{R} \times \mathbf{R} \to \mathbf{R}$$

is λ-measurable with respect to the first variable, then $\Phi(f)$ is λ-measurable for every generalized step-function $f \in L(\mathbf{R}, \mathbf{R})$.

We thus see (in view of Theorem 1) that the generalized step-functions are exactly those functions $f \in L(\mathbf{R}, \mathbf{R})$ for which every superposition operator Φ Lebesgue measurable with respect to the first variable yields Lebesgue measurable $\Phi(f)$.

Remark 2. If a superposition operator $\Phi : \mathbf{R} \times \mathbf{R} \to \mathbf{R}$ is lower semicontinuous (more generally, Borel) with respect to the first variable and continuous with respect to the second variable, then Φ is a Borel mapping from $\mathbf{R}^2$ into $\mathbf{R}$, hence Φ is also sup-measurable (see Exercise 3 of Chapter 15).

A much stronger result is contained in the following

Exercise 1. Let X be a topological space, Y be a separable metric space, Z be a metric space and let

$$\Phi : X \times Y \to Z$$

be a mapping satisfying these two conditions:
 (a) for each $x \in X$, the partial mapping $\Phi(x, \cdot)$ is continuous;
 (b) for each $y \in Y$, the partial mapping $\Phi(\cdot, y)$ is Borel.
Show that Φ is a Borel mapping.

Remark 3. Theorems 1 and 2 admit direct analogues for functions possessing the Baire property. Those analogues can be proved by the same scheme as for Lebesgue measurable functions. Only one essential moment should be mentioned. Namely, the proofs of Theorems 1 and 2 are based on the classical Luzin theorem concerning the structure of λ-measurable functions. Because we cannot apply the Luzin theorem to functions possessing the Baire property, we must replace this theorem by an appropriate similar statement. Such a statement is well known in general topology (see [120], [162] or Exercise 13 from the Introduction) and is formulated as follows.

Let E_1 be a topological space, E_2 be a topological space with a countable base and let $f : E_1 \to E_2$ be a mapping possessing the Baire property. Then there exists a first category set $Z \subset E_1$ such that the restriction $f|(E_1 \setminus Z)$ is continuous.

In this statement we may assume, without loss of generality, that Z is an F_σ-subset of E_1, hence $E_1 \setminus Z$ is a G_δ-set in E_1. If the original space E_1 is Polish, then $E_1 \setminus Z$ is also Polish (by virtue of the Alexandrov theorem). Consequently, if E_1 is a Polish space and $E_2 = \mathbf{R}$, we are able to apply Lemma 1 to the continuous function $f|(E_1 \setminus Z)$.

Under some additional set-theoretical axioms, Lemma 2 admits a significant generalization. Let us consider some abstract version of this lemma.

Fix an uncountable set E and a σ-ideal $\mathcal{I}$ of subsets of E, containing all singletons in E.

We shall say that $g : E \to \mathbf{R}$ is a step-function if

$$card(ran(g)) \le \omega.$$

We shall say that $f : E \to \mathbf{R}$ is a generalized step-function with respect to $\mathcal{I}$ if there exists at least one step-function $g : E \to \mathbf{R}$ for which we have

$$\{x \in E : f(x) \neq g(x)\} \in \mathcal{I},$$

i.e., f and g are $\mathcal{I}$-equivalent functions.

Recall that a family of sets $\mathcal{B} \subset \mathcal{I}$ forms a base of $\mathcal{I}$ if, for any set $Y \in \mathcal{I}$, there exists a set $Z \in \mathcal{B}$ such that $Y \subset Z$.

The following statement is valid.

Theorem 3. *Let $card(E) = \omega_1$, let $\mathcal{I}$ be a σ-ideal of subsets of E, containing all singletons in E and possessing a base whose cardinality does not exceed ω_1, and let*

$$f : E \to \mathbf{R}$$

be a function. Then these two assertions are equivalent:

(1) f is not a generalized step-function with respect to $\mathcal{I}$;

(2) there exists a set $X \subset E$ such that $X \notin \mathcal{I}$ and the restriction $f|X$ is injective.

Proof. The implication (2) $\Rightarrow$ (1) is evident. Let us establish the validity of the implication (1) $\Rightarrow$ (2). Suppose that f satisfies (1) and introduce the following two sets:

$$T_0 = \{t \in ran(f) : f^{-1}(t) \notin \mathcal{I}\},$$

$$T_1 = \{t \in ran(f) : f^{-1}(t) \in \mathcal{I}\}.$$

According to our assumption, there exists a base $\mathcal{B} = \{B_\xi : \xi < \omega_1\}$ of the given σ-ideal $\mathcal{I}$. Only two cases are possible.

1. $card(T_0) = \omega_1$. In this case we may write

$$T_0 = \{t_\xi : \xi < \omega_1\},$$

where $t_\xi \neq t_\zeta$ for all $\xi < \omega_1, \zeta < \omega_1, \xi \neq \zeta$.

Consider the family of sets

$$\{f^{-1}(t_\xi) \setminus B_\xi : \xi < \omega_1\}.$$

Obviously, we have

$$f^{-1}(t_\xi) \setminus B_\xi \neq \emptyset$$

for each ordinal $\xi < \omega_1$. Let us choose an element

$$x_\xi \in f^{-1}(t_\xi) \setminus B_\xi$$

for any $\xi < \omega_1$, and let us put

$$X = \{x_\xi : \xi < \omega_1\}.$$

From the definition of X it immediately follows that the restriction $f|X$ is an injection. Moreover, we have

$$X \setminus B_\xi \neq \emptyset$$

whenever $\xi < \omega_1$. The latter circumstance implies at once that the set X does not belong to $\mathcal{I}$.

2. $card(T_0) \leq \omega$. In this case we obtain

$$card(T_1) = \omega_1$$

and $f^{-1}(T_1) \notin \mathcal{I}$ (because our f is not a generalized step-function with respect to $\mathcal{I}$). Let us construct, by using the method of transfinite recursion, an ω_1-sequence $\{x_\xi : \xi < \omega_1\}$ of points of $f^{-1}(T_1)$. Suppose that, for an ordinal $\xi < \omega_1$, the partial family of points $\{x_\zeta : \zeta < \xi\} \subset f^{-1}(T_1)$ has already been defined. Clearly,

$$(\cup\{f^{-1}(f(x_\zeta)) : \zeta < \xi\}) \cup B_\xi \in \mathcal{I}.$$

Therefore,

$$f^{-1}(T_1) \setminus ((\cup\{f^{-1}(f(x_\zeta)) : \zeta < \xi\}) \cup B_\xi) \neq \emptyset.$$

Choose any element x from the above nonempty set and put $x_\xi = x$. Proceeding in this manner, we are able to construct the required ω_1-sequence $\{x_\xi : \xi < \omega_1\}$. Finally, put

$$X = \{x_\xi : \xi < \omega_1\}.$$

In view of our construction, X is a partial selector of the disjoint family of sets $\{f^{-1}(t) : t \in T_1\}$. Hence, the restriction of f to X is injective. Furthermore,

$$X \setminus B_\xi \neq \emptyset$$

for all ordinals $\xi < \omega_1$, whence it follows that X does not belong to $\mathcal{I}$.

This completes the proof of Theorem 3.

Remark 4. Assume the Continuum Hypothesis and take as $\mathcal{I}$ the σ-ideal $\mathcal{I}(\lambda)$ of all Lebesgue measure zero subsets of $\mathbf{R}$. Let $f : \mathbf{R} \to \mathbf{R}$ be a function distinct from all generalized step-functions with respect to $\mathcal{I}(\lambda)$.

Suppose also that the graph of f is of λ_2-measure zero. Then it is not hard to show that, for such an f, there always exists a superposition operator

$$\Phi : \mathbf{R} \times \mathbf{R} \to \mathbf{R}$$

satisfying relations (1)–(5) of Theorem 1. In this connection, let us underline that f does not need to be a λ-measurable function.

Remark 5. Let E be a nonempty set, $\mathcal{I}$ be a σ-ideal of subsets of E and let $\mathcal{S}$ be a σ-algebra of subsets of E such that $\mathcal{I} \subset \mathcal{S}$. Elements of $\mathcal{S}$ are usually called measurable sets in E and elements of $\mathcal{I}$ are called negligible sets in E. The triple $(E, \mathcal{S}, \mathcal{I})$ is called a measurable space with negligibles (see Chapter 14). If $X \subset E$ and $X \notin \mathcal{I}$, then, in general, we cannot assert that X contains at least one subset not belonging to $\mathcal{S}$. However, in some situations the specific features of a given σ-ideal $\mathcal{I}$ imply that any nonnegligible set in E includes a nonmeasurable set.

For example, assume again that $card(E) = \omega_1$ and that E is a topological space of second category, whose all singletons are of first category. Let $\mathcal{I} = \mathcal{K}(E)$ denote the σ-ideal of all first category subsets of E and suppose that this σ-ideal possesses a base whose cardinality does not exceed ω_1. Denote also by $\mathcal{Ba}(E)$ the σ-algebra of all subsets of E having the Baire property (obviously, $\mathcal{K}(E) \subset \mathcal{Ba}(E)$). Then, for any set $X \subset E$, the following two assertions are equivalent:

(a) $X \notin \mathcal{K}(E)$ (i.e., X is not of first category in E);

(b) there exists a set $Y \subset X$ such that $Y \notin \mathcal{Ba}(E)$ (i.e., Y does not have the Baire property in E).

The proof of the equivalence of (a) and (b) can be found in [100] where some related results are also presented (see also Exercise 9 from Chapter 14). Notice once more that this equivalence relies only on the inner properties of the σ-ideal $\mathcal{K}(E)$ and does not touch the structure of the σ-algebra $\mathcal{Ba}(E)$.

Let us continue our consideration of the question of measurability of functions obtained by using the superposition operator which is induced by a given function of two variables. In this connection, we would like to discuss some related measurability properties of functions of two variables.

Let $\Phi : \mathbf{R}^2 \to \mathbf{R}$ be again a function of two variables and let F be a class of functions acting from $\mathbf{R}$ into $\mathbf{R}$. As in Chapter 15, for any $f \in F$, we denote by Φ_f the function acting from $\mathbf{R}$ into $\mathbf{R}$ and defined by

$$\Phi_f(x) = \Phi(x, f(x)) \qquad (x \in \mathbf{R}).$$

We have already mentioned that, in some sense, Φ plays the role of a super-position operator whose domain coincides with the given class F of functions. A general problem arises to describe those conditions on Φ under which various nice properties of functions from F are preserved by Φ. For example, suppose that $F = L(\mathbf{R}, \mathbf{R})$ is the class of all real-valued Lebesgue measurable functions on $\mathbf{R}$. Then it is natural to try to characterize those operators Φ for which $L(\mathbf{R}, \mathbf{R})$ is preserved (i.e., the Lebesgue measurability of functions of one variable is preserved by Φ). We already know that, in general, the Lebesgue measurability of Φ (regarded as a function of two variables) does not guarantee the Lebesgue measurability of Φ_f for $f \in L(\mathbf{R}, \mathbf{R})$. Also, it is widely known that if Φ satisfies the so-called Carathéodory conditions, then it preserves the class $L(\mathbf{R}, \mathbf{R})$.

In this context, the next example is relevant.

Example 1. For any Lebesgue measurable function $f : \mathbf{R} \to \mathbf{R}$, there exists a function

$$\Phi : \mathbf{R}^2 \to \mathbf{R}$$

satisfying the Carathéodory conditions and there is a continuous function

$$g : \mathbf{R} \to \mathbf{R}$$

such that

$$f(x) = g'(x) = \Phi(x, g(x))$$

for almost all $x \in \mathbf{R}$. Indeed, according to the classical Luzin theorem (see [135] or [180]), there exists a continuous function $g : \mathbf{R} \to \mathbf{R}$ such that $g'(x) = f(x)$ for almost all $x \in \mathbf{R}$. Let us define

$$\Phi(x, y) = f(x) + y - g(x) \qquad (x \in \mathbf{R}, \ y \in \mathbf{R}).$$

Then Φ is measurable with respect to x and affine with respect to y (hence, Φ satisfies the Lipschitz condition with respect to y). Obviously, we also have

$$f(x) = g'(x) = \Phi(x, g(x))$$

for almost all $x \in \mathbf{R}$.

We thus conclude that any real-valued Lebesgue measurable function can be simultaneously regarded as the derivative (almost everywhere) of a continuous function and as the image of the same continuous function, under an appropriate superposition operator satisfying the Carathéodory conditions.

As mentioned earlier, several works were devoted to constructions of a non-Lebesgue measurable function Φ that, however, preserves the class

$L(\mathbf{R}, \mathbf{R})$. All those constructions were based on some extra set-theoretical axioms. In this connection, a problem was posed whether it is possible to construct analogous Φ within the theory **ZFC**. Recently, Roslanowski and Shelah [176] announced that the existence of such a function Φ cannot be established in **ZFC**.

A similar question for the Baire property was considered by Ciesielski and Shelah in [41].

Here some related topics will be discussed concerning measurability and sup-measurability properties of functions.

Let E be a nonempty set, $\mathcal{S}$ be a σ-algebra of subsets of E and let $\mathcal{I}$ be a σ-ideal of subsets of E, such that $\mathcal{I} \subset \mathcal{S}$. It will be assumed in our further considerations that $\mathcal{I}$ contains all one-element subsets of E and that the pair $(\mathcal{S}, \mathcal{I})$ satisfies the countable chain condition, i.e., every disjoint family of sets belonging to $\mathcal{S} \setminus \mathcal{I}$ is at most countable.

Let F be a class of real-valued $\mathcal{S}$-measurable functions on E. Suppose also that a function

$$\Phi : E \times \mathbf{R} \to \mathbf{R}$$

is given. In accordance with the above consideration, this Φ will be treated as a superposition operator for the class F, i.e., by using Φ, we obtain from any function $f \in F$ the function Φ_f defined by

$$\Phi_f(x) = \Phi(x, f(x)) \qquad (x \in E).$$

We shall say that Φ is sup-measurable with respect to F if, for each $f \in F$, the corresponding function Φ_f is $\mathcal{S}$-measurable.

Starting with the class F, it is reasonable to consider other classes F' of $\mathcal{S}$-measurable real-valued functions, containing F and such that each operator Φ sup-measurable with respect to F remains also sup-measurable with respect to F'. In this case, we shall say that F' extends F with preserving the sup-measurability property. It is also reasonable to try to characterize maximal extensions of F for which this property is still preserved.

It will be demonstrated below that, in some natural situations, it is possible to describe such maximal extensions in terms of $(\mathcal{S}, \mathcal{I})$ and F.

Fix a class F of $\mathcal{S}$-measurable real-valued functions. We shall say that $f \in F^*$ if there exist a countable disjoint covering $\{E_n : n < \omega\} \subset \mathcal{S}$ of E and a countable family $\{f_n : 1 \leq n < \omega\} \subset F$, such that

$$E_0 \in \mathcal{I}, \quad f|E_n = f_n|E_n \quad (1 \leq n < \omega).$$

Clearly, we have the inclusion $F \subset F^*$. In some cases, this inclusion is reduced to the equality. For instance, if F is the family of all $\mathcal{S}$-measurable functions, then $F^* = F$.

Example 2. Let F denote the class of all constant real-valued functions on E. Then it is easy to see that F^* coincides with the class of all those real-valued functions on E, which are $\mathcal{I}$-equivalent to step-functions (we recall that a step-function on E is any real-valued $\mathcal{S}$-measurable function whose range is at most countable).

In the sequel, we need the following simple auxiliary statement.

Lemma 4. *Let* $h : E \to \mathbf{R}$ *be an* $\mathcal{S}$-*measurable function and let* F *be some family of* $\mathcal{S}$-*measurable real-valued functions. Then* h *does not belong to* F^* *if and only if there exists a set* $A \in \mathcal{S} \setminus \mathcal{I}$ *possessing the following property: for any subset* B *of* A *belonging to* $\mathcal{S} \setminus \mathcal{I}$ *and for any function* $f \in F$, *the relation* $f|B \neq h|B$ *is fulfilled.*

The proof can easily be obtained by using the Zorn Lemma or the method of transfinite induction, taking into account the countable chain condition for the pair $(\mathcal{S}, \mathcal{I})$.

Exercise 2. Give a detailed proof of Lemma 4.

The next lemma is almost trivial.

Lemma 5. *Let* Φ *be a sup-measurable operator with respect to* F. *Then* Φ *is also sup-measurable with respect to* F^*.

We omit an easy proof of this lemma. Notice only that, for its validity, the countable chain condition is not necessary.

Let F be a family of $\mathcal{S}$-measurable real-valued functions.
We shall say that a family G of real-valued functions on E is fundamental for F if every function f from F is $\mathcal{I}$-equivalent to some function g from G.
It can easily be shown that the next auxiliary statement is true.

Lemma 6. *An operator* Φ *is sup-measurable with respect to a class* F *if and only if it is sup-measurable with respect to some class* G *fundamental for* F.

Exercise 3. Give a proof of Lemma 6 and verify that the countable chain condition is not needed here.

Example 3. The class of all Borel functions (acting from $\mathbf{R}$ into $\mathbf{R}$) is fundamental for the class of all Lebesgue measurable functions (acting from $\mathbf{R}$ into $\mathbf{R}$). The same class of Borel functions is also fundamental for the class of all those functions that act from $\mathbf{R}$ into $\mathbf{R}$ and possess the Baire property.

We recall that a family $\mathcal{B} \subset \mathcal{S} \setminus \mathcal{I}$ is a pseudo-base for a space $(E, \mathcal{S}, \mathcal{I})$ if every set $X \in \mathcal{S} \setminus \mathcal{I}$ contains at least one member of $\mathcal{B}$.

Because the given pair $(\mathcal{S}, \mathcal{I})$ satisfies the countable chain condition, every set $X \in \mathcal{S} \setminus \mathcal{I}$ contains a subset Y such that $X \setminus Y \in \mathcal{I}$ and Y is representable as the union of a countable disjoint family of members of a pseudo-base $\mathcal{B}$. This circumstance implies the validity of the next auxiliary proposition.

Lemma 7. *Let $\mathcal{B}$ be a pseudo-base for a space $(E, \mathcal{S}, \mathcal{I})$ with $card(\mathcal{B}) \geq 2$. Then there exists a family G of $\mathcal{S}$-measurable real-valued functions, fundamental for the family of all $\mathcal{S}$-measurable real-valued functions and satisfying the inequality*

$$card(G) \leq (card(\mathcal{B}))^{\omega}.$$

Exercise 4. Give a detailed proof of Lemma 7.

Lemma 8. *Suppose that the following relations are satisfied for a given space $(E, \mathcal{S}, \mathcal{I})$:*

(1) there exists a pseudo-base $\mathcal{B}$ for $(E, \mathcal{S}, \mathcal{I})$ containing at least two members and such that $(card(\mathcal{B}))^{\omega} \leq card(E)$;

(2) for any set $X \in \mathcal{S} \setminus \mathcal{I}$ and for any family $\{X_\theta : \theta \in \Theta\} \subset \mathcal{I}$ with $card(\Theta) < card(E)$, we have

$$X \setminus \cup\{X_\theta : \theta \in \Theta\} \neq \emptyset;$$

(3) each subset of E with cardinality strictly less than $card(E)$ belongs to $\mathcal{I}$.

Let F be a family of $\mathcal{S}$-measurable real-valued functions and let h be any $\mathcal{S}$-measurable real-valued function not belonging to F^*. Then there exists a superposition operator

$$\Phi : E \times \mathbf{R} \to \mathbf{R}$$

such that Φ is sup-measurable with respect to F, but is not sup-measurable with respect to the one-element class $\{h\}$.

Proof. Lemma 7 and relation (1) readily imply that there exists a family G of $\mathcal{S}$-measurable real-valued functions, fundamental for F and such that

$$card(G) \leq card(E).$$

We may also assume, without loss of generality, that every function from G is $\mathcal{I}$-equivalent to some function from F.

Let α denote the least ordinal number whose cardinality is equal to $card(E)$ and let $\{g_\xi : \xi < \alpha\}$ be an enumeration of all functions from G. Taking into account the relation $h \notin F^*$ and applying Lemma 4, we can find a set $A \in \mathcal{S} \setminus \mathcal{I}$ such that h differs from any $f \in F$ on each $\mathcal{S}$-measurable subset of A not belonging to the ideal $\mathcal{I}$. Obviously, the same is true for all functions from G, i.e., for any $g \in G$, the function h differs from g on each $\mathcal{S}$-measurable subset of A not belonging to $\mathcal{I}$. We may assume in the sequel (without loss of generality) that $A = E$.

Let $\{B_\xi : \xi < \alpha\}$ be an enumeration of all members from the pseudo-base $\mathcal{B}$. Applying the method of transfinite recursion, let us construct two injective disjoint α-sequences

$$\{x_\xi : \xi < \alpha\}, \quad \{x'_\xi : \xi < \alpha\}$$

of points of the given space E. Suppose that, for an ordinal $\xi < \alpha$, the partial families

$$\{x_\zeta : \zeta < \xi\}, \quad \{x'_\zeta : \zeta < \xi\}$$

have already been constructed. For any ordinal $\zeta < \xi$, denote

$$A_\zeta = \{z \in E : g_\zeta(z) = h(z)\}.$$

Then it is clear that $A_\zeta \in \mathcal{I}$. Consider the set

$$P_\xi = B_\xi \setminus ((\cup\{A_\zeta : \zeta < \xi\}) \cup \{x_\zeta : \zeta < \xi\} \cup \{x'_\zeta : \zeta < \xi\}).$$

In view of the relations (2) and (3), we have

$$card(P_\xi) = card(E).$$

Therefore, we can choose two distinct points x and x' from P_ξ. Finally, we put

$$x_\xi = x, \quad x'_\xi = x'.$$

Proceeding in this manner, we will be able to construct the required α-sequences of points. Now, we define a function

$$\Phi : E \times \mathbf{R} \to \mathbf{R}$$

as follows:

$\Phi(x_\xi, h(x_\xi)) = 1$ for each ordinal $\xi < \alpha$;
$\Phi(x'_\xi, h(x'_\xi)) = -1$ for each ordinal $\xi < \alpha$;
$\Phi(x, t) = 0$ for all other pairs $(x, t) \in E \times \mathbf{R}$.

Let us show that Φ is sup-measurable with respect to F and, at the same time, the function Φ_h is not $\mathcal{S}$-measurable. Indeed, take any $f \in F$ and find a function $g \in G$ which is $\mathcal{I}$-equivalent to f. Clearly, g coincides with some g_η where $\eta < \alpha$. Further, introduce the set

$$Z = \{z \in E : \Phi(z, g(z)) \neq 0\}.$$

For each $z \in Z$, the disjunction

$$\Phi(z, g(z)) = 1 \quad \vee \quad \Phi(z, g(z)) = -1$$

must be valid. This implies that either $z = x_\xi$ and $g(z) = h(z)$, or $z = x'_\xi$ and $g(z) = h(z)$. It follows directly from our construction that, in the both cases above, $\xi \leq \eta$. Consequently, the cardinality of Z must be strictly less than $card(E)$. Hence, in view of relation (3), the function Φ_g must be $\mathcal{S}$-measurable, and the same is true for Φ_f.

On the other hand, the definition of Φ easily yields that the function Φ_h cannot be $\mathcal{S}$-measurable. Indeed, for any set $B \in \mathcal{B}$, we have from our construction that

$$\{-1, 1\} \subset ran(\Phi_h|B).$$

Remembering that $\mathcal{B}$ is a pseudo-base for $(E, \mathcal{S}, \mathcal{I})$, we obtain that the sets $\Phi_h^{-1}(-1)$ and $\Phi_h^{-1}(1)$ are $\mathcal{S}$-thick in E, i.e., they intersect all members from $\mathcal{S} \setminus \mathcal{I}$. This fact immediately implies that both these sets are not $\mathcal{S}$-measurable and hence Φ_h is not $\mathcal{S}$-measurable, too. The proof is thus completed.

Taking into account the preceding lemmas, we are able to formulate the following statement.

Theorem 4. *Let a space $(E, \mathcal{S}, \mathcal{I})$ be given, let F be a class of $\mathcal{S}$-measurable real-valued functions and let the assumptions of Lemma 8 be fulfilled. Then the class F^* is the largest extension of F which preserves the sup-measurability property.*

Lemma 8 and hence Theorem 4 were proved under assumptions of somewhat set-theoretical flavor. These assumptions are known to be consistent for canonical measurable spaces with negligible sets, studied in real analysis (cf. Examples 5 and 6 below). However, we do not know whether the same assumptions are essential for the validity of the result in a general situation.

Now, let us give several examples illustrating the theorem obtained above. We begin with the following very simple example.

Example 4. Let $E = \mathbf{R}$, let $\mathcal{S} = dom(\lambda)$ be the σ-algebra of all Lebesgue measurable sets in $\mathbf{R}$ and let $\mathcal{I} = \mathcal{I}(\lambda)$ be the σ-ideal of all

Lebesgue measure zero subsets of $\mathbf{R}$. Denote by F the class of all real-valued constant functions on $\mathbf{R}$. Obviously, there are many real-valued functions h on $\mathbf{R}$ not belonging to F^* (for instance, any strictly monotone function acting from $\mathbf{R}$ into $\mathbf{R}$ can be taken as h). Thus, we obtain that there exists a superposition operator

$$\Phi : \mathbf{R} \times \mathbf{R} \to \mathbf{R},$$

which is Lebesgue measurable with respect to the first variable but produces non-Lebesgue measurable functions of type Φ_h. Actually, this result needs no additional set-theoretical assumptions.

The next example is less trivial.

Example 5. Again, let $E = \mathbf{R}$, let $S = dom(\lambda)$ be the σ-algebra of all Lebesgue measurable sets in $\mathbf{R}$ and let $\mathcal{I} = \mathcal{I}(\lambda)$ be the σ-ideal of all Lebesgue measure zero subsets of $\mathbf{R}$. We denote by F the family of all real-valued continuous functions on $\mathbf{R}$ differentiable almost everywhere (with respect to the Lebesgue measure). Let h be a real-valued continuous function on $\mathbf{R}$ such that it is nowhere approximately differentiable. Then $h \notin F^*$. Therefore, under the corresponding set-theoretical assumptions on $(E, S, \mathcal{I})$, there exists a superposition operator Φ sup-measurable with respect to F, for which the function Φ_h is not Lebesgue measurable (cf. Chapter 15 where a similar approach was suggested).

Example 6. Let $E = \mathbf{R}$, let $S = \mathcal{B}a(\mathbf{R})$ be the σ-algebra of all those sets in $\mathbf{R}$ which possess the Baire property and let $\mathcal{I} = \mathcal{K}(\mathbf{R})$ be the σ-ideal of all first category subsets of $\mathbf{R}$. We denote by F the family of all real-valued continuous functions f on $\mathbf{R}$ having the property that each nonempty open subinterval of $\mathbf{R}$ contains at least one point at which f is differentiable. Take any real-valued continuous function h on $\mathbf{R}$ that is nowhere differentiable. Then it is not hard to demonstrate that $h \notin F^*$. Therefore, under the corresponding set-theoretical assumptions on $(E, S, \mathcal{I})$, there exists a superposition operator Φ sup-measurable with respect to F, for which the function Φ_h does not possess the Baire property (cf. again Chapter 15).

Example 7. It is not difficult to show that the existence of a Sierpiński subset of the Euclidean plane $\mathbf{R}^2$ implies the existence of a superposition operator

$$\Phi : \mathbf{R}^2 \to \mathbf{R},$$

which is sup-measurable (with respect to the class $L(\mathbf{R}, \mathbf{R})$) but is not Lebesgue measurable as a function of two variables. Indeed, it suffices to take as Φ the characteristic function of a Sierpiński set on the plane (we shall say that such a Φ determines a Sierpiński superposition operator). Moreover, it can be observed that the same Φ yields Lebesgue measurable functions Φ_f for all those functions $f : \mathbf{R} \to \mathbf{R}$ whose graphs are sets of Lebesgue measure zero in $\mathbf{R}^2$. Obviously, there are many non-Lebesgue measurable functions among those f.

An analogous situation holds in terms of category and the Baire property. In this case, the existence of a Luzin subset of the plane is needed for constructing an appropriate example.

In connection with Example 7, the next statement is of some interest.

Theorem 5. *Let $h : \mathbf{R} \to \mathbf{R}$ be a function whose graph has a strictly positive outer Lebesgue measure in $\mathbf{R}^2$. Then, under the Continuum Hypothesis, there exists a Sierpiński superposition operator Φ such that Φ_h is not Lebesgue measurable.*

Proof. Let λ denote the standard one-dimensional Lebesgue measure on $\mathbf{R}$, let $\lambda_2 = \lambda \times \lambda$ denote the usual two-dimensional Lebesgue measure on $\mathbf{R}^2$ and let $\Gamma \subset \mathbf{R}^2$ be the graph of h. In view of the assumption of the theorem, Γ is not contained in a set of λ_2-measure zero.

Let $\{X_\xi : \xi < \omega_1\}$ be the family of all Borel sets in $\mathbf{R}$ of λ-measure zero and let $\{B_\xi : \xi < \omega_1\}$ be the family of all Borel sets in $\mathbf{R}^2$ of λ_2-measure zero. We shall construct, by applying the method of transfinite recursion, an injective family $\{x_\xi : \xi < \omega_1\}$ of points of $\mathbf{R}$. Suppose that, for an ordinal $\xi < \omega_1$, the partial family of points $\{x_\zeta : \zeta < \xi\}$ has already been defined. Consider the set

$$T_\xi = (\cup\{B_\zeta : \zeta < \xi\}) \cup (\cup\{X_\zeta \times \mathbf{R} : \zeta < \xi\}) \cup (\cup\{\{x_\zeta\} \times \mathbf{R} : \zeta < \xi\}).$$

Clearly, we have the equality

$$\lambda_2(T_\xi) = 0.$$

Hence, $\Gamma \setminus T_\xi \neq \emptyset$ and there exists a point

$$(x, y) \in \Gamma \setminus T_\xi.$$

We put $x_\xi = x$. Proceeding in this way, we will be able to construct the required family of points $\{x_\xi : \xi < \omega_1\}$. It immediately follows from our construction that:

(i) the set $\{x_\xi : \xi < \omega_1\}$ is a Sierpiński subset of the real line $\mathbf{R}$;

(ii) the set $\{(x_\xi, h(x_\xi)) : \xi < \omega_1\}$ is a Sierpiński subset of the plane $\mathbf{R}^2$.

Let Φ denote the characteristic function of the latter subset of $\mathbf{R}^2$. Then Φ is a Sierpiński superposition operator. At the same time, considering the function Φ_h, we easily observe that

$$\Phi_h^{-1}(1) = \{x_\xi : \xi < \omega_1\}.$$

Thus, Φ_h is not Lebesgue measurable since no Sierpiński subset of $\mathbf{R}$ is λ-measurable (see Theorem 5 from Chapter 10). This completes the proof.

Remark 6. Actually, the argument presented above yields (under **CH**) a more general result. Namely, for any set $\Gamma \subset \mathbf{R}^2$ of positive outer Lebesgue measure, there exists a partial function h acting from $\mathbf{R}$ into $\mathbf{R}$ and having the following properties:

(1) the graph of h is contained in Γ;
(2) the graph of h is a Sierpiński subset of $\mathbf{R}^2$;
(3) the domain of h is a Sierpiński subset of $\mathbf{R}$.

Exercise 5. Formulate and prove (under **CH**) an analogous result in terms of category, Baire property and Luzin sets.

Also, assuming Martin's Axiom, establish similar statements in terms of generalized Sierpiński sets and in terms of generalized Luzin sets.

17. Ordinary differential equations with bad right-hand sides

In this chapter we wish to consider some set-theoretical questions concerning the existence and uniqueness of solutions of ordinary differential equations. In particular, we deal here with those ordinary differential equations the right-hand sides of which are rather bad (even nonmeasurable in the Lebesgue sense), but for which we are able to establish the theorem on the existence and uniqueness of a solution with nice descriptive properties.

The results presented in this chapter are essentially based on the material of Chapter 15 (see also [101]).

The existence of a solution of an ordinary differential equation with a continuous right-hand side is stated by the famous Peano theorem (see, e.g., [166]). The following exercise shows that this classical theorem does not rely on any form of the Axiom of Choice and, in fact, is a result of **ZF** theory.

Exercise 1. Let W be a nonempty open subset of the plane $\mathbf{R} \times \mathbf{R}$ and let

$$\Phi : W \to \mathbf{R}$$

be a function. Fix a point $(x_0, y_0) \in W$. We recall that a differentiable function

$$f : \,]t_1, t_2[\,\to \mathbf{R}$$

is a (local) solution of the ordinary differential equation

$$y' = \Phi(x, y) \qquad (y(x_0) = y_0)$$

if $x_0 \in \,]t_1, t_2[$, the graph of f is contained in W and

$$f(x_0) = y_0, \quad f'(x) = \Phi(x, f(x)) \qquad (x \in \,]t_1, t_2[).$$

In this case, we also say that f is a solution of the Cauchy problem for the given function Φ and for the initial condition $(x_0, y_0) \in W$.

The Peano theorem mentioned above says that if Φ is continuous on W, then a solution always exists for any initial condition from W.

Demonstrate that this theorem is provable within **ZF**.

In connection with this exercise, see also the paper by Simpson [205] where some more precise results are presented.

There are many simple examples of ordinary differential equations whose right-hand sides are not so good as their solutions. For instance, let us take the ordinary differential equation

$$y' = |y|.$$

Then it is easy to check that:

(1) the right-hand side of this equation is continuous but not differentiable;

(2) all solutions of this equation are analytic;

(3) for any initial condition, there exists a unique solution of this equation, satisfying the condition.

In our further considerations we shall show that some differential equations

$$y' = \Phi(x, y)$$

are possible, for which Φ is nonmeasurable in the Lebesgue sense but relations (2) and (3) hold true.

We begin with an old remarkable result of Orlicz (see [160]) stating that, for almost each (in the category sense) function Φ from the Banach space $C_b(\mathbf{R} \times \mathbf{R})$ consisting of all bounded continuous real-valued functions defined on $\mathbf{R} \times \mathbf{R}$, the corresponding Cauchy problem

$$y' = \Phi(x, y) \qquad (y(x_0) = y_0, \ x_0 \in \mathbf{R}, \ y_0 \in \mathbf{R})$$

has a unique solution. In order to present this result we, first of all, want to recall the purely topological Kuratowski lemma on closed projections (see, e.g., the Introduction). Namely, if X and Y are some topological spaces and, in addition, Y is quasicompact, then the canonical projection

$$pr_1 \ : \ X \times Y \to X$$

is a closed mapping, i.e., for each closed subset A of $X \times Y$, the image $pr_1(A)$ is closed in X. Several applications of the Kuratowski lemma were discussed in the Introduction. Here we are going to consider an application of this lemma to the theory of ordinary differential equations. Actually, we need here a slightly more general version of the lemma.

Let us recall that a topological space E is σ-quasicompact if it can be represented in the form

$$E = \cup \{E_n \ : \ n < \omega\}$$

where all sets E_n $(n < \omega)$ are quasicompact subspaces of E. Now, the following slight generalization of the Kuratowski lemma is true.

Lemma 1. *Let X be a topological space and let Y be a σ-quasicompact space. Let, as above, pr_1 denote the canonical projection from $X \times Y$ into X. Then, for each F_σ-subset A of $X \times Y$, the image $pr_1(A)$ is an F_σ-subset of X.*

Proof. In fact, the Kuratowski lemma easily implies this result. Indeed, since Y is σ-quasicompact, we may write

$$Y = \cup \{Y_n \ : \ n < \omega\}$$

where all Y_n $(n < \omega)$ are quasicompact subspaces of Y. Then, for any set $A \subset X \times Y$, we have the equality

$$pr_1(A) = \cup \{pr_1(A \cap (X \times Y_n)) \ : \ n < \omega\}.$$

Suppose now that A is an F_σ-subset of $X \times Y$. Then A can be represented in the form

$$A = \cup \{A_m \ : \ m < \omega\}$$

where all sets A_m $(m < \omega)$ are closed in $X \times Y$. Therefore, we obtain

$$pr_1(A) = \cup \{pr_1(A_m \cap (X \times Y_n)) \ : \ m < \omega, \ n < \omega\}.$$

Now, every set

$$A_m \cap (X \times Y_n) \qquad (m < \omega, \ n < \omega)$$

is closed in the product space $X \times Y_n$ and every space Y_n is quasicompact. Hence, by the Kuratowski lemma, the set

$$pr_1(A_m \cap (X \times Y_n))$$

is closed in X. Consequently, $pr_1(A)$ is an F_σ-subset of X. This completes the proof of Lemma 1.

Now, let us return to the Banach space $C_b(\mathbf{R} \times \mathbf{R})$. For each function Φ from this space, we can consider the ordinary differential equation

$$y' = \Phi(x, y)$$

and, for any point $(x_0, y_0) \in \mathbf{R} \times \mathbf{R}$, we can investigate the corresponding Cauchy problem of finding a solution $y : \mathbf{R} \to \mathbf{R}$ of this equation, satisfying the initial condition $y(x_0) = y_0$. It is well known (see, e.g., [166] or Exercise 1 above) that such a solution does always exist and, since Φ is bounded, any solution is global, i.e., it is defined on the whole real line $\mathbf{R}$. On the other hand, we cannot assert, in general, the uniqueness of a solution. There are simple examples of continuous bounded real-valued functions Φ on $\mathbf{R} \times \mathbf{R}$ for which the corresponding Cauchy problem admits at least two distinct solutions (in this connection, let us mention the famous work by Lavrentieff [128] where a much stronger result was obtained).

Exercise 2. Give an example of a function Φ from the space $C_b(\mathbf{R} \times \mathbf{R})$, for which there exists an initial condition $(x_0, y_0) \in \mathbf{R} \times \mathbf{R}$ such that the corresponding Cauchy problem

$$y' = \Phi(x, y) \quad (y(x_0) = y_0)$$

possesses at least two distinct solutions.

Actually, we need some additional properties of the original function Φ in order to have the uniqueness of a solution of the differential equation

$$y' = \Phi(x, y) \quad (y(x_0) = y_0, \ x_0 \in \mathbf{R}, \ y_0 \in \mathbf{R}).$$

For instance, if Φ satisfies the so-called local Lipschitz condition with respect to the second variable y, then we have a unique solution for each Cauchy problem corresponding to Φ. It is reasonable to recall here that Φ satisfies the local Lipschitz condition with respect to y if, for any point $(x_0, y_0) \in \mathbf{R} \times \mathbf{R}$, there exist a neighborhood $V(x_0, y_0)$ of this point and a positive real number $M = M(\Phi, (x_0, y_0))$, such that

$$|\Phi(x, y_1) - \Phi(x, y_2)| \leq M \, |y_1 - y_2|$$

for all points (x, y_1) and (x, y_2) belonging to $V(x_0, y_0)$.

Let us denote by $Lip_l(\mathbf{R} \times \mathbf{R})$ the family of all those functions from $C_b(\mathbf{R} \times \mathbf{R})$ that satisfy the local Lipschitz condition with respect to y. Then, obviously, $Lip_l(\mathbf{R} \times \mathbf{R})$ is a vector subspace of $C_b(\mathbf{R} \times \mathbf{R})$. Notice also that $Lip_l(\mathbf{R} \times \mathbf{R})$ is an everywhere dense subset of $C_b(\mathbf{R} \times \mathbf{R})$. Indeed, this fact is almost trivial from the geometrical point of view. Thus, we can conclude that, for all functions Φ belonging to some everywhere dense subset of $C_b(\mathbf{R} \times \mathbf{R})$, the Cauchy problem

$$y' = \Phi(x, y) \quad (y(x_0) = y_0, \ x_0 \in \mathbf{R}, \ y_0 \in \mathbf{R})$$

has a unique solution. Orlicz essentially improved this result and showed that it holds true for almost all (in the category sense) functions from the Banach space $C_b(\mathbf{R} \times \mathbf{R})$.

More precisely, one can formulate the following statement.

Theorem 1. *The set of all those functions from $C_b(\mathbf{R} \times \mathbf{R})$ for which the corresponding Cauchy problem has a unique solution (for any point (x_0, y_0) from $\mathbf{R} \times \mathbf{R}$) is an everywhere dense G_δ-subset of $C_b(\mathbf{R} \times \mathbf{R})$.*

Proof. Let us denote by the symbol U the family of all those functions from $C_b(\mathbf{R} \times \mathbf{R})$ for which the corresponding Cauchy problem has a unique solution (for each point (x_0, y_0) belonging to $\mathbf{R} \times \mathbf{R}$). As mentioned above, the set U is everywhere dense in $C_b(\mathbf{R} \times \mathbf{R})$. Therefore it remains to prove that U is a G_δ-subset of $C_b(\mathbf{R} \times \mathbf{R})$. In order to show this, let us first rewrite the Cauchy problem in the equivalent integral form:

$$y(x) = \int_{x_0}^{x} \Phi(t, y(t))dt \; + \; y_0.$$

Further, for any two rational numbers $\varepsilon > 0$ and q, let us denote by $P(\varepsilon, q)$ the set of all those elements

$$(\Phi, x_0, y_0) \in C_b(\mathbf{R} \times \mathbf{R}) \times \mathbf{R} \times \mathbf{R}$$

for which there exist at least two real-valued continuous functions ϕ_1 and ϕ_2 such that:

$$dom(\phi_1) = dom(\phi_2) = \mathbf{R},$$

$$\phi_1(x) = \int_{x_0}^{x} \Phi(t, \phi_1(t))dt \; + \; y_0 \quad (x \in \mathbf{R}),$$

$$\phi_2(x) = \int_{x_0}^{x} \Phi(t, \phi_2(t))dt \; + \; y_0 \quad (x \in \mathbf{R}),$$

$$|\phi_1(q) - \phi_2(q)| \geq \varepsilon.$$

It is not difficult to establish that $P(\varepsilon, q)$ is a closed subset of the product space $C_b(\mathbf{R} \times \mathbf{R}) \times \mathbf{R} \times \mathbf{R}$. Indeed, suppose that a sequence

$$\{(\Phi^{(n)}, x_0^{(n)}, y_0^{(n)}) \; : \; n \in \mathbf{N}\}$$

of elements of $P(\varepsilon, q)$ converges to some element

$$(\Phi, x_0, y_0) \in C_b(\mathbf{R} \times \mathbf{R}) \times \mathbf{R} \times \mathbf{R}.$$

Then we obviously have

$$lim_{n \to +\infty} x_0^{(n)} = x_0, \quad lim_{n \to +\infty} y_0^{(n)} = y_0,$$

and the sequence of functions

$$\{\Phi^{(n)} \; : \; n \in \mathbf{N}\} \subset C_b(\mathbf{R} \times \mathbf{R})$$

converges uniformly to the function Φ. We may assume without loss of generality that

$$(\forall n \in \mathbf{N})(||\Phi^{(n)}|| \leq ||\Phi|| + 1).$$

For every natural number n, let $\phi_1^{(n)}$ and $\phi_2^{(n)}$ denote two real-valued continuous functions satisfying the following relations:

$$dom(\phi_1^{(n)}) = dom(\phi_2^{(n)}) = \mathbf{R},$$

$$\phi_1^{(n)}(x) = \int_{x_0^{(n)}}^{x} \Phi^{(n)}(t, \phi_1^{(n)}(t))dt \; + \; y_0^{(n)} \qquad (x \in \mathbf{R}),$$

$$\phi_2^{(n)}(x) = \int_{x_0^{(n)}}^{x} \Phi^{(n)}(t, \phi_2^{(n)}(t))dt \; + \; y_0^{(n)} \qquad (x \in \mathbf{R}),$$

$$|\phi_1^{(n)}(q) - \phi_2^{(n)}(q)| \geq \varepsilon.$$

Then it is not hard to verify that all functions from the family

$$\{\phi_1^{(n)} \; : \; n \in \mathbf{N}\} \cup \{\phi_2^{(n)} \; : \; n \in \mathbf{N}\}$$

are equicontinuous. More precisely, for each function ϕ from this family and for any two points $x' \in \mathbf{R}$ and $x'' \in \mathbf{R}$, we have the inequality

$$|\phi(x') - \phi(x'')| \leq (||\Phi|| + 1)|x' - x''|.$$

So, applying the classical Ascoli-Arzelá theorem (see, e.g., [166]), we can easily derive that there exists an infinite subset K of $\mathbf{N}$ for which the partial sequences of functions

$$\{\phi_1^{(n)} \; : \; n \in K\}, \quad \{\phi_2^{(n)} \; : \; n \in K\}$$

converge uniformly (on each bounded subinterval of $\mathbf{R}$) to some functions ϕ_1 and ϕ_2, respectively. Also, it can easily be checked that, for ϕ_1 and ϕ_2, we have the analogous relations

$$dom(\phi_1) = dom(\phi_2) = \mathbf{R},$$

$$\phi_1(x) = \int_{x_0}^{x} \Phi(t, \phi_1(t))dt \ + \ y_0 \qquad (x \in \mathbf{R}),$$

$$\phi_2(x) = \int_{x_0}^{x} \Phi(t, \phi_2(t))dt \ + \ y_0 \qquad (x \in \mathbf{R}),$$

$$|\phi_1(q) - \phi_2(q)| \geq \varepsilon.$$

Thus we see that

$$(\Phi, x_0, y_0) \in P(\varepsilon, q),$$

and hence $P(\varepsilon, q)$ is closed in the product space $C_b(\mathbf{R} \times \mathbf{R}) \times \mathbf{R} \times \mathbf{R}$.

Now, let us put

$$P = \cup \{P(\varepsilon, q) \ : \ \varepsilon > 0, \ \varepsilon \in \mathbf{Q}, \ q \in \mathbf{Q}\}.$$

Then it is clear that a function $\Psi \in C_b(\mathbf{R} \times \mathbf{R})$ does not belong to the set U if and only if there exist a rational number $\varepsilon > 0$, a rational number q and some points $x_0 \in \mathbf{R}$ and $y_0 \in \mathbf{R}$, such that (Ψ, x_0, y_0) belongs to the set $P(\varepsilon, q)$. In other words, we may write

$$C_b(\mathbf{R} \times \mathbf{R}) \setminus U = pr_1(P)$$

where

$$pr_1 \ : \ C_b(\mathbf{R} \times \mathbf{R}) \times \mathbf{R} \times \mathbf{R} \rightarrow C_b(\mathbf{R} \times \mathbf{R})$$

denotes the canonical projection. It immediately follows from the definition of the set P that P is an F_σ-subset of the product space $C_b(\mathbf{R} \times \mathbf{R}) \times \mathbf{R} \times \mathbf{R}$. In addition, the plane $\mathbf{R} \times \mathbf{R}$ is a σ-compact space. So, applying Lemma 1, we conclude that $pr_1(P)$ is an F_σ-subset of $C_b(\mathbf{R} \times \mathbf{R})$ and, consequently, U is a G_δ-subset of $C_b(\mathbf{R} \times \mathbf{R})$. This finishes the proof of Theorem 1.

Remark 1. Evidently, the Banach space $C_b(\mathbf{R} \times \mathbf{R})$ is not separable. Let E denote the subset of this space, consisting of all those functions that are constant at infinity. In other words, $\Phi \in E$ if and only if there exists a constant

$$M = M(\Phi) \in \mathbf{R}$$

such that, for any $\varepsilon > 0$, a positive real number $a = a(\Phi, \varepsilon)$ can be found for which we have

$$(\forall x)(\forall y)((x, y) \in \mathbf{R} \times \mathbf{R} \setminus [-a, a] \times [-a, a] \Rightarrow |\Phi(x, y) - M| < \varepsilon).$$

Notice that E is a closed vector subspace of $C_b(\mathbf{R} \times \mathbf{R})$ and hence E is a Banach space, as well. Moreover, one can easily verify that E is separable. Clearly, a direct analogue of Theorem 1 holds true for E. Actually, in [160]

Orlicz deals with the space E. A number of analogues of Theorem 1, for other spaces similar to $C_b(\mathbf{R} \times \mathbf{R})$ or E, are discussed in [2].

Remark 2. Unfortunately, the set U considered above has a bad algebraic structure. In particular, U is not a subgroup of the additive group of $C_b(\mathbf{R} \times \mathbf{R})$ and, consequently, U is not a vector subspace of $C_b(\mathbf{R} \times \mathbf{R})$. Indeed, suppose for a while that U is a subgroup of $C_b(\mathbf{R} \times \mathbf{R})$. Then U must be a proper subgroup of $C_b(\mathbf{R} \times \mathbf{R})$. Let us take a function

$$\Psi \in C_b(\mathbf{R} \times \mathbf{R}) \setminus U.$$

Obviously,

$$U \cap (\{\Psi\} + U) = \emptyset.$$

But both sets U and $\{\Psi\} + U$ are the complements of some first category subsets of $C_b(\mathbf{R} \times \mathbf{R})$. Therefore their intersection $U \cap (\{\Psi\} + U)$ must be the complement of a first category subset of $C_b(\mathbf{R} \times \mathbf{R})$, too, and hence

$$U \cap (\{\Psi\} + U) \neq \emptyset.$$

We have thus obtained a contradiction which yields that U cannot be a subgroup of $C_b(\mathbf{R} \times \mathbf{R})$.

For some other properties of U interesting from the set-theoretical and algebraic points of view, see, e.g., [2].

Theorem 1 proved above shows us that, for many functions from the space $C_b(\mathbf{R} \times \mathbf{R})$, we have the existence and uniqueness of a solution of the Cauchy problem. In fact, this is one of the most important results in the theory of ordinary differential equations. Naturally, we may consider a more general class of functions

$$\Phi \ : \ \mathbf{R} \times \mathbf{R} \to \mathbf{R}$$

not necessarily continuous or Lebesgue measurable and investigate for such functions the corresponding Cauchy problem from the point of view of the existence and uniqueness of a solution.

For this purpose, let us recall that, as shown in Chapter 15 of our book, there exists a subset Z of the plane $\mathbf{R} \times \mathbf{R}$, satisfying the following relations:

(1) no three distinct points of Z belong to a straight line;

(2) Z is the graph of some partial function acting from $\mathbf{R}$ into $\mathbf{R}$;

(3) Z is a λ_2-thick subset of the plane $\mathbf{R} \times \mathbf{R}$, where λ_2 denotes the standard two-dimensional Lebesgue measure on $\mathbf{R} \times \mathbf{R}$;

(4) for any Borel mapping $\phi \ : \ \mathbf{R} \to \mathbf{R}$, the intersection of Z with the graph of ϕ has cardinality strictly less than the cardinality of the continuum.

We denote by Φ the characteristic function of the above-mentioned set Z. Then, obviously, Φ is a Lebesgue nonmeasurable function and, furthermore, if $[\mathbf{R}]^{<\mathbf{c}} \subset dom(\lambda)$, then Φ is sup-measurable as well.

Now, starting with the function Φ described above, we wish to consider an ordinary differential equation

$$y' = \Psi(x, y)$$

with the Lebesgue nonmeasurable right-hand side Ψ, and we are going to show that, in some situations, it is possible to obtain the existence and uniqueness of a solution of this equation (for any initial condition).

First of all, we need to determine the class of functions to which a solution must belong. It is natural to take the class $AC_l(\mathbf{R})$ consisting of all locally absolutely continuous real-valued functions on $\mathbf{R}$. In other words, $\psi \in AC_l(\mathbf{R})$ if and only if, for each point $x \in \mathbf{R}$, there exists a neighborhood $V(x)$ of x such that the restriction $\psi|V(x)$ is absolutely continuous. Another characterization of locally absolutely continuous functions on $\mathbf{R}$ is the following one: a function ψ belongs to $AC_l(\mathbf{R})$ if and only if there exists a Lebesgue measurable function $f : \mathbf{R} \to \mathbf{R}$ such that f is locally integrable and

$$\psi(x) = \int_0^x f(t)dt + \psi(0)$$

for any point $x \in \mathbf{R}$.

Let Ψ be a mapping from $\mathbf{R} \times \mathbf{R}$ into $\mathbf{R}$ and let us fix a point

$$(x_0, y_0) \in \mathbf{R} \times \mathbf{R}.$$

We say that the corresponding Cauchy problem

$$y' = \Psi(x, y) \qquad (y(x_0) = y_0)$$

has a unique solution (in the class $AC_l(\mathbf{R})$) if there exists a unique function $\psi \in AC_l(\mathbf{R})$ satisfying the relations:

(a) $\psi'(x) = \Psi(x, \psi(x))$ for almost all (with respect to the Lebesgue measure λ) points $x \in \mathbf{R}$;

(b) $\psi(x_0) = y_0$.

For example, if our mapping Ψ is bounded, Lebesgue measurable with respect to x and satisfies locally the Lipschitz condition with respect to y, then, for each $(x_0, y_0) \in \mathbf{R} \times \mathbf{R}$, the corresponding Cauchy problem has a unique solution. The reader can easily verify this fact by using the standard argument. Notice that, in this example, Ψ is necessarily Lebesgue measurable and sup-measurable (cf. Exercise 2 from Chapter 15). Notice also that an analogue of Theorem 1 holds true for a certain class of Banach spaces consisting of mappings acting from $\mathbf{R} \times \mathbf{R}$ into $\mathbf{R}$, which are Lebesgue measurable with respect to x and continuous with respect to y.

Exercise 3. Prove that an analogue of Theorem 1 remains true for any Banach space E of bounded mappings acting from $\mathbf{R} \times \mathbf{R}$ into $\mathbf{R}$, for which there exists an everywhere dense set $D \subset E$ such that each function from D is Lebesgue measurable with respect to x and satisfies locally the Lipschitz condition with respect to y.

The next statement shows that the existence and uniqueness of a solution can be fulfilled even for some ordinary differential equations whose right-hand sides are extremely bad, e.g., nonmeasurable in the Lebesgue sense.

Theorem 2. *There exists a Lebesgue nonmeasurable mapping*

$$\Psi \; : \; \mathbf{R} \times \mathbf{R} \to \mathbf{R}$$

such that the Cauchy problem

$$y' = \Psi(x,y) \qquad (y(x_0) = y_0)$$

has a unique solution for any point $(x_0, y_0) \in \mathbf{R} \times \mathbf{R}$.

Proof. Let Z be a subset of the plane, constructed in Chapter 15 (see Theorem 2 therein). Denote again by Φ the characteristic function of Z and fix a real number t. Further, put

$$\Psi(x,y) = \Phi(x,y) + t \qquad (x \in \mathbf{R}, \; y \in \mathbf{R}).$$

We assert that Ψ is the required mapping. Indeed, Ψ is Lebesgue non-measurable because Φ is Lebesgue nonmeasurable. Let now (x_0, y_0) be an arbitrary point of the plane $\mathbf{R} \times \mathbf{R}$. Consider a function

$$\psi \; : \; \mathbf{R} \to \mathbf{R}$$

defined by the formula

$$\psi(x) = tx + (y_0 - tx_0) \qquad (x \in \mathbf{R}).$$

The graph of this function is a straight line, so it has at most two common points with the set Z. Consequently, the function

$$\Psi_\psi \; : \; \mathbf{R} \to \mathbf{R}$$

is equal to t for almost all (with respect to the Lebesgue measure λ) points from $\mathbf{R}$. We also have $\psi'(x) = t$ for all $x \in \mathbf{R}$. In other words, ψ is a solution of the Cauchy problem

$$y' = \Psi(x, y) \qquad (y(x_0) = y_0).$$

It remains to show that ψ is a unique solution from the class $AC_l(\mathbf{R})$. For this purpose, let us take an arbitrary solution ϕ of the same Cauchy problem, belonging to $AC_l(\mathbf{R})$. Then, for almost all points $x \in \mathbf{R}$, we have the equality

$$\phi'(x) = \Phi(x, \phi(x)) + t.$$

It immediately follows from this equality that the function Φ_ϕ is measurable in the Lebesgue sense. But, as we know,

$$card(\{x \in \mathbf{R} \ : \ \Phi_\phi(x) \neq 0\}) < \mathbf{c}.$$

So we obtain that Φ_ϕ is equivalent to zero and hence $\phi'(x) = t$ for almost all $x \in \mathbf{R}$. Therefore we can conclude that

$$\phi(x) = tx + (y_0 - tx_0) \qquad (x \in \mathbf{R}).$$

This completes the proof of Theorem 2.

Remark 3. The preceding theorem was proved within the theory **ZFC**. In this connection, let us stress once more that the function Ψ of Theorem 2 is Lebesgue nonmeasurable and, under a certain set-theoretical hypothesis, is also sup-measurable (hence weakly sup-measurable). At the same time, we already know that it is impossible to establish within the theory **ZFC** the existence of a sup-measurable mapping, which is not measurable in the Lebesgue sense (a recent result of Roslanowski and Shelah).

Exercise 4. Let n be a natural number and let

$$a_0 x^n + a_1 x^{n-1} + ... + a_{n-1} x + a_n$$

be a polynomial of degree n. Show that there exists a mapping

$$\Psi \ : \ \mathbf{R} \times \mathbf{R} \to \mathbf{R}$$

satisfying the following relations:
  (a) Ψ is nonmeasurable in the Lebesgue sense;
  (b) for any initial condition $(x_0, y_0) \in \mathbf{R} \times \mathbf{R}$, the differential equation

$$y' = \Psi(x, y)$$

has a unique solution ψ with $\psi(x_0) = y_0$;

(c) all solutions ψ of the above-mentioned differential equation are of the form

$$\psi(x) = a_0 x^n + a_1 x^{n-1} + ... + a_{n-1} x + a \qquad (x \in \mathbf{R}),$$

where $a \in \mathbf{R}$.

Now, starting with the same function Φ (constructed in Chapter 15 of the book), we shall show that, under the set-theoretical assumption

$$[\mathbf{R}]^{<\mathbf{c}} \subset dom(\lambda),$$

Theorem 1 of Orlicz can be generalized to Banach spaces of mappings acting from $\mathbf{R} \times \mathbf{R}$ into $\mathbf{R}$, essentially larger than the classical space $C_b(\mathbf{R} \times \mathbf{R})$ (notice that all spaces of real-valued bounded mappings, considered in this chapter, are assumed to be equipped with the norm of uniform convergence).

More precisely, we can formulate and prove the next result.

Theorem 3. *Suppose that* $[\mathbf{R}]^{<\mathbf{c}} \subset dom(\lambda)$. *Then there exists a Banach space* B_0 *of mappings acting from* $\mathbf{R} \times \mathbf{R}$ *into* $\mathbf{R}$, *satisfying the following relations:*

(1) $C_b(\mathbf{R} \times \mathbf{R}) \subset B_0$;

(2) there are sup-measurable but Lebesgue nonmeasurable functions belonging to B_0;

(3) an analogue of Theorem 1 holds true for B_0, *i.e., the family of all functions* $\Psi \in B_0$ *such that the ordinary differential equation*

$$y' = \Psi(x,y)$$

has a unique solution for any initial condition $y(x_0) = y_0$, *is an everywhere dense* G_δ-*subset of* B_0.

Proof. Let $\Phi = \Phi_0$ be again a mapping constructed in Chapter 15 of the book. Obviously, this mapping does not belong to the vector space $C_b(\mathbf{R} \times \mathbf{R})$. Denote by B_0 the vector space of functions, generated by

$$\{\Phi_0\} \cup C_b(\mathbf{R} \times \mathbf{R}).$$

Clearly, each function Ψ belonging to B_0 can be represented in the form

$$\Psi = \Psi_1 + t_1 \Phi_0,$$

where $\Psi_1 \in C_b(\mathbf{R} \times \mathbf{R})$ and $t_1 \in \mathbf{R}$. Moreover, because B_0 is the direct sum of the vector spaces $C_b(\mathbf{R} \times \mathbf{R})$ and $\{t\Phi_0 \ : \ t \in \mathbf{R}\}$, such a representation is unique. We equip B_0 with the norm of uniform convergence. Taking account of the fact that Φ_0 is Lebesgue nonmeasurable, we may write

$$dist(\Phi_0, C_b(\mathbf{R} \times \mathbf{R})) > 0.$$

In other words, B_0 can be regarded as a direct topological sum of the two Banach spaces $C_b(\mathbf{R} \times \mathbf{R})$ and $\{t\Phi_0 \ : \ t \in \mathbf{R}\}$. Consequently, we may identify B_0 with the product space $C_b(\mathbf{R} \times \mathbf{R}) \times \mathbf{R}$.

Let now Ψ_1 be an arbitrary function from $C_b(\mathbf{R} \times \mathbf{R})$ such that the corresponding ordinary differential equation

$$y' = \Psi_1(x, y)$$

has a unique solution for any initial condition $y(x_0) = y_0$. Then it is not difficult to check (by using the properties of our function Φ_0) that, for each real number t_1, the ordinary differential equation

$$y' = \Psi_1(x, y) + t_1 \Phi_0(x, y)$$

has also a unique solution for any initial condition $y(x_0) = y_0$. Conversely, if a function

$$\Psi = \Psi_1 + t_1 \Phi_0$$

from the space B_0 (where $\Psi_1 \in C_b(\mathbf{R} \times \mathbf{R})$) is such that the ordinary differential equation

$$y' = \Psi(x, y)$$

possesses a unique solution for every initial condition, then the ordinary differential equation

$$y' = \Psi_1(x, y)$$

possesses a unique solution for every initial condition, too.

Let us recall that the symbol U denotes (in this chapter) the family of all functions Ψ_1 from $C_b(\mathbf{R} \times \mathbf{R})$ such that the differential equation $y' = \Psi_1(x, y)$ has a unique solution for any initial condition. Denote now by V an analogous family for the space B_0, i.e., let V be the family of all functions Ψ from B_0 such that the differential equation $y' = \Psi(x, y)$ has a unique solution for any initial condition. Then, taking account of the preceding argument, we can assert that

$$V = U + \{t\Phi_0 \ : \ t \in \mathbf{R}\}.$$

Because, according to Theorem 1, U is an everywhere dense G_δ-subset of the Banach space $C_b(\mathbf{R} \times \mathbf{R})$, we easily conclude that V is an everywhere dense G_δ-subset of the Banach space B_0.

Theorem 3 has thus been proved.

Exercise 5. By assuming the same hypothesis $[\mathbf{R}]^{<\mathbf{c}} \subset dom(\lambda)$, give an example of a Banach space B_1 of functions acting from $\mathbf{R} \times \mathbf{R}$ into $\mathbf{R}$, satisfying the following relations:

(1) $C_b(\mathbf{R} \times \mathbf{R}) \subset B_1$;

(2) there are discontinuous Lebesgue measurable sup-measurable functions belonging to B_1;

(3) there are Lebesgue nonmeasurable sup-measurable functions belonging to B_1;

(4) an analogue of Theorem 1 holds true for B_1.

Remark 4. Let B be a Banach space of bounded sup-measurable mappings, for which an analogue of Theorem 1 is valid, i.e., the family of all $\Psi \in B$ such that the differential equation $y' = \Psi(x, y)$ has a unique solution for any initial condition $y(x_0) = y_0$, is an everywhere dense G_δ-subset of B. It is not difficult to see that the class of all such Banach spaces B is rather wide. In particular, it follows from Theorem 3 that there is a space B belonging to this class and containing a Lebesgue nonmeasurable mapping.

In this connection, it would be interesting to obtain a characterization (description) of the above-mentioned class of Banach spaces.

Finally, let us point out once more that some logical and set-theoretical aspects of the classical Cauchy-Peano theorem on the existence of solutions of ordinary differential equations are discussed in the paper by Simpson [205].

18. Nondifferentiable functions from the point of view of category and measure

Earlier we were concerned with various continuous but nondifferentiable functions acting from **R** into **R**. In this chapter, we wish to discuss one general approach to such functions from the viewpoint of category and measure. Roughly speaking, our goal is to demonstrate that, for a given generalized notion of derivative (introduced within the theory **ZF** & **DC**), the set of continuous nondifferentiable functions (with respect to this notion) turns out to be sufficiently large.

We begin with an approach based on the concept of Baire category. More precisely, it is based on the important theorem of Kuratowski and Ulam from general topology (for the formulation and proof of this theorem see, e.g., [120], [162], or Chapter 14 of the present book). We have already mentioned that the Kuratowski-Ulam theorem can be interpreted as a purely topological analogue of the classical Fubini theorem from measure theory. It is widely known that the Fubini theorem is fundamental for all of measure theory. Moreover, this theorem has many applications in analysis, probability theory and other domains of mathematics. Also, it is well known that the Kuratowski-Ulam theorem possesses a number of nontrivial applications in general topology and in modern mathematical analysis (some of them are presented in the books [120] and [162]).

In our further considerations, the main role is played by the following statement.

Theorem 1. *Let E_1 and E_2 be any two topological spaces with countable bases (or, more generally, with countable π–bases) and let E_3 be a topological space. Let Z be a subset of the product space $E_1 \times E_2$. Suppose that a certain mapping*

$$\Phi : Z \to E_3$$

is given, and that this mapping satisfies the conditions:

(1) the partial function Φ acting from the topological space $E_1 \times E_2$ into the topological space E_3 has the Baire property, i.e., for any open set V from

E_3, the preimage $\Phi^{-1}(V)$ has the Baire property in $E_1 \times E_2$;

(2) for almost all (in the sense of category) points $x \in E_1$, the domain of the partial mapping $\Phi(x, .)$ given by

$$\Phi(x, .)(y) = \Phi(x, y)$$

is a first category set in the space E_2.

Then the following two relations hold:

(a) Z is a first category subset of the product space $E_1 \times E_2$;

(b) for almost all (in the sense of category) points $y \in E_2$, the set

$$\{x : (x, y) \in dom(\Phi)\}$$

is of first category in the space E_1; roughly speaking, almost each point $y \in E_2$ is almost singular with respect to the partial mapping $\Phi(., y)$.

The proof of this general statement is very simple. Indeed, according to the Kuratowski-Ulam theorem, relation (a) implies relation (b). Therefore it is sufficient to establish relation (a) only. In virtue of condition (1), the partial function Φ has the Baire property, so the set

$$Z = \Phi^{-1}(E_3)$$

has the Baire property in the product space $E_1 \times E_2$. Using condition (2) and the Kuratowski-Ulam theorem once more, we get the required result.

In connection with Theorem 1, a natural question arises: how can condition (2) be checked for the given partial mapping Φ?

The following situation can be frequently met in analysis and it will be the most interesting for us in the sequel. Suppose that E_2 is a Polish topological vector space, E_3 is a topological vector space with a countable base and our partial mapping Φ satisfies condition (1) and the next condition:

(2') for almost each (in the sense of category) point $x \in E_1$, the partial mapping $\Phi(x, .)$ is linear and discontinuous on its domain.

Then it can be shown that Φ satisfies condition (2), as well. Indeed, for almost all points $x \in E_1$, the function $\Phi(x, .)$ has the Baire property and is linear and discontinuous on the vector space

$$Z(x) = \{y : (x, y) \in Z\}.$$

Let us prove that, for the points x mentioned above, the set $Z(x)$ is of first category in the space E_2. Suppose otherwise, i.e., suppose that $Z(x)$ is a second category set with the Baire property. Then we may apply to

$Z(x)$ the well-known Banach-Kuratowski-Pettis theorem from the theory of topological groups (see, for example, [85], [120], or Exercise 1 of Chapter 8). This theorem is a topological analogue of the classical Steinhaus property of Lebesgue measurable sets with a strictly positive measure. Namely, according to this theorem, the set

$$Z(x) - Z(x) = \{y - z \; : \; y \in Z(x), \; z \in Z(x)\}$$

contains a nonempty open subset of the topological vector space E_2 (more precisely, the set $Z(x) - Z(x)$ is a neighborhood of zero of E_2). But since the set $Z(x)$ is a vector space, too, we come to the equality

$$Z(x) - Z(x) = Z(x)$$

and, finally, we obtain

$$Z(x) = E_2.$$

Hence the function $\Phi(x, .)$ is defined on the whole Polish topological vector space E_2 and is linear on this space. Now, by taking account of the fact that the function $\Phi(x, .)$ has the Baire property, it is not difficult to prove (by using the same Banach-Kuratowski-Pettis theorem) that $\Phi(x, .)$ is a continuous mapping. But this contradicts the choice of the point x. The contradiction obtained shows us that the set $Z(x)$ must be of first category in the space E_2. Therefore condition (2) is satisfied for our partial mapping Φ.

Remark 1. Theorem 1 may be considered as one of possible formalizations of a well-known principle in mathematical analysis that is frequently called "the principle of condensation of singularities." Among various works devoted to this principle, the most famous is the classical paper of Banach and Steinhaus [14]. It is easy to see that the Banach-Steinhaus principle of condensation of singularities is closely connected with Theorem 1 and can also be obtained as a consequence of the Kuratowski-Ulam theorem. Indeed, let us take

$$E_1 = \mathbf{N}$$

where the set $\mathbf{N}$ of all natural numbers is equipped with the discrete topology, and let E_2 be an arbitrary Banach space. Suppose that E_3 is another Banach space and a double sequence of continuous linear operators

$$L_{m,n} : E_2 \to E_3 \quad (m, n \in \mathbf{N})$$

is given, such that, for any $m \in \mathbf{N}$, we have

$$sup_{n \in \mathbf{N}} ||L_{m,n}|| = +\infty.$$

Let us define a partial mapping Φ from the product space $E_1 \times E_2$ into the space E_3 by the following formula:

$$\Phi(m, x) = lim_{n \to +\infty} L_{m,n}(x).$$

It is clear that this partial mapping has the Baire property and, for each $m \in \mathbf{N}$, the partial mapping $\Phi(m, .)$ is defined on a first category subset of the space E_2. Hence the domain of the partial mapping Φ is also a first category set in the product space $E_1 \times E_2$. Now, we may apply the Kuratowski-Ulam theorem and, evidently, we obtain that, for almost all elements $x \in E_2$, the set

$$\{m \; : \; (m, x) \in dom(\Phi)\}$$

is empty. But, actually, this is the Banach-Steinhaus principle of condensation of singularities.

Remark 2. The general scheme of applications of Theorem 1 is as follows. First of all, we must check that a given partial mapping Φ has the Baire property. Obviously, Φ has this property if it is a Borel mapping or, more generally, if it is a measurable mapping with respect to the $\sigma-$ algebra generated by a family of analytic sets (such situations are typical in modern analysis). Now, suppose that our partial mapping Φ of two variables has the Baire property. Then the second step is to check that the corresponding partial mappings of one variable are defined on the first category sets. This will be valid if E_2 and E_3 are Polish topological vector spaces and if, for almost all elements $x \in E_1$, the corresponding mappings $\Phi(x, .)$ are linear and discontinuous on their domains (notice that if the given space E_3 is a normed vector space, then we need to check the linearity and the unboundedness of the corresponding partial mappings). Finally, we can apply Theorem 1.

Now, we wish to present an application of Theorem 1 in a concrete situation. Namely, we will be interested in a certain type of generalized derivative.

Let c_0 denote the separable Banach space consisting of all real-valued sequences converging to zero. Let $\mathbf{R}$ denote the real line and let $[0, 1]$ be the closed unit interval in $\mathbf{R}$. Suppose that a mapping

$$\phi : [0, 1] \to c_0$$

is given. Evidently, we may write

$$\phi = \{\phi_n \; : \; n \in \mathbf{N}\}$$

where
$$\phi_n : [0,1] \to \mathbf{R} \qquad (n \in \mathbf{N}).$$

Let us assume that the mapping ϕ satisfies the following condition: for each point $x \in [0,1]$ and for each index $n \in \mathbf{N}$, the value $\phi_n(x)$ is not equal to zero. Moreover, let us assume (without loss of generality) that

$$0 < \phi_n(0) \leq 1,$$

$$0 > \phi_n(1) \geq -1$$

for all natural numbers n. If f is a real-valued function defined on the segment $[0,1]$ and a point x belongs to this segment, then the real number

$$lim_{n \to +\infty} \frac{f(x + \phi_n(x)) - f(x)}{\phi_n(x)}$$

is called the ϕ–derivative of f at x (if this limit exists, of course). In our further considerations, we denote the limit mentioned above by the symbol $f'_\phi(x)$.

Let us put
$$E_1 = [0,1], \quad E_2 = C[0,1], \quad E_3 = \mathbf{R}$$

and consider a partial mapping Φ acting from the product space $E_1 \times E_2$ into the space E_3 and defined by the formula

$$\Phi(x, f) = f'_\phi(x).$$

Suppose that the original function ϕ has the Baire property. We assert that, in such a case, the partial mapping Φ has the Baire property, too. Indeed, it suffices to observe that, for every natural number n, the mappings

$$(x, f) \to f(x + \phi_n(x)) - f(x),$$

$$(x, f) \to \phi_n(x)$$

have the Baire property. For the second mapping, this is obvious since the function ϕ_n has the Baire property. Further, the mapping

$$(x, f) \to f(x)$$

is continuous and the mapping

$$(x, f) \to f(x + \phi_n(x))$$

can be represented as the following superposition:

$$(x, f) \to (x, \phi_n(x), f) \to (x + \phi_n(x), f) \to f(x + \phi_n(x)).$$

In this superposition the first mapping has the Baire property and the two other mappings are continuous. Therefore we conclude that the superposition also has the Baire property. Let us notice, by the way, that the same result can be established in a different manner. Namely, if in the function of two variables

$$(x, f) \to f(x + \phi_n(x)) - f(x)$$

we fix a point x, then we obtain a continuous function of one variable, and if in the same function of two variables we fix a second variable f, then we obtain a function of one variable having the Baire property. So, we see that, for our function of two variables, the conditions similar to the classical Carathéodory conditions (i.e., the measurability with respect to one of the variables and the continuity with respect to another one) are fulfilled. From this fact it immediately follows that our function of two variables has the Baire property (in this connection, see also [141] where a general problem concerning the measurability of functions of two or more variables is investigated in detail).

Taking the above remarks into account, we conclude that the partial mapping

$$(x, f) \to f'_\phi(x)$$

has the Baire property. Moreover, it is easy to see that if a point x is fixed, then this partial mapping yields a linear discontinuous function of one variable f. Consequently, we can apply Theorem 1 and formulate the following statement.

Theorem 2. *If a mapping*

$$\phi : [0, 1] \to c_0$$

has the Baire property, then almost each function from the Banach space $C[0, 1]$ does not possess a ϕ-derivative almost everywhere on the segment $[0, 1]$.

We want to point out that the basic operations used in classical mathematical analysis are, as a rule, of the projective type, i.e., these operations are described completely by some projective sets lying in certain Polish topological spaces. In many natural situations, it can happen that the graph of our partial mapping Φ from Theorem 1 is a projective subset of the corresponding Polish product space. Then, according to the important results of Solovay, Martin, and others, we must apply some additional set–theoretical axioms for the validity of the corresponding version of Theorem 1. For example, suppose that Φ satisfies only condition (2) of Theorem 1, the graph of Φ lies in a Polish product space $E_1 \times E_2 \times E_3$ and this graph is a continuous image of the complement of an analytic subset of a Polish topological space. Then if we wish to preserve the assertion of Theorem 1 for Φ, we need the existence of a two-valued measurable cardinal or Martin's Axiom with the negation of the Continuum Hypothesis. Analogously, if the graph of our partial mapping Φ is a projective subset of a Polish product space, belonging to a higher projective class, then we need the Axiom of Projective Determinacy or a similar set–theoretical axiom (for more details, see [71] and [72]). Actually, suppose that we work in the following theory:

$$\textbf{ZF} \ \& \ \textbf{DC} \ \& \ (\textit{each subset of } \textbf{R} \textit{ has the Baire property}).$$

Then the assertion of Theorem 1 will be true for all Polish topological spaces E_1, E_2, E_3 and for all partial mappings Φ acting from $E_1 \times E_2$ into E_3 and satisfying condition (2) of this theorem. See, e.g., [86] where the theory mentioned above is applied to some questions connected with the existence of generalized derivatives of various types. In particular, it is established in [86] that if we work in the above-mentioned theory, then almost each function from the space $C[0,1]$ does not possess a generalized derivative almost everywhere on the segment $[0,1]$. Obviously, such an approach can also be applied to special types of generalized derivatives, for instance, to the so-called path derivatives (for the definition and basic properties of path derivatives, see, e.g., [29]).

In addition, let us stress that the direct analogue of the classical Banach-Mazurkiewicz theorem (which was considered in the Introduction) cannot be established for all generalized derivatives, since there is (in the theory $\textbf{ZF} \ \& \ \textbf{DC}$) a certain notion of a generalized derivative having the property that, for any continuous function

$$f \ : \ [0,1] \to \textbf{R},$$

there exists at least one point x from the segment $[0,1]$, such that f is differentiable at x in the sense of this generalized derivative (cf. [86]).

Further, the following natural question arises: does there exist an analogue of the above-mentioned result in terms of measure theory? In other words, does there exist a Borel diffused probability measure μ on the space $C[0,1]$ such that, for any generalized derivative introduced in the theory $\textbf{ZF} \ \& \ \textbf{DC}$, almost all (with respect to μ) functions from $C[0,1]$ are not

differentiable, in the sense of this derivative, at almost all (with respect to
λ) points of $[0, 1]$?

At the present time, this question remains open.

Here we give a construction of the classical Wiener measure μ_w on $C[0, 1]$
and demonstrate that, for the derivative in the usual sense, μ_w yields a
positive answer to this question. We recall that historically the Wiener
measure appeared as a certain interpretation (mathematical model) of the
Brownian motion (for an interesting survey of this phenomenon, see, e.g.,
[21] and, especially, [131]).

Note that the construction of the Wiener measure is not easy and needs
a number of auxiliary facts and statements (however, those facts and statements turn out to be useful for the general theory of stochastic processes).

To begin, we first of all wish to recall some simple notions from probability theory and the famous Kolmogorov theorem on mutually consistent
finite-dimensional probability distributions.

Let E be a set, let $\mathcal{S}$ be a σ-algebra of subsets of E and let μ be a
probability measure on $\mathcal{S}$. So we are dealing with the basic probability
space

$$(E, \mathcal{S}, \mu).$$

In our further constructions we assume, as a rule, that μ is a complete
measure. This does not restrict the generality of our considerations because
we can always replace μ by its completion.

Let f be a partial function acting from E into $\mathbf{R}$. We say that f is a
random variable if f is measurable with respect to the σ-algebra $\mathcal{S}$ (i.e., for
any open set $U \subset \mathbf{R}$, the preimage $f^{-1}(U)$ belongs to $\mathcal{S}$) and

$$\mu(E \setminus dom(f)) = 0.$$

For any random variable f, we may define the Borel probability measure
μ_f on $\mathbf{R}$, putting

$$\mu_f(X) = \mu(f^{-1}(X)) \qquad (X \in B(\mathbf{R})).$$

The measure μ_f is usually called the distribution of a random variable f.
Actually, the measure μ_f is defined in such a way that it becomes the
homomorphic image of the measure μ under the homomorphism f, so we
may write

$$\mu_f = \mu \circ f^{-1}.$$

Obviously, μ_f is uniquely determined by the function

$$F_f \ : \ \mathbf{R} \to [0, 1]$$

such that
$$F_f(x) = \mu(\{e \in E \ : \ f(e) < x\}) \qquad (x \in \mathbf{R}).$$
This function is also called the distribution of f. It is increasing and satisfies the relations:

(a) $lim_{t \to -\infty} F_f(t) = 0$;

(b) $lim_{t \to +\infty} F_f(t) = 1$;

(c) $(\forall x \in \mathbf{R})(lim_{t \to x-} F_f(t) = F_f(x))$, i.e., the function F_f is continuous from the left.

Exercise 1. Let F be an increasing function acting from $\mathbf{R}$ into $\mathbf{R}$ and satisfying the relations analogous to (a), (b), and (c). Show that there exist a probability space $(E, \mathcal{S}, \mu)$ and a random variable

$$f \ : \ E \to \mathbf{R},$$

such that $F = F_f$.

Exercise 2. We recall that a probability measure μ is separable if the topological weight of the metric space canonically associated with μ is less than or equal to ω (in other words, the above-mentioned metric space is separable). For instance, the classical Lebesgue measure on the unit segment $[0, 1]$ is separable. Check that this fact is a trivial consequence of the following statement: the completion of any probability measure given on a countably generated σ-algebra of sets is separable. Show the validity of this statement.

Check that any homomorphic image of a separable measure is separable, too.

Give an example of a topological space T and of a Borel probability measure on T which is not separable (take as T the commutative compact topological group $\{0, 1\}^{\omega_1}$).

Remark 3. In connection with the result of Exercise 2, let us note that there exist nonseparable measures on the segment $[0, 1]$ extending the classical Lebesgue measure on $[0, 1]$. Moreover, there are nonseparable extensions of the standard Lebesgue measure on the unit circumference, which are invariant under the group of all rotations of this circumference about its centre (for more information, see, e.g., [73], [82], [88], and [111]).

Exercise 3. Let $(E, \mathcal{S}, \mu)$ be a basic probability space and let T be a topological space (equipped with its Borel σ-algebra $\mathcal{B}(T)$). Any μ-measurable partial mapping

$$f \ : \ E \to T$$

satisfying the condition
$$\mu(E \setminus dom(f)) = 0$$
is usually called a T-valued random variable on E. The Borel probability measure μ_f on T defined by the formula

$$\mu_f(X) = \mu(f^{-1}(X)) \qquad (X \in \mathcal{B}(T))$$

is called the distribution of f in T, and we write

$$\mu_f = \mu \circ f^{-1}.$$

Show that there exist a probability space $(E, \mathcal{S}, \mu)$, a topological space T and a Borel probability measure ν on T, such that there is no T-valued random variable f on E for which $\mu_f = \nu$.

Let $(E, \mathcal{S}, \mu)$ be again a basic probability space and let

$$f \; : \; E \to \mathbf{R}$$

be a random variable. We recall that $\int_E f(e)d\mu(e)$ denotes the mathematical expectation of f (of course, under the assumption that this integral exists). We also recall the simple formula

$$\int_E f(e)d\mu(e) = \int_{\mathbf{R}} x dF_f(x).$$

More generally, for any Borel function

$$\phi \; : \; \mathbf{R} \to \mathbf{R},$$

we have the equality

$$\int_E \phi(f(e))d\mu(e) = \int_{\mathbf{R}} \phi(x)dF_f(x)$$

under the assumption that the corresponding integrals exist.

Exercise 4. Prove the formula presented above. Deduce, in particular, that, for each natural number n, the equality

$$\int_E (f(e))^n d\mu(e) = \int_{\mathbf{R}} x^n dF_f(x)$$

holds true (if these integrals exist).

In many cases, it may happen that the distribution μ_f of a random variable f can be defined with the aid of its density. We recall that a Lebesgue measurable function

$$p_f \; : \; \mathbf{R} \to [0, +\infty[$$

is a density of μ_f (of F_f) if, for each Borel set $X \subset \mathbf{R}$, we have

$$\mu_f(X) = \int_X p_f(x)dx.$$

This means that the measure μ_f is absolutely continuous with respect to the Lebesgue measure λ on $\mathbf{R}$. Evidently, any two densities of μ_f are equivalent with respect to λ. In addition, if p_f exists, then we can write

$$\int_E f(e)d\mu(e) = \int_{\mathbf{R}} x p_f(x)dx$$

and, more generally,

$$\int_E \phi(f(e))d\mu(e) = \int_{\mathbf{R}} \phi(x) p_f(x)dx$$

for every Borel function

$$\phi \; : \; \mathbf{R} \to \mathbf{R}$$

such that the corresponding integrals exist.

The classical example of a probability distribution is the normal (or Gaussian) distribution. For the real line $\mathbf{R}$, the density of the so-called centered normal distribution is given by the formula

$$p_f(x) = (2\pi)^{-1/2}(1/\sigma)exp(-x^2/2\sigma^2) \qquad (x \in \mathbf{R})$$

where $\sigma > 0$ is a fixed constant. It can easily be checked in this case that

$$\int_E f^2(e)d\mu(e) = \int_{\mathbf{R}} x^2 p_f(x)dx = \sigma^2.$$

Taking a derivative (with respect to a parameter σ) in the last equality, we obtain

$$\int_E f^4(e)d\mu(e) = \int_{\mathbf{R}} x^4 p_f(x)dx = c\sigma^4,$$

where c is some strictly positive constant whose precise value is not interesting for us.

We now wish to recall the Kolmogorov theorem on the existence of a probability measure with given finite-dimensional distributions (see, e.g., [21], [49], [156], [171]). This theorem plays the fundamental role in the contemporary theory of stochastic processes.

Let T be an arbitrary set of indices. Consider a family $\{R_t \ : \ t \in T\}$ where, for each index $t \in T$, the set R_t coincides with $\mathbf{R}$. Suppose that, for any finite set

$$\tau = \{t_1, ..., t_n\} \subset T,$$

a Borel probability measure μ_τ on the product space

$$R_\tau = R_{t_1} \times ... \times R_{t_n}$$

is given in such a way that the whole family

$$\{\mu_\tau \ : \ \tau \in [T]^{<\omega}\}$$

of probability measures is consistent, i.e., for any two finite subsets τ and τ' of T such that $\tau \subset \tau'$, we have

$$\mu_\tau = \mu_{\tau'} \circ pr_{\tau',\tau}^{-1},$$

where

$$pr_{\tau',\tau} \ : \ R_{\tau'} \to R_\tau$$

denotes the canonical projection from $R_{\tau'}$ onto R_τ. Further, consider the product space

$$\mathbf{R}^T = \prod_{t \in T} R_t$$

with the σ-algebra $\mathcal{S}$ generated by the family of mappings

$$\{pr_t \ : \ t \in T\}$$

where, for each index $t \in T$, the mapping

$$pr_t \ : \ \mathbf{R}^T \to R_t$$

coincides with the canonical projection from $\mathbf{R}^T$ onto R_t. In other words, we may define $\mathcal{S}$ as the smallest σ-algebra of subsets of $\mathbf{R}^T$, such that all mappings pr_t $(t \in T)$ are measurable with respect to $\mathcal{S}$ ($\mathcal{S}$ is also frequently called the cylindrical σ-algebra in the space $\mathbf{R}^T$, generated by the family of linear functionals $\{pr_t : t \in T\}$).

Exercise 5. Show that the cylindrical σ-algebra $\mathcal{S}$ of the topological product space $\mathbf{R}^T$ coincides with its Borel σ-algebra if and only if

$$card(T) \leq \omega.$$

Exercise 6. Let X be a set and let $\{f_i \ : \ i \in I\}$ be a family of real-valued functions defined on X. We recall that this family separates the points of X if, for any two distinct points x and y from X, there exists an index $i \in I$ such that

$$f_i(x) \neq f_i(y).$$

Let now X be a Polish topological space and let $\{f_i \ : \ i \in I\}$ be a countable family of Borel real-valued functions on X, separating the points of X. Denote by

$$\mathcal{A} = \mathcal{S}(\{f_i \ : \ i \in I\})$$

the smallest σ-algebra of subsets of X, for which all functions f_i $(i \in I)$ become measurable. Consider a mapping

$$f \ : \ X \to \mathbf{R}^I$$

defined by the formula

$$f(x) = (f_i(x))_{i \in I} \qquad (x \in X).$$

Note that, since $card(I) \leq \omega$, the space $\mathbf{R}^I$ is isomorphic to one of the spaces $\mathbf{R}^\omega$, $\mathbf{R}^n$ $(n \in \mathbf{N})$. Check that:

(a) f is injective and Borel;

(b) $\mathcal{A} = \{f^{-1}(Z) \ : \ Z \in \mathcal{B}(\mathbf{R}^I)\}$.

By using the classical theorem from descriptive set theory, stating that the image of a Borel subset of a Polish space under an injective Borel mapping into a Polish space is also Borel (see [120] or the Introduction), infer from (a) and (b) the equality

$$\mathcal{A} = \mathcal{B}(X).$$

In particular, consider the separable Banach space $C[0, 1]$ of all continuous real-valued functions on the segment $[0, 1]$ and take as I a countable subset of $[0, 1]$ everywhere dense in $[0, 1]$. For each $i \in I$, let

$$f_i \ : \ C[0, 1] \to \mathbf{R}$$

be the mapping defined by

$$f_i(\phi) = \phi(i) \qquad (\phi \in C[0, 1]).$$

Conclude from the result presented above that

$$\mathcal{A} = \mathcal{B}(C[0,1]).$$

Give also a direct proof of this equality, without the aid of the mentioned result.

The Kolmogorov extension theorem states that there exists a unique probability measure μ_T defined on the cylindrical σ-algebra $\mathcal{S}$ of $\mathbf{R}^T$ and satisfying the relations

$$\mu_\tau = \mu_T \circ pr_{T,\tau}^{-1} \qquad (\tau \in [T]^{<\omega}),$$

where, for each finite set $\tau \subset T$, the mapping

$$pr_{T,\tau} : \mathbf{R}^T \to R_\tau$$

is the canonical projection from $\mathbf{R}^T$ onto R_τ. The original measures μ_τ are usually called the finite-dimensional distributions of μ_T.

The proof of the Kolmogorov theorem is not very difficult. Indeed, using the consistency conditions, we first define the functional μ_T on the cylindrical algebra (consisting of all finite unions of elementary subsets of $\mathbf{R}^T$) in such a way that the equalities

$$\mu_\tau = \mu_T \circ pr_{T,\tau}^{-1}$$

will be fulfilled for all finite sets $\tau \subset T$. Then we have to show that this functional is countably additive on the above-mentioned algebra. This is not hard because all finite-dimensional spaces R_τ are Radon, i.e., for any Borel set $X \subset R_\tau$ and for each $\varepsilon > 0$, there exists a compact set $K \subset X$ such that

$$\mu_\tau(X \setminus K) < \varepsilon.$$

Finally, utilizing the classical Carathéodory theorem, we can extend our functional onto the whole cylindrical σ-algebra $\mathcal{S}$ (for details, see, e.g., [21], [156] or [171]).

Exercise 7. With the previous notation, show that, in the formulation of the Kolmogorov theorem, it suffices to assume only the consistency conditions of the form

$$\mu_\tau = \mu_{\tau'} \circ pr_{\tau',\tau}^{-1},$$

where τ and τ' are any finite subsets of T for which

$$\tau \subset \tau', \quad card(\tau' \setminus \tau) = 1.$$

Remark 4. There are various generalizations of the Kolmogorov theorem. For example, this theorem may be regarded as a particular case of the statement asserting the existence of a projective limit of a given projective system of Radon probability measures. Furthermore, there are some abstract versions of the Kolmogorov theorem in terms of the so-called compact classes of sets introduced by Marczewski. For more details, see again [21], [156], or [171].

It is interesting to note that, by using some generalized version of the Kolmogorov theorem, the well-known Riesz theorem about representations of all continuous linear functionals on the space $C(K)$, where K is a compact topological space, can be deduced.

For our further purposes, we need only that special case of the Kolmogorov theorem when

$$T = [0, 1].$$

Let us fix a finite set

$$\tau = \{t_1, ..., t_n\} \subset [0, 1] \setminus \{0\}.$$

Clearly, we may suppose that

$$0 < t_1 < ... < t_n.$$

Define a Borel probability measure μ_τ on R_τ by the formula

$$\mu_\tau(X) = \int_X p_\tau(x_1, ..., x_n) dx_1...dx_n \quad (X \in B(R_\tau)),$$

where the density p_τ satisfies the relation

$$p_\tau(x_1, ..., x_n) = (2\pi)^{-n/2}(t_1(t_2 - t_1)...(t_n - t_{n-1}))^{-1/2}.$$

$$exp((-1/2)(x_1^2/t_1 + (x_2 - x_1)^2/(t_2 - t_1) + ... + (x_n - x_{n-1})^2/(t_n - t_{n-1})))$$

for all points

$$(x_1, ..., x_n) \in R_\tau.$$

If τ is a finite subset of $[0, 1]$ whose minimal element coincides with 0, then we put

$$\mu_\tau = \mu_0 \times \mu_{\tau \setminus \{0\}}$$

where μ_0 is the Borel probability measure on R_0 concentrated at the origin of R_0 (i.e., the so-called Dirac measure). It is not difficult to check the consistency of the family of probability measures

$$\{\mu_\tau : \tau \text{ is a finite subset of } [0, 1]\}.$$

Exercise 8. By starting with the equality

$$\int_{-\infty}^{+\infty} exp(-ax^2/2)dx = (2\pi/a)^{1/2} \qquad (a > 0),$$

show that

$$\int_{-\infty}^{+\infty} exp((-1/2)((a-x)^2/c + (x-b)^2/d))dx$$

$$= (2\pi cd/(c+d))^{1/2} \cdot exp((-1/2)((a-b)^2/(c+d))),$$

where a, b, c, d are strictly positive real numbers.

Exercise 9. By using the results of Exercises 7 and 8, demonstrate the consistency of the above-mentioned family of measures

$$\{\mu_\tau \; : \; \tau \; is \; a \; finite \; subset \; of \; [0,1]\}.$$

Applying the Kolmogorov extension theorem to this family of measures, we get the probability measure μ_w on the product space $\mathbf{R}^{[0,1]}$. We shall demonstrate below that the latter measure canonically induces the required Wiener measure on the space $C[0,1] \subset \mathbf{R}^{[0,1]}$ (in this connection, note that the initial measure μ_w also is called the Wiener measure on the product space $\mathbf{R}^{[0,1]}$).

In order to obtain the main result of this chapter, we need some simple but important notions from the general theory of stochastic processes.

Let $(E, \mathcal{S}, \mu)$ be a space endowed with a probability measure and let T be a set of indices (parameters). We shall say that a partial function of two variables

$$H \; : \; E \times T \to \mathbf{R}$$

is a stochastic process if, for each $t \in T$, the partial function

$$H(\cdot, t) \; : \; E \to \mathbf{R}$$

is a random variable on the basic probability space $(E, \mathcal{S}, \mu)$. In this case, for any fixed $e \in E$, the partial function

$$H(e, \cdot) \; : \; T \to \mathbf{R}$$

is called the trajectory of a given process H, corresponding to e.

Suppose that T is equipped with a σ-algebra $\mathcal{S}'$ of its subsets, i.e., the pair $(T, \mathcal{S}')$ turns out to be a measurable space. We say that a stochastic process H is measurable if it (regarded as a partial function on $E \times T$) is measurable with respect to the product σ-algebra of $\mathcal{S}$ and $\mathcal{S}'$.

Exercise 10. Let us put $E = T = [0, 1]$ and equip $[0, 1]$ with the standard Lebesgue measure λ. Give an example of a nonmeasurable stochastic process H such that

$$dom(H) = E \times T$$

and all trajectories $H(e, \cdot)$ $(e \in E)$ and all random variables $H(\cdot, t)$ $(t \in T)$ belong to the first Baire class (see Chapter 2).

Suppose that some two stochastic processes H and G are given on $E \times T$. We say that they are stochastically equivalent if, for each $t \in T$, the random variables $H(\cdot, t)$ and $G(\cdot, t)$ are equivalent (i.e., coincide almost everywhere with respect to μ).

Stochastically equivalent processes have very similar properties and, as a rule, are identified. However, in certain problems of probability theory (e.g., in those where special features of trajectories of a given process play an essential role) such an identification cannot be done.

Assume now that a set T of parameters is a topological space. We say that a stochastic process

$$H \;:\; E \times T \to \mathbf{R}$$

is stochastically continuous at a point $t_0 \in T$ if, for each $\varepsilon > 0$, we have

$$lim_{t \to t_0} \mu(\{e \in E \;:\; |H(e, t) - H(e, t_0)| > \varepsilon\}) = 0.$$

Further, we say that a process H is stochastically continuous if H is stochastically continuous at all points $t \in T$.

Note that if H_1 and H_2 are any two stochastically equivalent processes, then H_1 is stochastically continuous if and only if H_2 is stochastically continuous.

Exercise 11. Suppose that the unit segment $[0, 1]$ is equipped with the Lebesgue measure λ. Give an example of a measurable stochastic process H with

$$dom(H) = [0, 1] \times [0, 1],$$

which is stochastically continuous but almost all its trajectories are discontinuous.

Lemma 1. Let $T = [0, 1]$ with the usual Euclidean topology and let

$$H \;:\; E \times T \to \mathbf{R}$$

be a stochastic process. Then the following two conditions are equivalent:
  (1) H is stochastically continuous;
  (2) for any $\varepsilon > 0$, we have

$$lim_{d \to 0+} sup_{t \in T, t' \in T, |t-t'| < d} \; \mu(\{e \in E \; : \; |H(e,t) - H(e,t')| > \varepsilon\}) = 0.$$

Proof. Suppose that condition (1) is fulfilled. Fix $\varepsilon > 0$ and $\delta > 0$. For each $t \in T$, there exists an open neighborhood $V(t)$ of t such that

$$sup_{t' \in V(t)} \mu(\{e \in E \; : \; |H(e,t') - H(e,t)| > \varepsilon/2\}) < \delta/2.$$

The family $\{V(t) \; : \; t \in T\}$ forms an open covering of $T = [0,1]$. Because $[0,1]$ is compact, there exists a Lebesgue number $d > 0$ for this covering, i.e., d has the property that any subinterval of $[0,1]$ with diameter $2d$ is contained in one of the sets of the covering. Consequently, if

$$t \in T, \quad t' \in T, \quad |t - t'| < d,$$

then $t' \in \;]t - d, t + d[$ and, for some $r \in T$, we get

$$]t - d, t + d[\; \subset V(r), \quad t \in V(r), \quad t' \in V(r).$$

Thus, for almost all $e \in E$, we may write

$$\{e \; : \; |H(e,t') - H(e,t)| > \varepsilon\} \subset$$

$$\{e \; : \; |H(e,t') - H(e,r)| > \varepsilon/2\} \cup \{e \; : \; |H(e,t) - H(e,r)| > \varepsilon/2\}$$

and, taking into account the definition of $V(r)$, we obtain

$$\mu(\{e \in E \; : \; |H(e,t') - H(e,t)| > \varepsilon\}) < \delta/2 + \delta/2 = \delta.$$

This establishes the implication (1) $\Rightarrow$ (2). The converse implication (2) $\Rightarrow$ (1) is trivial, and the lemma has thus been proved.

Exercise 12. Show that Lemma 1 holds true in a more general situation when T is an arbitrary nonempty compact metric space.

Exercise 13. Let $T = [0,1]$ and let

$$H \; : \; E \times T \to \mathbf{R}$$

be a stochastic process. Suppose also that, for some real number $\alpha > 0$, there exists a function

$$\phi \; : \; [0,1] \to \; [0,+\infty[$$

satisfying the following two conditions:
  (1) $\lim_{d \to 0+} \phi(d) = 0$;
  (2) for all t and t' from $[0, 1]$, we have

$$\int_E |H(e, t) - H(e, t')|^\alpha d\mu(e) \leq \phi(|t - t'|).$$

Show that the process H is stochastically continuous.

  The simple result presented in Exercise 13 can directly be applied to the Wiener measure μ_w introduced above. Indeed, we have the basic probability space

$$(\mathbf{R}^{[0,1]}, \mathcal{S}, \mu_w)$$

and the stochastic process

$$W \ : \ \mathbf{R}^{[0,1]} \times [0, 1] \to \mathbf{R}$$

canonically associated with μ_w, which is defined by the formula

$$W(\cdot, t) = pr_t \qquad (t \in [0, 1]).$$

In particular, we see that

$$dom(W) = \mathbf{R}^{[0,1]} \times [0, 1].$$

Choose any two points t_1 and t_2 from $[0, 1]$ such that

$$0 < t_1 < t_2.$$

According to the definition of μ_w, the two-dimensional distribution of the random vector

$$(W(\cdot, t_1), W(\cdot, t_2))$$

is given by the corresponding density

$$p_{t_1, t_2} \ : \ \mathbf{R}^2 \to \mathbf{R}$$

where, for all $(x_1, x_2) \in \mathbf{R}^2$, we have

$$p_{t_1, t_2}(x_1, x_2) =$$

$$(1/2\pi)(t_1(t_2 - t_1))^{-1/2} exp((-1/2)(x_1^2/t_1 + (x_2 - x_1)^2/(t_2 - t_1))).$$

Consider the random variable

$$W(\cdot, t_1) - W(\cdot, t_2).$$

It is easy to see that the density

$$p \: : \: \mathbf{R} \to \mathbf{R}$$

of this variable is defined by the formula

$$p(x) = (2\pi(t_2 - t_1))^{-1/2} exp(-x^2/2(t_2 - t_1)) \qquad (x \in \mathbf{R}).$$

Indeed, this immediately follows from the general fact stating that if (f_1, f_2) is a random vector whose density of distribution is

$$q_{(f_1, f_2)} \: : \: \mathbf{R}^2 \to \mathbf{R},$$

then the density of distribution of $f_1 - f_2$ is

$$q_{f_1 - f_2} \: : \: \mathbf{R} \to \mathbf{R}$$

where

$$q_{f_1 - f_2}(x) = \int_{\mathbf{R}} q_{(f_1, f_2)}(x + y, y) dy \qquad (x \in \mathbf{R}).$$

Exercise 14. Prove the fact mentioned above.

Now, if t and t' are any two points from $[0, 1]$, we may write

$$\mu_w(\{e \in \mathbf{R}^{[0,1]} \: : \: |W(e, t) - W(e, t')| > \varepsilon\}) \le$$

$$(1/\varepsilon^2) \int (W(e, t) - W(e, t'))^2 d\mu_w(e) = |t - t'|/\varepsilon^2.$$

This shows us that the process W is stochastically continuous. W is usually called the standard Wiener process. Let us remark that W may be regarded as a canonical example of a Gaussian process (for information about Gaussian processes, see, e.g., [49], [156], and [171]).

Let us return to a general probability space $(E, \mathcal{S}, \mu)$ and assume that T is a set of parameters equipped with some σ-algebra $\mathcal{S}'$ of its subsets. Consider two stochastic processes

$$H \: : \: E \times T \to \mathbf{R}, \qquad G \: : \: E \times T \to \mathbf{R}.$$

We shall say that G is a measurable modification of H if the following conditions are fulfilled:

(a) H and G are stochastically equivalent;

(b) G is a measurable process, i.e., G regarded as a partial function acting from $E \times T$ into $\mathbf{R}$ is measurable with respect to the product σ-algebra of $\mathcal{S}$ and $\mathcal{S}'$.

In particular, if

$$(T, \mathcal{S}', \nu)$$

is a probability space and G is a measurable modification of H, then we also say that G is a $(\mu \times \nu)$-measurable modification of H. But, sometimes, it is more convenient to define a $(\mu \times \nu)$-measurable modification of H as a stochastically equivalent process measurable with respect to the completion of the product measure $\mu \times \nu$.

Suppose that our set T of parameters is a topological space. We shall say that a stochastic process

$$H \; : \; E \times T \to \mathbf{R}$$

is separable if there are a μ-measure zero set $A \subset E$ and a countable set $Q \subset T$, such that, for any element $e \in E \setminus A$ and for any point $t \in dom(H(e, \cdot))$, there exists a sequence

$$\{t_n \; : \; n \in \mathbf{N}\} \subset Q \cap dom(H(e, \cdot))$$

converging to t and having the property

$$lim_{n \to +\infty} H(e, t_n) = H(e, t).$$

From this definition follows at once that Q is an everywhere dense in T, so T is separable (as a topological space). The above-mentioned set Q is usually called a set of separability of H.

Lemma 2. *Let T coincide with the unit segment $[0, 1]$ equipped with the standard Lebesgue measure λ, and let*

$$H = \{H(\cdot, t) \; : \; t \in T\}$$

be an arbitrary stochastically continuous process. Then there exists a process

$$G = \{G(\cdot, t) \; : \; t \in T\}$$

satisfying the relations:
(1) H and G are stochastically equivalent;
(2) G is measurable;
(3) G is separable and one of its sets of separability coincides with

$$Q = \{k/2^m \; : \; k \in \mathbf{N}, \; m \in \mathbf{N}, \; k/2^m \leq 1\};$$

(4) there exists a μ-measurable set E' with $\mu(E') = 1$ such that, for any point $t \in Q$, we have

$$H(\cdot, t)|E' = G(\cdot, t)|E'.$$

In particular, G turns out to be a measurable separable modification of H.

Proof. In view of Lemma 1, for each $\varepsilon > 0$, we can write

$$lim_{d \to 0+} sup_{t \in T, t' \in T, |t - t'| < d} \, \mu(\{e \in E \ : \ |H(e, t) - H(e, t')| > \varepsilon\}) = 0.$$

Consequently, for any integer $n > 0$, there exists a finite family of reals

$$0 = t_0^n < t_1^n < \, \ldots \, < t_{k(n)}^n = 1$$

belonging to Q and satisfying the conditions:
(a) the length of each segment $[t_i^n, t_{i+1}^n]$ is less than $1/n$;
(b) if t and t' belong to some segment $[t_i^n, t_{i+1}^n]$, then

$$\mu(\{e \in E \ : \ |H(e, t) - H(e, t')| > 1/n\}) < 1/2^n.$$

Moreover, we may choose the above-mentioned families

$$Q_n = \{t_i^n \ : \ i = 0, 1, ..., k(n)\}$$

in such a way that the following conditions will be fulfilled, too:
(c) for any $n \in \mathbf{N} \setminus \{0\}$, the set Q_n is contained in the set Q_{n+1};
(d) $Q = \cup\{Q_n \ : \ n \in \mathbf{N}, \ n > 0\}$.
Now, let us put

$$E' = \cap\{dom(H(\cdot, t)) \ : \ t \in Q\}.$$

Obviously, we have the equality

$$\mu(E') = 1.$$

Further, for each integer $n > 0$, define a function

$$G_n \ : \ E' \times [0, 1] \to \mathbf{R}$$

by the relations

$$G_n(e, t) = H(e, t_i^n) \qquad (t \in [t_i^n, t_{i+1}^n[),$$

$$G_n(e, 1) = H(e, 1).$$

Evidently, the partial function G_n is measurable with respect to the product σ-algebra of $\mathcal{S}$ and $\mathcal{B}([0,1])$. Furthermore, the series

$$\sum_{n>0} \mu(\{e \in E' \cap dom(H(\cdot,t)) \ : \ |H(e,t) - G_n(e,t)| > 1/n\})$$

is convergent for any point $t \in [0,1]$. Hence, for each $t \in [0,1]$, we get

$$lim_{n \to +\infty} G_n(\cdot,t) = H(\cdot,t)$$

almost everywhere in E (with respect to μ, of course). Let us put

$$G(e,t) = limsup_{n \to +\infty} G_n(e,t)$$

for all those pairs $(e,t) \in E' \times T$ for which the above-mentioned $limsup$ exists. In this way, we obtain a partial mapping

$$G \ : \ E \times T \to \mathbf{R}.$$

The definition of G implies at once that G is a measurable stochastic process stochastically equivalent to H and, for any point $t \in Q$, we have

$$G(\cdot,t)|E' = H(\cdot,t)|E'.$$

Let now t be an arbitrary point from $[0,1] \setminus Q$. Then there exists an increasing sequence

$$\{t_{i(n)}^n \ : \ n \in \mathbf{N}, \ n > 0\} \subset [0,1]$$

such that

$$t_{i(n)}^n \in Q_n, \quad t \in [t_{i(n)}^n, t_{i(n)+1}^n[, \quad lim_{n \to +\infty} t_{i(n)}^n = t.$$

In virtue of the definition of G, we easily obtain

$$G(e,t) \in cl(\{G(e,t_{i(n)}^n) \ : \ n \in \mathbf{N} \setminus \{0\}\})$$

for any point $e \in E' \cap dom(G(\cdot,t))$. This completes the proof of the lemma.

Remark 5. The process G of Lemma 2 is usually called a separable modification of the original process H. Note that the existence of a separable modification of a given process can be established in a much more general situation than in that described by Lemma 2. For our further purposes, this lemma is completely sufficient. More deep results may be found in [156] and

[171]. It is interesting to mention here that the general theorem concerning the existence of a separable modification of a stochastic process essentially relies on the notion of a von Neumann topology (multiplicative lifting). For details, see, e.g., [171] where such an approach is developed.

Lemma 3. *Let $(\alpha_n)_{n\in\mathbf{N}}$ and $(\beta_n)_{n\in\mathbf{N}}$ be two sequences of strictly positive real numbers, such that*

$$\sum_{n\in\mathbf{N}} \alpha_n < +\infty, \quad \sum_{n\in\mathbf{N}} \beta_n < +\infty,$$

and let $\{f_n : n \in \mathbf{N}\}$ be a sequence of random variables on $(E, \mathcal{S}, \mu)$ satisfying the relations

$$\mu(\{e \in E : |f_n(e)| > \alpha_n\}) < \beta_n \quad (n \in \mathbf{N}).$$

Then there exists a μ-measure zero set $A \subset E$ such that, for any point $e \in E \setminus A$, the series

$$\sum_{n\in\mathbf{N}} |f_n(e)|$$

is convergent.

Proof. For each $n \in \mathbf{N}$, let us denote

$$A_n = \{e \in E : |f_n(e)| > \alpha_n\}.$$

Then, according to our assumption,

$$\mu(A_n) < \beta_n \quad (n \in \mathbf{N}).$$

Let us put

$$A = \cap_{n\in\mathbf{N}}(\cup_{m\in\mathbf{N},\ m>n} A_m).$$

Then we obviously have $\mu(A) = 0$. Take any point e from $E \setminus A$. There exists a natural number k for which

$$e \notin \cup_{m\in\mathbf{N},\ m>k} A_m.$$

This means that, for every integer $m > k$, the inequality

$$|f_m(e)| \leq \alpha_m$$

is fulfilled. Hence the series $\sum_{n\in\mathbf{N}} |f_n(e)|$ is convergent, and the proof is completed.

Lemma 4. *Let H be a stochastic process such that*

$$dom(H) = E \times [0,1]$$

and

$$\int_E |H(\cdot, t+r) - H(\cdot, t)|^4 d\mu \le d \cdot r^2$$

for all $t \in [0,1]$ and $t+r \in [0,1]$, where $d > 0$ is some constant. Then there exists a stochastic process G satisfying the relations:
(1) G and H are stochastically equivalent;
(2) G is measurable;
(3) G is separable with a set of separability

$$Q = \{k/2^m \ : \ k \in \mathbf{N}, \ m \in \mathbf{N}, \ k/2^m \le 1\};$$

(4) for any point $t \in Q$, we have

$$H(\cdot, t) = G(\cdot, t);$$

(5) almost all (with respect to μ) trajectories of G are continuous real-valued functions defined on the whole segment $[0,1]$.

Proof. First of all, we may write

$$\mu(\{e \in E \ : \ |H(e, t+r) - H(e,t)| > |r|^{1/5}\}) \le (|r|^{-4/5}) dr^2 = d|r|^{6/5},$$

for any $t \in [0,1]$ and $t + r \in [0,1]$. This immediately implies that H is stochastically continuous. Applying Lemma 2, we can find a process

$$G \ : \ E \times [0,1] \to \mathbf{R}$$

satisfying relations (1)–(4). Indeed, relations (1)–(3) are satisfied in virtue of Lemma 2, and relation (4) is valid because

$$dom(H) = E \times [0,1].$$

Let us denote

$$\Phi_m = sup_{0 \le k < 2^m} |G(\cdot, (k+1)/2^m) - G(\cdot, k/2^m)|,$$

where k and m are assumed to be natural numbers. Obviously, Φ_m is a random variable. Furthermore, we have

$$\mu(\{e \in E \ : \ \Phi_m(e) > 2^{-m/5}\}) \le$$

$$\sum_{0 \leq k < 2^m} \mu(\{e \in E \ : \ |G(e, (k+1)/2^m) - G(e, k/2^m)| > 2^{-m/5}\}) \leq$$

$$\leq 2^m d2^{-6m/5} = d2^{-m/5}.$$

In view of Lemma 3, the series

$$\sum_{m \in \mathbf{N}} \Phi_m$$

is convergent almost everywhere in E, i.e., there exists a μ-measure zero set A such that

$$\sum_{m \in \mathbf{N}} \Phi_m(e) < +\infty$$

for all elements $e \in E \setminus A$. Now, we fix $n \in \mathbf{N}$ and easily observe that if $t \in [0,1]$, $t' \in [0,1]$ and $|t - t'| < 2^{-n}$, then, for some integer $k \geq 0$, the number $k/2^n$ is less than or equal to 1 and

$$|t - k/2^n| < 1/2^n, \quad |t' - k/2^n| < 1/2^n.$$

Evidently,

$$|G(\cdot, t) - G(\cdot, t')| \leq |G(\cdot, t) - G(\cdot, k/2^n)| + |G(\cdot, t') - G(\cdot, k/2^n)|.$$

But if, in addition, $t \in Q$ and $t' \in Q$, then it can directly be checked that

$$|G(\cdot, t) - G(\cdot, k/2^n)| \leq \sum_{m \in \mathbf{N}, \ m > n} \Phi_m,$$

$$|G(\cdot, t') - G(\cdot, k/2^n)| \leq \sum_{m \in \mathbf{N}, \ m > n} \Phi_m,$$

which yields the relation

$$|G(\cdot, t) - G(\cdot, t')| \leq 2(\sum_{m \in \mathbf{N}, \ m > n} \Phi_m).$$

Utilizing the separability of G, we infer that there exists a μ-measure zero set B having the following property: if e is an arbitrary element from $E \setminus (A \cup B)$ and t and t' are any two points such that

$$t \in dom(G(e, \cdot)), \quad t' \in dom(G(e, \cdot)), \quad |t - t'| < 1/2^n,$$

then

$$|G(e, t) - G(e, t')| \leq 2(\sum_{m \in \mathbf{N}, \ m > n} \Phi_m(e)).$$

But we know that, for $e \in E \backslash (A \cup B)$, the series $\sum_{m \in \mathbf{N}} \Phi_m(e)$ is convergent. Thus, we conclude that the trajectory $G(e, \cdot)$ is uniformly continuous. This immediately implies that $G(e, \cdot)$ is a restriction of a continuous real-valued function defined on $[0, 1]$. So we may extend G to a new process in such a way that all trajectories of this process, corresponding to the elements from $E \backslash (A \cup B)$, turn out to be continuous on $[0, 1]$. It can easily be seen that the new process (denoted by the same symbol G) is separable and measurable as well. Indeed, the separability of G holds trivially and the measurability of G follows from the fact that G is measurable with respect to $e \in E$ and is continuous with respect to $t \in [0, 1]$. Lemma 4 has thus been proved.

We now are ready to establish the following result.

Theorem 3. *The Wiener measure μ_w induces a Borel probability measure μ on the space $C[0, 1]$, with properties analogous to the corresponding properties of μ_w.*

Proof. Indeed, we have the probability measure space

$$(\mathbf{R}^{[0,1]}, \mathcal{S}, \mu_w)$$

and the standard Wiener process

$$W = (pr_t)_{t \in [0,1]}$$

for this space. In view of the preceding lemma, there exists a process G for the same space, such that:

(1) W and G are stochastically equivalent;

(2) G is measurable;

(3) G is separable with a set of separability

$$Q = \{k/2^m \; : \; k \in \mathbf{N}, \; m \in \mathbf{N}, \; k/2^m \leq 1\};$$

(4) for any point $t \in Q$, we have $W(\cdot, t) = G(\cdot, t)$;

(5) almost all trajectories of G are continuous real-valued functions on $[0, 1]$.

Let E' denote the set of all those elements $e \in E = \mathbf{R}^{[0,1]}$ for which the trajectory $G(e, \cdot)$ is continuous on $[0, 1]$. Obviously,

$$\mu_w(E') = 1.$$

Define a mapping

$$\phi \; : \; E' \to C[0, 1]$$

by the formula

$$\phi(e) = G(e, \cdot) \qquad (e \in E').$$

Observe that ϕ is measurable with respect to μ_w (this fact easily follows from the result of Exercise 6). So we can put

$$\mu = \mu_w \circ \phi^{-1}.$$

Because μ is a homomorphic image of μ_w, we have

$$\mu(X) = \mu_w(\{e \in E \ : \ G(e, \cdot) \in X\})$$

for each Borel subset X of $C[0,1]$. In particular, if $a > 0$ and t and t' are any two points of $[0,1]$, then

$$\mu(\{f \in C[0,1] \ : \ |f(t) - f(t')| < a\}) = \mu_w(\{e \in E \ : \ |G(e,t) - G(e,t')| < a\})$$

$$= \mu_w(\{e \in E \ : \ |W(e,t) - W(e,t')| < a\}).$$

In a certain sense, we may identify μ and μ_w. So it will be convenient to preserve the same notation μ_w for the obtained measure μ. In other words, we consider μ_w as a Borel probability measure on the space $C[0,1]$.

At last, we are able to return to the question of the differentiability of continuous real-valued functions on $[0,1]$ (from the point of view of μ_w). Namely, the following statement is true.

Theorem 4. *Almost all (with respect to μ_w) functions from $C[0,1]$ are nondifferentiable almost everywhere on $[0,1]$ (with respect to λ).*

Proof. Let us introduce the set

$$D = \{(f,t) \in C[0,1] \times [0,1] \ : \ f \text{ is differentiable at } t\}.$$

It can easily be checked that the set D is $(\mu_w \times \lambda)$-measurable in the product space $C[0,1] \times [0,1]$. So, taking into account the Fubini theorem, it suffices to show that, for each $t \in [0,1]$, the set

$$D_t = \{f \in C[0,1] \ : \ f \text{ is differentiable at } t\}$$

is of μ_w-measure zero. In order to do this, we first observe that the inclusion

$$D_t \subset \cup_{n \in \mathbf{N}}\{f \in C[0,1] \ : \ limsup_{|r| \to 0+}|f(t+r) - f(t)|/|r| < n\}$$

is satisfied. Hence, it suffices to prove, for each $n \in \mathbf{N}$, that

$$\mu_w(D_{t,n}) = 0,$$

where

$$D_{t,n} = \{f \in C[0,1] \ : \ limsup_{|r|\to 0+}|f(t+r) - f(t)|/|r| < n\}.$$

Further, one can easily verify that

$$D_{t,n} \subset \cup_{\delta \in \mathbf{Q}, \ \delta>0}D_{t,n,\delta},$$

where

$$D_{t,n,\delta} = \cap_{r \in \mathbf{Q}, \ 0<|r|<\delta}D_{t,n,\delta,r}$$

and

$$D_{t,n,\delta,r} = \{f \in C[0,1] \ : \ 0 < |r| < \delta, \ |f(t+r) - f(t)|/|r| < n\}.$$

Thus, it remains to demonstrate that

$$\mu_w(D_{t,n,\delta}) = 0.$$

But, for any r satisfying $0 < |r| < \delta$, we may write

$$\mu_w(D_{t,n,\delta,r}) = \mu_w(\{f \in C[0,1] \ : \ |f(t+r) - f(t)| < n|r|\}) \le$$

$$(2\pi|r|)^{-1/2} \int_{-n|r|}^{n|r|} exp(-x^2/2|r|)dx =$$

$$(2\pi)^{-1/2} \int_{-n|r|^{1/2}}^{n|r|^{1/2}} exp(-y^2/2)dy = O(|r|^{1/2}).$$

This immediately implies the desired result, since $|r| > 0$ can be chosen arbitrarily small.

Remark 6. A more general result obtained by Wiener and Lévy holds true; namely, they proved that almost all (with respect to μ_w) functions from $C[0,1]$ are nowhere differentiable on $[0,1]$. Briefly speaking, almost all trajectories of the modified Wiener process are nowhere differentiable on $[0,1]$. For extensive information concerning the relationships between stochastic processes and Brownian motion, we refer the reader to the fundamental monograph by Lévy [131].

Remark 7. As mentioned earlier, the standard Wiener process is a very particular case of a Gaussian stochastic process. Gaussian processes form a natural class of stochastic processes, which have many interesting properties (see, e.g., [31], [156], [171], [206]) and are important from the point of view of numerous applications.

Exercise 15. Let E be an infinite-dimensional separable Hilbert space over $\mathbf{R}$. Check that E is isomorphic to the space $L_2[0,1]$ of all Lebesgue measurable square integrable real-valued functions on the unit segment $[0,1]$, equipped with the canonical scalar product

$$< \phi, \psi > \; = \int_0^1 \phi(x)\psi(x)dx \qquad (\phi \in L_2[0,1], \; \psi \in L_2[0,1]).$$

As usual, we identify all those functions from $L_2[0,1]$, which coincide almost everywhere on $[0,1]$.

For each point $t \in [0,1]$, denote by f_t the characteristic function (i.e., indicator) of the interval $[0,t] \subset [0,1]$. Verify that $f_t \in L_2[0,1]$ and that a mapping

$$f : [0,1] \to L_2[0,1]$$

defined by

$$f(t) = f_t \qquad (t \in [0,1])$$

satisfies the following relations:

(a) f is injective and continuous;

(b) for any four points t_1, t_2, t_3, t_4 from $[0,1]$ such that

$$t_1 < t_2 \le t_3 < t_4,$$

the line segments $[f(t_1), f(t_2)]$ and $[f(t_3), f(t_4)]$ in $L_2[0,1]$ are perpendicular to each other;

(c) f is nowhere differentiable on $[0,1]$.

Note that, in view of relation (a), the set $f([0,1])$ is a curve in $L_2[0,1]$ homeomorphic to $[0,1]$ (it is called the Wiener curve).

We thus conclude that, for an infinite-dimensional separable Hilbert space E, the construction of a continuous nowhere differentiable function acting from $[0,1]$ into E is much easier than the classical construction of a continuous nowhere differentiable function acting from $[0,1]$ into $\mathbf{R}$ (cf. also Theorem 2 of the Introduction).

Bibliography

[1] S.I. Adian, P.S. Novikov. On a semicontinuous function. *Uchen. Zap. Moskov. Gos. Ped. Inst.*, vol. 138, no. 3, pp. 3–10, 1958 (in Russian).

[2] A. Andretta, A. Marcone. Ordinary differential equations and descriptive set theory: uniqueness and globality of solutions of Cauchy problems in one dimension. *Fund. Math.*, vol. 153, no. 2, pp. 157–190, 1997.

[3] J. Appell, P.P. Zabrejko. *Nonlinear Superposition Operators*. Cambridge: Cambridge University Press, 1990.

[4] R. Baire. Sur les fonctions de variables réelles. *Ann. di Math.*, vol. 3, no. 3, 1899.

[5] R. Baire. Sur la représentation des fonctions discontinues. *Acta Math.*, vol. 30, 1905.

[6] R. Baire. *Leçons sur les Fonctions Discontinues*. Paris: Gauthier-Villars, 1905.

[7] M. Balcerzak. Another nonmeasurable set with property (s^0). *Real Analysis Exchange*, vol. 17, no. 2, pp. 781–784, 1991–1992.

[8] M. Balcerzak. Some remarks on sup-measurability. *Real Analysis Exchange*, vol. 17, no. 2, pp. 597–607, 1991–1992.

[9] M. Balcerzak. Typical properties of continuous functions via the Vietoris topology. *Real Analysis Exchange*, vol. 18, no. 2, pp. 532–536, 1992–1993.

[10] S. Baldwin. Martin's axiom implies a stronger version of Blumberg's theorem. *Real Analysis Exchange*, vol. 16, pp. 67–73, 1990–1991.

[11] S. Banach. Sur l'équation fonctionnelle $f(x+y) = f(x)+f(y)$. *Fund. Math.*, vol. 1, p. 123, 1920.

[12] S. Banach. Über die Baire'sche Kategorie gewisser Funktionenmengen. *Studia Math.*, vol. 3, pp. 174–179, 1931.

[13] S. Banach, C. Kuratowski. Sur une généralisation du probléme de la mesure. *Fund. Math.*, vol. 14, pp. 127–131, 1929.

[14] S. Banach, H. Steinhaus. Sur le principe de la condensation de singularités. *Fund. Math.*, vol. 9, pp. 50–61, 1927.

[15] F. Bernstein. Zur Theorie der trigonometrischen Reihen, *Sitzungsber. Sächs. Akad. Wiss. Leipzig. Math.-Natur. Kl.*, 60, pp. 325–338, 1908.

[16] A. Blass. A partition theorem for perfect sets. *Proc. Amer. Math. Soc.*, vol. 82, pp. 271–277, 1981.

[17] J. Blazek, E. Borák, J. Malý. On Köpcke and Pompeiu functions. *Čas. pro pest. mat.*, vol. 103, pp. 53–61, 1978.

[18] H. Blumberg. New properties of all real functions. *Trans. Amer. Math. Soc.*, vol. 24, pp. 113–128, 1922.

[19] V.G. Boltjanskii. *The Hilbert Third Problem.* Moscow: Izd. Nauka, 1977 (in Russian).

[20] N. Bourbaki. *Set Theory.* Moscow: Izd. Mir, 1965 (in Russian, translation from French).

[21] N. Bourbaki. *Integration.* Moscow: Izd. Mir, 1977 (in Russian; translation from French).

[22] J.B. Brown. Restriction theorems in real analysis. *Real Analysis Exchange*, vol. 20, no. 2, pp. 510–526, 1994–1995.

[23] J.B. Brown. Variations on Blumberg's theorem. *Real Analysis Exchange*, vol. 9, pp. 123–137, 1983–1984.

[24] J.B. Brown, G.V. Cox. Classical theory of totally imperfect spaces. *Real Analysis Exchange*, vol. 7, pp. 1–39, 1982.

[25] J.B. Brown, K. Prikry. Variations on Lusin's theorem. *Trans. Amer. Math. Soc.*, vol. 302, pp. 77–86, 1987.

[26] A. Bruckner. *Differentiation of Real Functions.* Berlin: Springer-Verlag, 1978.

[27] A. Bruckner, J. Haussermann. Strong porosity features of typical continuous functions. *Acta Math. Hung.*, vol. 45, no. 1–2, pp. 7–13, 1985.

[28] A. Bruckner, J.L. Leonard. Derivatives. *Amer. Math. Monthly*, vol. 73, no. 4, part II, pp. 24–56, 1966.

[29] A.M. Bruckner, R.J. O'Malley, B.S. Thomson. Path derivatives: a unified view of certain generalized derivatives. *Trans. Amer. Math. Soc.*, vol. 283, no. 1, pp. 97–125, 1984.

[30] L. Bukovský, N.N. Kholshchevnikova, M. Repický. Thin sets of harmonic analysis and infinite combinatorics. *Real Analysis Exchange*, vol. 20, no. 2, pp. 454–509, 1994–1995.

[31] V.V. Buldygin, A.B. Kharazishvili. *Geometric Aspects of Probability Theory and Mathematical Statistics.* Dordrecht: Kluwer Academic Publishers, 2000.

[32] J. Cichoń, A. Kharazishvili, B. Węglorz. On selectors associated with some subgroups of the real line. *Bull. Acad. Sci. of Georgian SSR*, vol. 144, no. 2, 1991.

[33] J. Cichoń, A. Kharazishvili, B. Węglorz. On sets of Vitali's type. *Proc. Amer. Math. Soc.*, vol. 118, no. 4, 1993.

[34] J. Cichoń, A. Kharazishvili, B. Węglorz. *Subsets of the Real Line.* Łódź: Łódź University Press, 1995.

[35] J. Cichoń, M. Morayne. On differentiability of Peano type functions, III. *Proc. Amer. Math. Soc.*, vol. 92, no. 3, pp. 432–438, 1984.

[36] J. Cichoń, M. Morayne. Universal functions and generalized classes of functions. *Proc. Amer. Math. Soc.*, vol. 102, no. 1, pp. 83–89, 1988.

[37] J. Cichoń, M. Morayne, J. Pawlikowski, S. Solecki. Decomposing Baire functions. *The Journal of Symbolic Logic*, vol. 56, no. 4, pp. 1273–1283, 1991.

[38] K. Ciesielski. Set-theoretic real analysis. *Journal of Applied Analysis*, vol. 3, no. 2, pp. 143–190, 1997.

[39] K. Ciesielski, L. Larson, K. Ostaszewski. Differentiability and density continuity. *Real Analysis Exchange*, vol. 15, no. 1, pp. 239–247, 1989–1990.

[40] K. Ciesielski, L. Larson, K. Ostaszewski. I-density continuous functions. *Memoirs of the Amer. Math. Soc.*, vol. 107, no. 515, pp. 1–132, 1994.

[41] K. Ciesielski, S. Shelah. Category analogue of sup-measurability problem. *Journal of Applied Analysis*, vol. 6, no. 2, pp. 159–172, 2000.

[42] P.M. Cohn. *Universal Algebra*. London: Harper & Row, 1965.

[43] P. Corazza. Ramsey sets, the Ramsey ideal, and other classes over **R**. *The Journal of Symbolic Logic*, vol. 57, pp. 1441–1468, 1992.

[44] U.B. Darji. Decomposition of functions. *Real Analysis Exchange*, vol. 21, no. 1, pp. 19–25, 1995–1996.

[45] U.B. Darji. Countable decomposition of derivatives and Baire 1 functions. *Journal of Applied Analysis*, vol. 2, no. 2, pp. 119–124, 1996.

[46] M. Dehn. Uber den Rauminhalt. *Math. Ann.*, vol. 55, pp. 465–478, 1902.

[47] C. Dellacherie. *Capacités et Processus Stochastiques*. Berlin: Springer-Verlag, 1972.

[48] A. Denjoy. Sur les fonctions dérivées sommables, *Bull. Soc. Math. France*, vol. 43, pp. 161–248, 1915.

[49] J.L. Doob. *Stochastic Processes*. New York: Wiley & Sons, Inc., 1953.

[50] D. Egoroff. Sur les suites de fonctions mesurables. *C. R. Acad. Sci. Paris*, vol. CLII, 1911.

[51] I. Ekeland. *Elements d'Economie Mathematique*. Paris: Hermann, 1979.

[52] P. Erdös, S. Kakutani. On nondenumerable graphs. *Bull. Amer. Math. Soc.*, vol. 49, pp. 457–461, 1943.

[53] P. Erdös, K. Kunen, R.D. Mauldin. Some additive properties of sets of real numbers. *Fund. Math.*, vol. CXIII, pp. 187–199, 1981.

[54] P. Erdös, R.D. Mauldin. The nonexistence of certain invariant measures. *Proc. Amer. Math. Soc.*, vol. 59, pp. 321–322, 1976.

[55] P. Erdös, A.H. Stone. On the sum of two Borel sets. *Proc. Amer. Math. Soc.*, vol. 25, no. 2, pp. 304–306, 1970.

[56] M.P. Ershov. Measure extensions and stochastic equations. *Probability Theory and its Applications*, vol. 19, no. 3, 1974 (in Russian).

[57] M. Evans. On continuous functions and the approximate symmetric derivatives. *Colloq. Math.*, vol. 31, pp. 129–136, 1974.

[58] G. Fichtenholz. Sur une fonction de deux variables sans intégrale double. *Fund. Math.*, vol. 6, pp. 30–36, 1924.

[59] C. Freiling. A converse to a theorem of Sierpiński on almost symmetric sets. *Real Analysis Exchange*, vol. 15, no. 2, pp. 760–767, 1989–1990.

[60] C. Freiling. Axioms of symmetry: throwing darts at the real number line. *The Journal of Symbolic Logic*, vol. 51, pp. 190–200, 1986.

[61] D.H. Fremlin. Measure additive coverings and measurable selectors. *Dissertationes Mathematicae*, vol. CCLX, 1987.

[62] H. Friedman. A consistent Fubini-Tonelli theorem for non-measurable functions. *Illinois Journal of Math.*, vol. 24, pp. 390–395, 1980.

[63] S. Fuchino, Sz. Plewik. On a theorem of E. Helly. *Proc. Amer. Math. Soc.*, vol. 127, no. 2, pp. 491–497, 1999.

[64] B.R. Gelbaum, J.M.H. Olmsted. *Counterexamples in Analysis.* San Francisco: Holden-Day, 1964.

[65] C. Goffman. Everywhere differentiable functions and the density topology. *Proc. Amer. Math. Soc.*, vol. 51, pp. 250–251, 1975.

[66] C. Goffman, C.J. Neugebauer, T. Nishiura. Density topology and approximate continuity. *Duke Mathem. Journal*, vol. 28, pp. 497–505, 1961.

[67] Z. Grande, J. Lipiński. Un exemple d'une fonction sup-mesurable qui n'est pas mesurable. *Colloq. Math.*, vol. 39, pp. 77–79, 1978.

[68] E. Grzegorek. Remarks on σ-fields without continuous measures. *Colloq. Math.*, vol. 39, pp. 73–75, 1978.

[69] P.R. Halmos. *Measure Theory.* New York: D. Van Nostrand, 1950.

[70] G. Hamel. Eine Basis aller Zahlen und die unstetigen Lösungen der Funktionalgleichung: $f(x + y) = f(x) + f(y)$. *Math. Ann.*, vol. 60, pp. 459–462, 1905.

[71] *Handbook of Mathematical Logic* (edited by J. Barwise). Amsterdam: North-Holland Publishing Comp., 1977.

[72] *Handbook of Mathematical Logic*, Part 2. Moscow: Izd. Nauka, 1982 (in Russian, translation from English).

[73] E. Hewitt, K. Ross. *Abstract Harmonic Analysis*, vol. 1. Berlin: Springer-Verlag, 1963.

[74] E.W. Hobson. *Theory of Functions of a Real Variable*, II. New York: Dover, 1957.

[75] D.H. Hyers, S.M. Ulam. Approximately convex functions. *Proc. Amer. Math. Soc.*, vol. 3, pp. 821–828, 1952.

[76] S. Jackson, R.D. Mauldin. Some complexity results in topology and analysis. *Fund. Math.*, vol. 141, pp. 75–83, 1992.

[77] V. Jarník. Sur la dérivabilité des fonctions continues. *Spisy Privodov, Fak. Univ. Karlovy*, vol. 129, pp. 3–9, 1934.

[78] J. Jasiński, I. Reclaw. Restrictions to continuous and pointwise discontinuous functions. *Real Analysis Exchange*, vol. 23, no. 1, pp. 161–174, 1997–1998.

[79] T. Jech. *Set Theory*. New York-London: Academic Press, 1978.

[80] B. Jessen. The algebra of polyhedra and the Dehn-Sydler theorem. *Math. Scand.*, vol. 22, pp. 241–256, 1968.

[81] S. Kaczmarz, H. Steinhaus. *Theorie der Orthogonalreihen*. Warszawa-Lwow: Monogr. Matem., 1935.

[82] S. Kakutani, J.C. Oxtoby. Construction of a non-separable invariant extension of the Lebesgue measure space. *Ann. Math.*, vol. 52, pp. 580–590, 1950.

[83] Y. Katznelson, K. Stromberg. Everywhere differentiable, nowhere monotone functions. *Amer. Math. Monthly*, vol. 81, pp. 349–354, 1974.

[84] A.S. Kechris. *Classical Descriptive Set Theory*. New York: Springer-Verlag, 1995.

[85] J.L. Kelley. *General Topology*. New York: D. Van Nostrand, 1955.

[86] A.B. Kharazishvili. *Applications of Set Theory*. Tbilisi: Izd. Tbil. Gos. Univ., 1989 (in Russian).

[87] A.B. Kharazishvili. Certain types of invariant measures. *Dokl. Akad. Nauk SSSR*, vol. 222, no. 3, pp. 538–540, 1975 (in Russian).

[88] A.B. Kharazishvili. *Invariant Extensions of Lebesgue Measure*. Tbilisi: Izd. Tbil. Gos. Univ., 1983 (in Russian).

[89] A.B. Kharazishvili. Martin's axiom and Γ-selectors. *Bull. Acad. Sci. of Georgian SSR*, vol. 137, no. 2, 1990 (in Russian).

[90] A.B. Kharazishvili. On sections of analytic sets. *Bull. Acad. Sci. of Georgia*, vol. 147, no. 2, pp. 223–226, 1993 (in Russian).

[91] A.B. Kharazishvili. On translations of sets and functions. *Journal of Applied Analysis*, vol. 1, no. 2, pp. 145–158, 1995.

[92] A.B. Kharazishvili. *Selected Topics of Point Set Theory*. Łódź: Łódź University Press, 1996.

[93] A.B. Kharazishvili. Some applications of Hamel bases. *Bull. Acad. Sci. Georgian SSR*, vol. 85, no. 1, pp. 17–20, 1977 (in Russian).

[94] A.B. Kharazishvili. Some questions concerning invariant extensions of Lebesgue measure. *Real Analysis Exchange*, vol. 20, no. 2, pp. 580–592, 1994–1995.

[95] A.B. Kharazishvili. *Some Questions of Set Theory and Measure Theory*. Tbilisi: Izd. Tbil. Gos. Univ., 1978 (in Russian).

[96] A.B. Kharazishvili. Some questions of the theory of invariant measures. *Bull. Acad. Sci. of Georgian SSR*, vol. 100, no. 3, 1980 (in Russian).

[97] A.B. Kharazishvili. Some remarks on density points and the uniqueness property for invariant extensions of the Lebesgue measure. *Acta Universitatis Carolinae - Mathematica et Physica*, vol. 35, no. 2, pp. 33–39, 1994.

[98] A.B. Kharazishvili. Some remarks on the property (N) of Luzin. *Annales Mathematicae Silesianae*, vol. 9, pp. 33–42, 1995.

[99] A.B. Kharazishvili. *Vitali's Theorem and Its Generalizations*. Tbilisi: Izd. Tbil. Gos. Univ. 1991 (in Russian).

[100] A.B. Kharazishvili. Baire property and its applications. *Proc. I. Vekua Inst. of Appl. Math.*, vol. 43, pp. 5–113, 1992 (in Russian).

[101] A.B. Kharazishvili. Sup-measurable and weakly sup-measurable mappings in the theory of ordinary differential equations. *Journal of Applied Analysis*, vol. 3, no. 2, pp. 211–224, 1997.

[102] A.B. Kharazishvili. On countably generated invariant σ-algebras that do not admit measure type functionals. *Real Analysis Exchange*, vol. 23, no. 1, pp. 287–294, 1997–1998.

[103] A.B. Kharazishvili. *Transformation Groups and Invariant Measures: Set-theoretical Aspects*. Singapore: World Scientific Publ. Co., 1998.

[104] A.B. Kharazishvili. On measurability properties connected with the superposition operator. *Real Analysis Exchange*, vol. 28, no. 1, pp. 205–214, 2002–2003.

[105] A.B. Kharazishvili. On absolutely nonmeasurable additive functions. *Georgian Mathematical Journal*, vol. 11, no. 2, pp. 301–306, 2004.

[106] A.B. Kharazishvili. On generalized step-functions and superposition operators. *Georgian Mathematical Journal*, vol. 11, no. 4, pp. 753–758, 2004.

[107] A.B. Kharazishvili. *Nonmeasurable Sets and Functions*. North-Holland Mathematics Studies, 195, Amsterdam: Elsevier, 2004.

[108] A.B. Kharazishvili, A.P. Kirtadze. On the measurability of functions with respect to certain classes of measures. *Georgian Mathematical Journal*, vol. 11, no. 3, pp. 489–494, 2004.

[109] B. King. Some remarks on difference sets of Bernstein sets. *Real Analysis Exchange*, vol. 19, no. 2, pp. 478–490, 1993–1994.

[110] E.M. Kleinberg. *Infinitary Combinatorics and the Axiom of Determinateness*. Berlin: Springer-Verlag, 1977.

[111] K. Kodaira, S. Kakutani. A non-separable translation-invariant extension of the Lebesgue measure space. *Ann. Math.*, vol. 52, pp. 574–579, 1950.

[112] A.N. Kolmogoroff. Sur la possibilité de la définition générale de la dérivée, de l'intégrale et de la sommation des séries divergentes. *C. R. Acad. Sci. Paris*, vol. 180, pp. 362–364, 1925.

[113] P. Komjáth. A note on set mappings with meager images. *Studia Scientiarum Mathematicarum Hungarica*, vol. 30, pp. 461–467, 1995.

[114] M.A. Krasnoselskii. *Topological Methods in the Theory of Nonlinear Integral Equations*. Moscow: GITTL, 1956 (in Russian).

[115] M.A. Krasnoselskii, A.V. Pokrovskii. On a discontinuous superposition operator. *Uspekhi Mat. Nauk*, vol. 32, no. 1, pp. 169–170, 1977 (in Russian).

[116] M.A. Krasnoselskii, A.V. Pokrovskii. *Systems with Hysteresis*. Berlin: Springer-Verlag, 1988.

[117] M. Kuczma. *An Introduction to the Theory of Functional Equations and Inequalities: Cauchy's Equation and Jensen's Inequality*. Katowice: PWN, 1985.

[118] K. Kunen. Random and Cohen reals. In: *Handbook of Set-Theoretic Topology*, editors K.Kunen and J.E.Vaughan, Amsterdam: North-Holland Publishing Comp., 1984.

[119] K. Kunen. *Set Theory*. Amsterdam: North–Holland Publishing Comp., 1980.

[120] K. Kuratowski. *Topology*, vol. 1. New York: Academic Press, 1966.

[121] K. Kuratowski. *Topology*, vol. 2. New York: Academic Press, 1968.

[122] K. Kuratowski. On the concept of strongly transitive systems in topology. *Annali di Matematica Pura ed Applicata*, vol. 98, no. 4, 1974.

[123] K. Kuratowski. A theorem on ideals and its application to the Baire property in polish spaces. *Uspekhi Mat. Nauk*, vol. 31, 5 (191), 1976.

[124] K. Kuratowski, A. Mostowski. *Set Theory*. Amsterdam: North-Holland Publishing Comp., 1967.

[125] K. Kuratowski, C. Ryll–Nardzewski. A general theorem on selectors. *Bull. Acad. Polon. Sci.*, Ser. Math., vol. 13, no. 6, pp. 397–402, 1965.

[126] M. Laczkovich, G. Petruska. On the transformers of derivatives. *Fund. Math.*, vol. 100, pp. 179–199, 1978.

[127] M.A. Lavrentieff. Contribution á la théorie des ensembles homéomorphes. *Fund. Math.*, vol. 6, 1924.

[128] M.A. Lavrentieff. Sur une équation différentielle du premiér ordre. *Math. Zeitschrift*, vol. 23, pp. 197–209, 1925.

[129] H. Lebesgue. Sur les fonctions représentables analytiquement. *C.R. Acad. Sci. Paris*, vol. 139, pp. 29–31, 1904.

[130] H. Lebesgue. Sur les fonctions représentables analytiquement. *Journal de Math. Pures et Appl.*, 1, Ser. 6, pp. 139–216, 1905.

[131] P. Lévy. *Processus Stochastiques et Mouvement Brownien*. Paris: Gauthier-Villars, 1965.

[132] A. Liapunoff. Sur quelques propriétés des cribles rectilignes et des ensembles *C*. *C.r. Soc. sci. Varsovie*, vol. 29, pp. 1–8, 1936.

[133] N. Lusin. *Leçons sur les Ensembles Analytiques et leurs Applications*. Paris: Gauthier - Villars, 1930.

[134] N. Lusin. Sur une probléme de M. Baire. *C.R. Acad. Sci. Paris*, vol. 158, p. 1259, 1914.

[135] N.N. Lusin. *Integral and Trigonometric Series*. Moscow: Izd. GITTL, 1956 (in Russian).

[136] N.N. Lusin. *Collected Works*, vol. 2, Moscow: Izd. Akad. Nauk SSSR, 1958 (in Russian).

[137] N. Lusin, W. Sierpiński. Sur une décomposition du continu. *C.R. Acad. Sci. Paris*, vol. 175, pp. 357–359, 1922.

[138] D. Maharam. On a theorem of von Neumann. *Proc. Amer. Math. Soc.*, vol. 9, pp. 987–994, 1958.

[139] D. Maharam, A.H. Stone. Expressing measurable functions by one-one ones. *Advances in Mathematics*, vol. 46, no. 2, pp. 151–161, 1982.

[140] J. Malý. The Peano curve and the density topology. *Real Analysis Exchange*, vol. 5, pp. 326–329, 1979–1980.

[141] E. Marczewski, C. Ryll-Nardzewski. Sur la mesurabilité des fonctions de plusieurs variables. *Annales de la Societé Polonaise de Mathématique*, vol. 25, pp. 145–155, 1952.

[142] R.D. Mauldin. The set of continuous nowhere differentiable functions. *Pacific Journal of Mathematics*, vol. 83, no. 1, pp. 199–205, 1979; vol. 121, no. 1, pp. 119–120, 1986.

[143] S. Mazurkiewicz. Sur les fonctions non-derivables. *Studia Math.*, vol. 3, pp. 92–94, 1931.

[144] S. Mazurkiewicz. Sur les suites de fonctions continues. *Fund. Math.*, vol. 18, pp. 114–117, 1932.

[145] E. Michael. Continuous selections, I. *Ann. Math.*, vol. 63, no. 2, pp. 361–382, 1956.

[146] E. Michael. Selected selection theorems. *Amer. Math. Monthly*, vol. 63, pp. 233–238, 1956.

[147] J. van Mill. *Infinite-dimensional Topology: Prerequisites and Introduction.* Amsterdam: North-Holland Publishing Comp., 1989.

[148] J. van Mill, R. Pol. Baire 1 functions which are not countable unions of continuous functions. *Acta Math. Hungar.*, vol. 66, no. 4, pp. 289–300, 1995.

[149] A.W. Miller. Special subsets of the real line. In: *Handbook of Set-Theoretic Topology*, editors K. Kunen and J. Vaughan, Amsterdam: North-Holland Publishing Comp., 1984.

[150] G. Mokobodzki. Ensembles á coupes dénombrables et capacités dominées par une mesure. *Seminaire de Probabilites, Universite de Strasbourg*, 1977–1978.

[151] M. Morayne. On differentiability of Peano type functions. *Colloq. Math.*, vol. 53, no. 1, pp. 129–132, 1987.

[152] M. Morayne. On differentiability of Peano type functions, II. *Colloq. Math.*, vol. 53, no. 1, pp. 133–135, 1987.

[153] J.C. Morgan II. *Point Set Theory.* New York: Marcel Dekker, Inc., 1990.

[154] I.P. Natanson. *The Theory of Functions of a Real Variable.* Moscow: Gos. Izd. Techn.-Teor. Lit., 1957 (in Russian).

[155] T. Natkaniec. *Almost Continuity.* Bydgoszcz: WSP, 1992.

[156] J. Neveu. *Bases Mathématiques du Calcul des Probabilités.* Paris: Masson et Cie, 1964.

[157] P.S. Novikov. Separation of C-sets. *Izv. Akad. Nauk SSSR*, Ser. Mat., vol. 1, no. 2, pp. 253–264, 1937.

[158] P.S. Novikov. *Selected Works.* Moscow: Izd. Nauka, 1979 (in Russian).

[159] W. Orlicz. *Collected Papers*, vols. 1 and 2. Warszawa: PWN, 1988.

[160] W. Orlicz. Zur Theorie der Differentialgleichung $y' = f(x, y)$. *Polska Akademia Umiejetnosci*, Krakow, pp. 221–228, 1932.

[161] J.C. Oxtoby. Cartesian products of Baire spaces. *Fund. Math.*, vol. XLIX, pp. 157–166, 1961.

[162] J.C. Oxtoby. *Measure and Category.* Berlin: Springer-Verlag, 1971.

[163] G.R. Pantsulaia. Density points and invariant extensions of the Lebesgue measure. *Bull. Acad. Sci. of Georgia*, vol. 151, no. 2, 1995 (in Russian).

[164] J. Pawlikowski. The Hahn-Banach theorem implies the Banach-Tarski paradox. *Fund. Math.*, vol. 138, pp. 21–22, 1991.

[165] A. Pelc, K. Prikry. On a problem of Banach. *Proc. Amer. Math. Soc.*, vol. 89, no. 4, pp. 608–610, 1983.

[166] I.G. Petrovskii. *Lectures on the Theory of Ordinary Differential Equations.* Moscow: GITTL, 1949 (in Russian).

[167] W.F. Pfeffer, K. Prikry. Small spaces. *Proc. London Math. Soc.*, vol. 58, no. 3, pp. 417–438, 1989.

[168] Z. Piotrowski. Separate and joint continuity. *Real Analysis Exchange*, vol. 11, 1985–1986.

[169] J. Raisonnier. A mathematical proof of S.Shelah's theorem on the measure problem and related results. *Israel Journal of Mathematics*, vol. 48, no. 1, 1984.

[170] F.P. Ramsey. On a problem of formal logic. *Proc. London Math. Soc.*, ser. 2, vol. 30, pp. 264–286, 1930.

[171] M.M. Rao. *Stochastic Processes and Integration.* Sijthoff and Noord-
       hoff, Alphen aan den Rijn, 1979.

[172] I. Reclaw. Restrictions to continuous functions and boolean algebras.
       Proc. Amer. Math. Soc., vol. 118, no. 3, pp. 791–796, 1993.

[173] D. Repovs, P.V. Semenov. E.Michael's theory of continuous selec-
       tions; development and applications. *Uspekhi Mat. Nauk*, vol. 49, no.
       6 (300), 1994 (in Russian).

[174] C.A. Rogers. A linear Borel set whose difference set is not a Borel
       set. *Bull. London Math. Soc.*, vol. 2, pp. 41–42, 1970.

[175] A. Rosental. On the continuity of functions of several variables. *Math.
       Zeitschrift*, vol. 63, no. 1, pp. 31–38, 1955.

[176] A. Roslanowski, S. Shelah. Measured Creatures. *Israel Journal of
       Mathematics* (to appear).

[177] F. Rothberger. Eine Äquivalenz zwischen der Kontinuumhypothese
       und der Existenz der Lusinschen und Sierpińskischen Mengen. *Fund.
       Math.*, vol. 30, pp. 215–217, 1938.

[178] S. Ruziewicz. Sur une proprieté de la base hamelienne. *Fund. Math.*,
       vol. 26, pp. 56–58, 1936.

[179] C. Ryll-Nardzewski, R. Telgarský. The nonexistence of universal in-
       variant measures. *Proc. Amer. Math. Soc.*, vol. 69, pp. 240–242,
       1978.

[180] S. Saks. *Theory of the Integral.* Warszawa-Lwów, 1937.

[181] E.A. Selivanovskii. On a class of effective sets. *Mat. Sbor.*, vol. 35,
       pp. 379–413, 1928 (in Russian).

[182] S. Shelah. Can you take Solovay's inaccessible away? *Israel Journal
       of Mathematics*, vol. 48, no. 1, pp. 1–47, 1984.

[183] S. Shelah. Possibly every real function is continuous on a non-meagre
       set. *Publ. Inst. Mat. Beograd (N.S.)*, vol. 57, no. 71, pp. 47–60,
       1995.

[184] J. Shinoda. Some consequences of Martin's axiom and the negation
       of the continuum hypothesis. *Nagoya Math. Journal*, vol. 49, pp.
       117–125, 1973.

[185] J. Shipman. Cardinal conditions for strong Fubini theorems. *Trans. Amer. Math. Soc.*, vol. 321, pp. 465–481, 1990.

[186] J.R. Shoenfield. Measurable cardinals. In: *Logic Colloquium'69*, Amsterdam: North-Holland Publishing Comp., 1971.

[187] J.W. Shragin. Conditions for measurability of superpositions. *Dokl. Akad. Nauk SSSR*, vol. 197, pp. 295–298, 1971, (in Russian).

[188] J.W. Shragin. On representation of a locally defined operator in the form of the Nemytskii operator. *Functional Differential Equations*, vol. 3, no. 3-4, pp. 447–452, 1996.

[189] W. Sierpiński. L'axiome de M.Zermelo et son role dans la théorie des ensembles et l'analyse. *Bull. Intern. Acad. Sci. Cracovie*, Ser. A, pp. 97–152, 1918.

[190] W. Sierpiński. *Cardinal and Ordinal Numbers*. Warszawa: PWN, 1958.

[191] W. Sierpiński. *Hypothése du Continu* (2-nd Edition). New York: Chelsea Publ. Co., 1956.

[192] W. Sierpiński. Les correspondances multivoques et l'axiome du choix. *Fund. Math.*, vol. 34, pp. 39–44, 1947.

[193] W. Sierpiński. *Oeuvres Choisies*, Volumes I-III. Warszawa: PWN, 1972-1976.

[194] W. Sierpiński. Sur la question de la mesurabilité de la base de M. Hamel. *Fund. Math.*, vol. 1, pp. 105–111, 1920.

[195] W. Sierpiński. Sur l'équation fonctionnelle $f(x + y) = f(x) + f(y)$. *Fund. Math.*, vol. 1, pp. 116–122, 1920.

[196] W. Sierpiński. Sur les fonctions convexes mesurables. *Fund. Math.*, vol. 1, pp. 125–128, 1920.

[197] W. Sierpiński. Sur l'ensemble de distances entre les points d'un ensemble. *Fund. Math.*, vol. 7, pp. 144–148, 1925.

[198] W. Sierpiński. Sur le produit combinatoire de deux ensembles jouissant de la propriété C. *Fund. Math.*, vol. 24, pp. 48–50, 1935.

[199] W. Sierpiński. Sur une fonction non mesurable partout presque symetrique. *Acta Litt. Scient. Szeged*, vol. 8, pp. 1–6, 1936.

[200] W. Sierpiński. Sur un théoréme équivalent á l'hypothése du continu. *Bull. Internat. Acad. Sci. Cracovie*, Ser. A, pp. 1–3, 1919.

[201] W. Sierpiński. Sur un probleme concernant les fonctions semi-continues. *Fund. Math.*, vol. 28, pp. 1–6, 1937.

[202] W. Sierpiński. *Introduction to Set Theory and Topology*. Warszawa: PZWS, 1947 (in Polish).

[203] W. Sierpiński, A. Zygmund. Sur une fonction qui est discontinue sur tout ensemble de puissance du continu. *Fund. Math.*, vol. 4, pp. 316–318, 1923.

[204] J.C. Simms. Sierpiński's theorem. *Simon Stevin*, vol. 65, no. 1–2, pp. 69–163, 1991.

[205] S.G. Simpson. Which set existence axioms are needed to prove the Cauchy-Peano theorem for ordinary differential equations? *The Journal of Symbolic Logic*, vol. 49, pp. 783–802, 1984.

[206] A.V. Skorokhod. *Integration in Hilbert Space*. Moscow: Izd. Nauka, 1975 (in Russian).

[207] B.S. Sodnomov. An example of two G_δ-sets whose arithmetical sum is not Borel measurable. *Dokl. Akad. Nauk SSSR*, vol. 99, pp. 507–510, 1954 (in Russian).

[208] B.S. Sodnomov. On arithmetical sums of sets. *Dokl. Akad. Nauk SSSR*, vol. 80, no. 2, pp. 173–175, 1951 (in Russian).

[209] S. Solecki. Measurability properties of sets of Vitali's type. *Proc. Amer. Math. Soc.*, vol. 119, no. 3, pp. 897–902, 1993.

[210] S. Solecki. On sets nonmeasurable with respect to invariant measures. *Proc. Amer. Math. Soc.*, vol. 119, no. 1, pp. 115–124, 1993.

[211] R.M. Solovay. A model of set theory in which every set of reals is Lebesgue measurable. *Ann. Math.*, vol. 92, pp. 1–56, 1970.

[212] R.M. Solovay. Real-valued measurable cardinals. *Proceedings of Symposia in Pure Mathematics*, vol. XII, pp. 397–428, Axiomatic set theory, Part 1, Providence 1971.

[213] J. Stallings. Fixed point theorems for connectivity maps. *Fund. Math.*, vol. 47, pp. 249–263, 1959.

[214] H. Steinhaus. Sur les distances des points de mesure positive. *Fund. Math.*, vol. 1, pp. 93–104, 1920.

[215] J.-P. Sydler. Conditions nécessaires et suffisantes pour l'équivalence des polyédres de l'espace euclidien á trois dimensions. *Comment. Math. Helv.*, vol. 40, pp. 43–80, 1965.

[216] E. Szpilrajn (E. Marczewski). Sur une classe de fonctions de M. Sierpiński et la classe correspondante d'ensembles. *Fund. Math.*, vol. 24, pp. 17–34, 1935.

[217] E. Szpilrajn (E. Marczewski). The characteristic function of a sequence of sets and some of its applications. *Fund. Math.*, vol. 31, pp. 207–223, 1938.

[218] M. Talagrand. Sommes vectorielles d'ensembles de mesure nulle, *C.R. Acad. Sci. Paris*, v. 280, pp. 853–855, 1975.

[219] F. Tall. The density topology. *Pacific Journal of Mathematics*, vol. 62, pp. 275–284, 1976.

[220] G.P. Tolstov. A note on D.F. Egorov's theorem. *Dokl. Akad. Nauk SSSR*, vol. 22, pp. 309–311, 1939.

[221] G.P. Tolstov. On partial derivatives. *Izv. Akad. Nauk SSSR*, Ser. Mat., vol. 13, no. 5, pp. 425–446, 1949 (in Russian).

[222] T. Traynor. An elementary proof of the lifting theorem. *Pacific Journal of Mathematics*, vol. 53, pp. 267–272, 1974.

[223] S. Ulam. *A Collection of Mathematical Problems*. New York: Interscience, 1960.

[224] S. Ulam. Zur Masstheorie in der allgemeinen Mengenlehre. *Fund. Math.*, vol. 16, pp. 140–150, 1930.

[225] S. Ulam, K. Kuratowski. Quelques propriétés topologiques du produit combinatoire. *Fund. Math.*, vol. 19, 1932.

[226] G. Vitali. *Sul problema della misura dei gruppi di punti di una retta.* Nota, Bologna, 1905.

[227] S. Wagon. *The Banach-Tarski Paradox.* Cambridge: Cambridge University Press, 1985.

[228] J.T. Walsh. Marczewski sets, measure and the Baire property II. *Proc. Amer. Math. Soc.*, vol. 106, no. 4, pp. 1027–1030, 1989.

[229] C. Weil. On nowhere monotone functions. *Proc. Amer. Math. Soc.*, vol. 56, pp. 388–389, 1976.

[230] H.E. White Jr. Topological spaces in which Blumberg's theorem holds. *Proc. Amer. Math. Soc.*, vol. 44, pp. 454–462, 1974.

[231] W. Wilczyński. A category analogue of the density topology, approximate continuity and the approximate derivative. *Real Analysis Exchange*, vol. 10, no. 2, pp. 241–265, 1984–1985.

[232] W. Wilczyński, A. Kharazishvili. On translations of measurable sets and of sets with the Baire property. *Bull. Acad. Sci. of Georgia*, vol. 145, no. 1, pp. 43–46, 1992 (in Russian).

[233] Z. Zahorski. Sur la premiére dérivée. *Trans. Amer. Math. Soc.*, vol. 69, pp. 1–54, 1950.

[234] P. Zakrzewski. Extensions of isometrically invariant measures on Euclidean spaces. *Proc. Amer. Math. Soc.*, vol. 110, pp. 325–331, 1990.

[235] P. Zakrzewski. On a construction of universally small sets. *Real Analysis Exchange*, vol. 28, no. 1, pp. 221–226, 2002–2003.

Subject Index